Farm Animal Management

NIPA® GENX ELECTRONIC RESOURCES & SOLUTIONS P. LTD.
New Delhi-110 034

Farm Animal Management

Principles and Practices

Rana Ranjeet Singh
Assistant Professor,
Department of LPM,
Vanbandhu College of Veterinary Sciences & Animal Husbandry
Navsari Agricultural University
Navsari - 396 450, Gujarat

and

Md. Manzarul Islam
Assistant Professor,
Department of LPM,
College of Veterinary Science & Animal Husbandry
Anand Agricultural University
Anand – 388 110, Gujarat

NIPA® GENX ELECTRONIC RESOURCES & SOLUTIONS P. LTD.
New Delhi-110 034

NIPA® GENX ELECTRONIC RESOURCES & SOLUTIONS P. LTD.

101,103, Vikas Surya Plaza, CU Block
L.S.C. Market, Pitam Pura, New Delhi-110 034
Ph : +91 11 27341616, 27341717, 27341718
E-mail: newindiapublishingagency@gmail.com
www: www.nipabooks.com

For customer assistance, please contact
Phone: + 91-11-27 34 17 17
Fax: + 91-11- 27 34 16 16
E-Mail: feedbacks@nipabooks.com

ISBN: 978-81-19103-50-8

Composed and Designed by NIPA®.

Dedicated to

Our Family, Teachers & Alma Mater

Preface

Livestock rearing is an integral part of socio-economic framework of India since time immemorial. Livestock sub-sector being a vital component of agriculture sector, plays a multidimensional role and acts as a tool in achieving nutritional security, employment generation and socio-economic development of rural sector, particularly among the landless, small, marginal farmers and women. India possesses huge livestock population of varying production potentials, distributed across different ago-ecological zones under different operational and livestock holding size. Hence, it requires different package of practices for their management. Further, factors like drivers of development change and climate change pose many challenges to this sector. Farm animal management encompasses integrated and precise application of basic scientific principles of breeding, feeding, heeding and weeding in general as well as in times of specific need. Therefore, an attempt has been made in this book to cover all these basic and applied aspects of livestock management in detail. Additionally, in the perspective of reorganization of the syllabus of veterinary science and animal husbandry programme, we felt the need of a suitable text book to cater the need of the students. It has also been noticed that students find it difficult to go through many standard but voluminous books on different aspects of veterinary science and animal husbandry. Hence, the attempt has been made to cover the whole syllabus with most updated information available in a very concise form. This book is appropriately divided into fifteen chapters covering almost each and every aspect of livestock production management. We believe that this book will not only help to augment the professional acumen of students of B. V. Sc. &

A. H., animal science, dairy science but will also act as an adjunct to the wisdom of teachers, farm managers, extension and dairy workers, farmers etc in the subject of livestock production and management.

Authors do not claim credits for any original work discoursed in this book but expects this book to be considered as a concise assortment of the diverse key topics in this field. We are indebted to the various authors, researchers, extension workers who dedicated their life to the science, enlightened this field with their knowledge. We express our gratitude to various organizations/institutions like FAO, Purebred Dairy Cattle Association, NASCO, ILRI, ICAR, USBGA, GOI and others for their valuable resources which are cited in this book.

We thank the contributors of this book because without their help this endeavour would never have been accomplished. We sincerely express our deepest sense of gratitude and sincere thanks to Dr. H. N. Pandey, Dr. Triveni Dutt, Dr. P. M. Desai, Dr. L. H. Saiyed, Dr. A. B. Fulsoundar, Dr. V. B. Kharadi, Dr. B. P. Brahmkshatriya, Dr. Sandhya Choudhary and Dr. K. N. Wadhwani for their benevolent guidance and erudite suggestions. We thank Dr. N. H. Kelawala (Dean, VCVS&AH, NAU, Navsari) for his adoring empathy, touching generosity and decisive appreciations which made our task easier. We are also thankful to our colleagues (Dr. Navin, Sabapara, Rao, Sorathiya, Mahipal, Vipul) for their support and to Mr. Vishal for his assistance for drawing various figures. Further, we are also thankful to our critics whose words kept us busy and helped immensely in successful completion of the job. We thank our publisher NIPA for supporting our efforts commendably.

Finally we express our sincere thanks to our family members for their support and encouragements.

Even though utmost care has been taken during conceptualising, drafting and printing of this book to avoid any kind of errors but we believe that this attempt may still be refined so we welcome any suggestion for the further improvement of the book.

Editors

List of Contributors

H N Pandey
Principal Scientist (Retired)
LPM Section. IVRI, Izatnagar,
Bareilly-243122, U. P.

Prem Prakash Dubey
Assistant Scientist,
Department of Animal Genetics & Breeding,
GADVASU, Ludhiana, Punjab,
email: prakashagb@gmail.com

K B Kore
Assistant Professor, Department of Animal Nutrition, VCVS & AH.,
NAU, Navsari-396450, Gujarat,
email: kumarkore@gmail.com

Amit Kumar
Scientist (LAR Section)
Animal Genetics Division, IVRI,
Izatnagar, Bareilly-243122, U. P.
email: vetamitchandan07@gmail .com

Pushpendra Kumar
Principal Scientist,
Animal Genetics Division, IVRI,
Izatnagar, Bareilly-243122, U. P.
email:pushpendraivri@rediffmail. com

Subodh Kumar
Senior Scientist.
Animal Genetics Division, IVRI.
Izatnagar, Bareilly-243122, U. P.
email: subkum@gmail.com

Triveni Dutt
Joint Director,
Extension Education,
IVRI. Izatnagar, Bareilly-243122,
U. P.

B. H. M. Patel
Senior Scientist,
LPM Section, IVRI, Izatnagar,
Bareilly-243122, U. P.
email: mpatellpm@gmail.com

Mukesh Singh
Principal Scientist,
LPM Section, IVRI. Izatnagar,
Bareilly-243122, U. P.
email: drmsingh9@gmail.com

K N Wadhwani
Professor & Head,
Department of LPM,
CVS & A. H.,AAU, Anand, Gujarat
email: knwadhwani@yahoo.co.in

Virendra Kumar Singh
Assistant Professor,
Department of Physiology &
Biochemistry,
VCVS & AH, NAU,
Navsari-396450, Gujarat,
email: drvksinghl98l@gmail.com

Contents

Chapter - 1

An Introduction to Animal Husbandry

Rana Ranjeet Singh

Taxonomic Classification of Common Livestock Species

A series of terms (kingdom, phylum, class, order, family, genus, species etc) is used to denote the member of a kingdom and each division of the hierarchy contains one or more groups from the next lower division like subphyla, subclasses, suborders, subfamilies, or subgenera etc if required. Further, species may be subdivided into varieties, breeds, or strains. Domestic farm animals belong to either class Mammalia or Aves. The scientific name of an animal comprise of its genus and species names. The first letter of genus name is written in capital letter but not the species name. While writing the scientific name of an animal the genus and species names are underlined when handwritten and always italicized when printed. General characteristics and taxonomic arrangements of common domestic animals have been given below.

Kingdom : Animalia

Members of kingdom Animalia are multicellular, eukaryotic heterotrophs. Adult animals develop from embryos. Most animals ingest and digest their food in some kind of internal cavity.

Phylum : Chordata

All chordates have a dorsal hollow nerve tube, a notochord, post-anal tail and pharyngeal gill slits at some stage of their life cycle.

Subphylum : Vertebrata

All vertebrates have a backbone (spinal column) and normally a closed circulatory system.

A. Class: Mammalia

Members of class Mammalia has four-chambered heart, a diaphragm which separates thoracic and abdominal cavity, body remains covered with hair and their embryo develops in uterus of the mother. The mother secretes milk for the young ones through the mammary glands. They are homoeothermic animals.

1. Order : Perissodactyla

Members of this order are odd-toed ungulates, weight is distributed on central digits 2, 3, 4. First digit remains absent, has simple stomach and is hindgut fermenters (colon with microorganisms for the digestion of cellulose).

- 1 toe per foot = horses
- 3 toes per foot = tapirs, rhinos

- **Family: Equidae**
- **Genus: Equus**
    - Equus caballus -- Horse
    - Equus asinus -- Ass or donkey

2. Order : Artiodactyla

Members of this order are even-toed ungulates (2 toes or 4 toes, or 2 toes with dew hooves), have compound stomach and also known as cloven hoofed animals.

- **Suborder: Ruminantia**
- **Family: Bovidae**

One of the defining characteristics of this family is the presence of unbranched horns. Horns are present in males of all bovid species and in females of some genera. When horns are present in both sexes, horn of males are always thicker at the base and are more complex. The horns are permanently attached to the frontal bones of the skull and are composed of a bone core covered with a keratin sheath (which is never shed). An air space separates these two layers; hence they are often called as "hollow-horned ungulates".

- *Bos taurus* – European cattle
- *Bos indicus* – Indian or zebu or Afrikander cattle (humped cattle)
- *Bos grunniens* – Yak
- *Bos sondaicus* – Banteng
- *Bos frontalis* – Gayal or Mithun
- *Bubalus bubalis* – Indian buffalo
- *Bison bison* – Bison
- *Oribos monoschatus* – Musk ox
- *Ovis aries* – Sheep
- *Capra hircus* – Goat
- **Suborder: Suiformes**
- **Family: Suidae**
    - *Sus scrofa* -pig
- **Suborder: Tylopoda**
- **Family: Camelidae**
    - *Camelus bactrianus* – Bactrian camel (two humped)
    - *Camelus dromedaries* –Arabian camel (one humped)
    - *Lama pacos* – Alpaca
    - *Lama glama* – Llama
- **Family: Cervidae**
    - Rangifer tarandus -- Reindeer

3. Order : Carnivora

These are carnivorous and have large pointed canine teeth.

- **Family: Canidae**
    - *Canis familiaris* –Dog

- **Family: Felidae**
  - *Felis catus* – Cat

4. Order : Proboscidae

They have long, muscular trunks and thick, loose skin. Their upper incisors are elongated as tusks.

- **Family: Elephantidae**
  - *Elephas maximus* – Asiatic elephant
  - *Loxodonta africana* – African elephant

5. Order : Rodentia

These have chisel-like, constantly-growing incisors.

- **Family: Muridae**
  - *Rattus novregicus* – Rat
- **Family: Caviidae**
  - *Caria porcellus* – Guinea pig

6. Order : Lagomorpha

They have chisel-like incisors and hind legs modified for jumping.

- **Family: Leporidae**
  - *Orictolagus cuniculus* – Rabbit

B. Class- Aves

Members of class Aves also have four-chambered heart and are homeotherm. However, their bodies are covered with feathers, lack diaphragm, have beaks and gizzards instead of teeth, have light, hollow bones and an air-sac system attached to the lungs and lay eggs in which their young ones develop.

1. Order : Anseriformes

They are web-footed swimmers and divers.

- **Family: Anatidae**
  - *Anas platyrhynchos* – Duck
  - *Cairina moschata* – Muscovey duck
  - *Anser anser* – Goose

2. Order : Galliformes

They are clawed, runner and scratchers.

- **Family: Phasianidae**
    - *Gallus gallus* – Chicken
    - *Coturnix coturnix japanoica* – Japanese quail
    - *Pavo cristatus* – Peafowl
- **Family: Numididae**
    - *Numida meleagris* – Guinea fowl
- **Family: Meleagrididae**
    - *Meleagris gallopova* -- Turkey

3. Order : Columbiformes

- **Family: Columbidae**
    - *Columba livia* – Pigeaon

Domestication of Farm Animals

Domestication is defined as the condition where the breeding, feeding and care of animals are more or less controlled by the man. In other words, domestication is the process by which captive animals adapt to man and the environment provided by him. Taming is defined as elimination of tendencies to flee from man and changes to an animal through taming are primarily behavioural (reduction of its flight distances). Feral animals are the animals which have reverted to the wild state. Our present civilization has its roots in the domestication of animals. The process of domestication started when early humans came in contact with the wild animals and recognized worthiness of animals to them (supply of food, skins, fibre, draft, and hunting etc). After certain period of time they began to confine some of these animals and bred them in confinement. This practice resulted in some changes in their morphological, physiological and behavioural characteristics. After generations of continuous selection for desirable behavioural and morphological features, animals differed from their ancestors and their genotype also differed. Original or early domesticates were mainly ungulates and gallinaceous birds. The characteristics which favoured the domestication of a species were flexible diet, reasonably fast growth rate, ability to be bred in captivity, pleasant disposition, temperament and modifiable social hierarchy. The actual timing of original prehistoric domesticates has always been contentious and in some cases may differ by a millennium or even more. As excavation of an ancient hidden or burial site could possibly change the timing of their domestication presently known to us. Time and area of domestication of some of the domesticated animals have been mentioned below.

1. Dog : First animal domesticated by man, around 12,000 BC, during the time of the hunter-gatherers (the cavemen).
2. Goat : First food animal to be domesticated probably about 10,000 years ago (8000 BC) in Iran, at the edge of the Fertile Crescent.
3. Sheep : Domesticated probably about 8000 BC, around the neighbourhood area where goats were domesticated.
4. Pigs : At least 7000 BC, separately in the Europe and Asia.
5. Cattle : Humpless cattle around 6000 BC in Asia Minor, Greece, or Turkey and zebu cattle domestication began first in India about 4000 BC.
6. Horse (tarpan) : Around 4000 BC by nomadic tribes on the steppes of eastern Asia.
7. Buffalo : About 300 BC around India, Ceylon, and Burma.
8. Yak : About 5,000 years ago on the Qinghai-Tibetan plateau.

Source : Roots (2007).

History of Livestock Rearing in India

History of animal husbandry is as old as our civilization and there are ample evidences which indicate that animals were domesticated before the commencement of the written history. Association of livestock to Indian society during ancient times can be traced from zoomorphics, excavation findings, old scriptures and edicts etc. The ancient Indian literatures (holy Vedas, Puranas, Brahmanas and epics etc) are inundated with information on various aspects of animal husbandry. In the following paragraphs we will discuss association of livestock to Indian society during different periods.

Information on the animal husbandry practices in the Vedic period (1500-1000 BC) can be traced in the Rigveda (the oldest holy book of the Aryans). The Vedic society in India was dominated by cow culture and it was considered as the source of good fortune, happiness and good health. Aryans maintained their cattle on pastures near to their dwellings. They milked them thrice a day and cows which gave plentiful milk with ease were prayed for. Castration of males were practiced and bullocks were used for ploughing, farm transport and for irrigation purposes. Buffalo was not a common animal for Aryans. Goat, sheep and dogs were kept primarily for milk, wool and guarding houses, respectively. Barley, sugarcane and sesame (til) cake have been mentioned as animal feed (Somvanshi, 2006).

Information about animal husbandry practices followed in Later Vedic period (1000-600 BC) can be obtained from Yajurveda and two epics -Mahabharata (900 BC) and Ramayana (200 BC). Keeping of tamed elephants has also been referred in the Yajurveda. Horses were kept for riding purpose. Idea about the existence various species of livestock, treatment of many diseases with medicinal herbs & surgical methods are available in these epics. Beside the other uses of animals, dried cow dung was suggested to be better manure. During Budhist and Magadh period use of horses for riding was common. In Gupta dynasty and Kannauj empire horses were main war animals.

Animal husbandry achieved new heights during Mauryan age (322-232 BC) as this period preceded with Buddha and Mahavir's period. 'Suttanipata' (earliest Buddhist text) expresses cattle as a giver of food, beauty and happiness (annada, vannada and sukhada) and advocate their safety. Kautilya's 'Arthasastra' illustrates cow as worshipped animal and killing of cow was the deadly sin. It was one of the first duties of the king to do parikrama of cow with her calf and a bull. Buffalo become a familiar dairy animal in this period. In a Sanchi sculpture, two buffaloes are depicted swimming in a pool in front of a hermitage. Concept of green, dry fodder, rations for different species of animals were known during this period. Cows were milked once or twice a day. Sheep, goat, horses and donkeys were reared for wool, milk, riding and war purposes and carrying loads, respectively. Mules were also present in the Mauryan period.

The emblem of Veterinary Council of India was adopted from the period of Emperor Ashoka (around 300 BC), which projected the veterinary profession as its "best heritage" (Singh, 2002). Ashoka gave impetus to the development of veterinary science in India. The first veterinary hospital (Baniyan Hospital' of Suratis) existed in Ashoka's regime. Further, in ancient literatures citation of in excess of 2000 drugs of plant origin, knowledge about various diseases and the treatments and various management practices are available.

By the seventh century the veterinary sciences reached its peak and then started declining due to the attack of foreign invaders. The revival of veterinary sciences started in the eighteenth century by the arrival of British who started many research institutions entirely devoted to the animal husbandry and veterinary sciences. Some of the eminent veterinarians of ancient India are mentioned in Table 1.1.

Table 1.1 : Eminent veterinarians of ancient India.

Eminent veterinarians of ancient India	Known for/Specialization
1. Salihotra	: Known as 'father of Indian veterinary sciences'. He was an equine expert composed three books namely: Haya-Ayurveda, Asvaprasnsa and Asvalaksanashastram.
2. Palkapya	: Composed a voluminous work known as Hasti-Ayurveda or Gaj-Ayurveda or Palkapya Samhita.
3. Nal	: Composed manuscript on equine medicine.
4. Lord Krishna	: Famous for his love with cow & also cured horses.
5. Gautama	: Composed Gautam samhita (treatment & management of cow).
6. Nakula	: Composed a book "Asva Chikitsa" or "Equine Medicine" in Sanskrit.
7. Sahdeva	: Expert in cattle husbandry.
8. Jivak	: Identified various diseases of animals and did pregnancy diagnosis.
9. Charaka	: Expert of medicine.
10. Sushruta	: Expert of surgery and identified rabies like disease which occurs when a mad dog bites.

Livestock Production System

According to Seré *et al.* (1995), livestock production system is defined as a subset of the farming systems, where livestock contribute more than 10 percent to the total farm production in value terms or where intermediate contributions (value of animal traction and/or manure) represent more than 10 percent of the total value of purchased inputs.

Classification of world livestock production systems

1. **Solely Livestock Systems :** Livestock system in which rangelands, pastures, annual forages and purchased feeds supplies more than 90 percent of dry matter fed to animals and less than 10 percent of the total value of output comes from non-livestock farming activities. This system is divided into landless livestock production system and grassland based systems.

a) **Landless livestock production systems:** In this system, less than 10 percent of the dry matter fed to animals is farm produced and annual average stocking rates are above ten livestock units per hectare of agricultural land. This system is further divided into landless monogastric systems (predominantly in Organization for Economic Cooperation and Development (OECD) member countries and mainly chicken and pigs are raised) and landless ruminant systems (principally cattle and marginally sheep). Cattle are almost exclusively found in Eastern Europe and the Commonwealth of Independent States (CIS) and in a few OECD member countries. Landless sheep production systems are only found in western Asia and Northern Africa.

b) **Grassland based systems :** In this system, annual average stocking rate remains less than ten livestock units per hectare of agricultural land and more than 10 percent of the dry matter fed to animals from farm. This system is further divided into temperate (southern Australia, New Zealand, and parts of the United States, China and Mongolia) and tropical highland (parts of the highlands of South America and eastern Africa), humid/sub-humid tropics & sub-tropics (mostly in lowlands of South America: the llanos of Colombia and Venezuela as well as the cerrados of Brazil, lowlands of Mexico and estancias in Argentina) and arid/semi-arid tropics and sub-tropics (sub-Saharan Africa, Near East, North Africa regions, Australia, parts of western United States and southern Africa).

2. **Mixed Farming Systems :** Livestock system where crop by-products, stubble supplies more than 10 percent of the dry matter fed to animals or non-livestock farming activities contributes more than 10 percent of the total value of production. This system is divided into

   a. **Rainfed mixed farming systems :** A systems in which more than 90 percent of the value of non-livestock farm production comes from rain fed land use. This system includes-temperate (North America, Europe and north-eastern Asia) and tropical highland (eastern Africa : Ethiopia, Burundi, Rwanda) and the Andean region of

Latin America (Ecuador, Mexico), humid/sub-humid tropics and sub-tropics (Typical cases are smallholder rice-buffalo systems in southeast Asia or soybean-maize-pasture operations in the Brazilian cerrados of the large-scale commercial type) and arid/semi-arid tropics & sub-tropics (typical cases are dry land farming-sheep systems in northern Africa and in the Indian subcontinent, and also the small ruminant-cassava systems in north-eastern Brazil).

b. **Irrigated mixed farming systems:** In contrast to the previous system, in this system more than 10 percent of the value of non-livestock farm production comes from irrigated land uses. This system is further divided into-temperate (particularly in the Mediterranean region; Portugal, Italy, Greece, Albania, Bulgaria and in the Far East (North and South Korea, Japan, and parts of China) & tropical highland, humid/sub-humid tropics & sub-tropics (Typical cases are irrigated rice-buffalo systems of the Philippines, Vietnam, etc.) and arid/semi-arid tropics & sub-tropics (typical cases are lucerne/maize-based intensive dairy systems in California, Israel and Mexico; small-scale buffalo milk production in Pakistan; and animal traction based cash crop production in Egypt and Afghanistan).

India is rich in agro-ecological diversity and a range of distinct livestock production systems has evolved in each of these regions in relation to preference for different livestock species, native breed, prevailing management & marketing systems and existing natural flora and fauna. Landless (rural, urban & peri-urban industrial landless system), mixed farming (crop-animal system) and rangeland based livestock production systems are found across the various agro-ecological zones in India.

Agro-Climatic Zones

According to FAO (1983), agro-climatic zones is a land unit, in terms of major climate and growing period, that is climatically suitable for a certain range of crops and cultivars. Agro-climatic conditions chiefly refer to soil types, temperature, rainfall and water availability which influences the type of vegetations. India has been divided into 15 agro-climatic regions (Planning Commission) out of which 14 regions are in the main land and the remaining one in the islands of Bay of Bengal and the Arabian Sea (Figure 1.1). These zones were further divided into agro-ecological zones. An agro-ecological zone or region is a land resources mapping unit, defined in terms of climate, landform and soils, and/or land cover with a specific range of potentials and constraints with respect to land use. Agro-Ecological Zoning (AEZ) is one of the most important foundations for sustainable development of agricultural planning of a region. According to National Bureau of Soil Survey & Land Use Planning (NBSS&LUP) there are 20 agro-ecological zones in the country which is further subdivided into 60 sub-zones. Agro-climatic zones have a characteristic inter-relationship between farming systems and climate whereas agro-ecological regions have a characteristic inter-relationship between farming systems and various environmental features, apart from climate. The agro-climatic zones/ regions of India are

1. **Western Himalayan (WH):** States (total 39 districts) under this zone are Himachal Pradesh, Jammu & Kashmir and Uttarakhand.
2. **Eastern Himalayan (EH):** States (total 79 districts) under this zone are Arunachal Pradesh, Assam, Manipur, Meghalaya, Mizoram, Nagaland, Sikkim, Tripura and West Bengal (3 districts).
3. **Lower Gangetic Plains (LGP):** Major portion of West Bengal (15 districts) comes under this zone.
4. **Middle Gangetic Plains (MGP) :** States (total 60 districts) under this zone are Uttar Pradesh (23 districts) and Bihar.
5. **Upper Gangetic Plains (UGP):** Major portion of Uttar Pradesh (total 40 districts) comes under this zone.
6. **Trans Gangetic Plains (TGP):** States/UT (total 48 districts) under this zone are Chandigarh, Delhi, Haryana, Punjab and Rajasthan (2 districts).
7. **Eastern Plateau and Hills (EPH):** States (total 57 districts) under this zone are Chhattisgarh, Jharkhand, Madhya Pradesh (3 districts), Maharashtra (4 districts), Orissa (15 districts) and West Bengal (1 districts).
8. **Central Plateau and Hills (CPH) :** States (total 57 districts) under this zone are Madhya Pradesh (30 districts), Rajasthan (20 districts) and Uttar Pradesh (7 districts).

9. **Western Plateau and Hills (WPH):** States (total 41 districts) under this zone are Madhya Pradesh (15 districts), Maharashtra (25 districts) and Rajasthan (1 district).
10. **Southern Plateau and Hills (SPH) :** States (total 48 districts) under this zone are Andhra Pradesh (14 districts), Karnataka (21 districts) and Tamil Nadu (13 districts).
11. **East Coast Plains and Hills (ECPH):** States (total 43 districts) under this zone are Andhra Pradesh (9 districts), Orissa (15 districts), Pondicherry (4 districts) and Tamil Nadu (15 districts).
12. **West Coast Plains & Ghats (WCPG):** States (total 30 districts) under this zone are Goa, Karnataka (6 districts), Kerala, Maharashtra (6 districts) and Tamil Nadu (2 districts).
13. **Gujarat Plains and Hills (GPH) :** States (total 28 districts) under this zone are Gujarat, Dadra & Nagar Haveli and Daman & Diu.
14. **Western Dry (WD) :** State (total 9 districts) under this zone is Rajasthan. The zone has all the characteristics of a hot desert.
15. **Island Region (I):** Andaman & Nicobar Islands and Lakshdweep (total 3 districts) comes under this zone.

Source : Planning Commission (Khanna, 1989 & Singh, 2006).

Comparative Highest and Lowest Priority Agro-climatic Zones According to the Density of the Different Livestock Species (Sastry, 1994)

Cattle

a. Highest cattle densities are found in LGP, MGP and WCPH region.
b. The farmers of LGP and MGP are more resource poor and they rear cattle for draught animal power whereas in WCPH more crossbred cattle are reared especially in Kerala.

Figure 1.1 : Agro-climatic zones of India

c. Lowest cattle densities are found in WH, UGP, WD, GPH and ECPH region mainly because WH, WD and GPH are traditionally a sheep rearing area.

d. In UGP and TGP with adoption of more modern methods of agriculture dependence on cattle for DAP is decreasing so the cattle number whereas in ECPH region buffalo is more important animal for both work and milk production.

Buffalo

a. Highest buffalo densities are found in majority of Gangetic plain regions except LGP other regions with highest densities are CHP, SHP and ECPH regions as in these zones more conducive climate, terrain and more developed agriculture.

b. Lowest buffalo densities are found in WH, EH, WD, GPH, EPH, WHP and WCH regions.

c. It is evident that in the areas where agriculture is less developed, extreme climate, undulating terrain and greater economic importance of other species densities of buffalo population is low.

d. Moreover, MGP region is the only one where both densities of cattle and buffalo are higher.

Sheep & Goat

a. SHP is the only region in the country where highest densities of sheep and goat is found.

b. In CPH region both densities of sheep and goat are uniform.

c. WD region is the area where sheep density is also highest where it is reared for wool production in contrast to SHP region where it is reared for mutton.

d. Next highest sheep densities are found in TGP and ECPH.

e. Lowest sheep densities are found in WCPG, GPH, EPH and EH region. In WCPG, EPH and EH region cattle rearing is more important.

Goat

a. Highest goat densities are found in CPH, LGP and MGP regions.

b. Lowest densities are found in SPH, EPH, ECPH and EH regions.

Pigs

a. Highest pig densities are found in LGP, MGP, UGP, EH, TGP and WCPG regions and lowest in other part of the country.

b. (EH, WCPG- more Christian population and eastern Gangetic plains weaker section of society).

Horse

a. Kathiawari and Marwari- north western plains

b. Manipuri and Bhutia- mountainous region

Camel

Rajasthan, Gujarat, Haryana and Punjab.

Yak

WH & EH region

Mithun

EH region

Distribution of Land and Livestock Holdings Across Different Operational Holding in India

Distribution of livestock and size of livestock holdings across different operational holding has been shown in Table 1.2. Around 32 per cent households didn't have access to land for agriculture and classified as landless households. Share of landless households increased while share of other categories in the total household decreased over a period of ten years. Further, the ownership of livestock by different operational holding decreased except for marginal household in the total household in 2002-03 national sample survey report as compared to 1992-93 report. Size of livestock holding in landless category decreased significantly and in general size of livestock holding decreased in all categories except few cases in between 1992-2002.

Table 1.2 : Distribution of land and livestock holdings in India (2002-03).

Category	Landless < 0.002 ha	Marginal 0.002-1.0 ha	Small 1.0-2.0 ha	Medium 2.0-4.0 ha	Large > 4.0 ha	All
% households	31.9	47.1	11.2	6.2	3.4	100.0
Distribution of livestock (%)						
Bovine	0.6	51.3	21.2	15.0	11.9	100.0
Ovine	2.1	61.5	15.7	9.6	11.0	100.0
Poultry	4.4	62.7	17.4	6.8	8.6	100.0
Pigs	3.2	76.2	12.0	5.5	3.0	100.0
Size of livestock holdings, no./100 households						
Bovine	3	169	293	374	535	156
Ovine	4	84	90	99	203	64
Poultry	17	164	191	136	306	123
Pigs	0.3	5.3	3.5	2.9	2.9	3.3

Source: NSS Report No. 493, Livestock Ownership Across Operational Land Holding Classes in India 2002-03, Ministry of Statistics and Program Implementation, GOI.

Role of Livestock in Indian Economy

Livestock sub-sector is an important component of agricultural sector all over the world. In India, livestock rearing is an integral part of socio-economic fabric since time immemorial. About 73 percent of household in the rural area possess some kind of livestock and three quarters of these households belongs to small and marginal farmers. Livestock is believed as the oldest form of wealth and was once a symbol of prosperity in the society. Livestock sector plays a multidimensional role and helps in achieving nutritional security, employment generation and socio-economic development of rural sector, particularly among the landless, small, marginal farmers and women. In brief, livestock sector provides

- Draught animal power for various purposes (ploughing, transport, drawing of water etc)
- Quality animal protein
- Employment opportunities
- Source of income
- Opportunities of farm diversification
- It acts as a supplementary and complementary enterprise
- Acts as a dependable "bank on hooves" at the times of need
- Organic manure
- Domestic fuel
- Utilizes crop residues etc
- Skin & hides for leather; bones for bone meal and glands for pharmaceutical industry etc
- Acts as a tool to alleviate poverty etc

In the year of 2010-11, livestock sector contributed around 3.64 per cent (at current prices) in the total gross domestic product (GDP) of the country and about 23.80 per cent to agricultural GDP. Role of livestock in arid and semi-arid regions is of special importance as it is the main source of family income and the contribution of livestock to agricultural GDP is as high as 70 per cent in arid areas whereas in semi-arid areas it is over 40 per cent. Total value of output from livestock sector in the year 2008-09, 2009-10 and 2010-11 were Rs. 292146, 341491 and 388370 crores at current prices, respectively. Among different livestock products, milk group was largest contributor in total value of output from livestock sector (Figure 1.2). The value of export of livestock and livestock products was about Rs. 25, 540,886 lakh and it contributed around 2.2 per cent of total export earning of India during 2010-11 (GOI, 2012). The contribution of draught animal power is roughly estimated to be around Rs.10, 000 crores per year besides saving approximately six million tones of petroleum worth Rs. 6,000 crores. On an average, 800 million tones of manure is produced out of which about 300 million tones are used as fuel and remaining as manure. As many as 75 million women &15 million men are engaged in the livestock sector and milk production only involves more than 30 million small producers (GOI,2006).

Figure 1.2 : Value of output from livestock sector in India (at current prices).

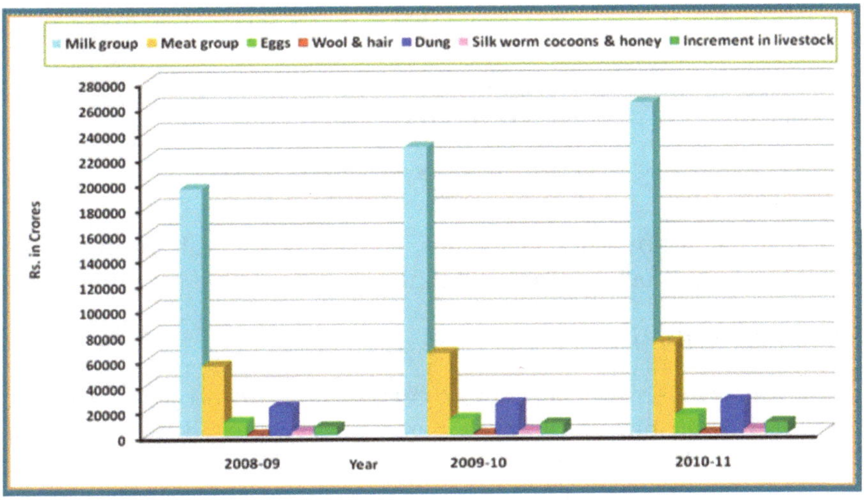

Source : GOI (2012).

Livestock Population

Livestock population of India and their annual growth rates has been given in Table 1.3.1. Number of pigs, camels, mules, donkeys, horses and ponies decreased while number of other species increased between the years 2003-2007. During this period highest per cent increase in number was observed for exotic & crossbred cattle followed by yak, sheep and goat while highest decrease in number was observed for donkeys, mules, horses & ponies, camels and pigs.

Table 1.3.1 : Livestock population of India.

Species	Population (million)		Annual Growth Rates (%), 2003-2007
	2003	2007	
1. Cattle	185.18	199.08	1.83
2. Buffalo	97.92	105.34	1.84
3. Sheep	61.47	71.56	3.87
4. Goats	124.36	140.54	3.10
5. Pigs	13.52	11.13	-4.74
6. Camels	0.63	0.52	-4.83
7. Horses & ponies	0.75	0.61	-4.98
8. Mules	0.18	0.14	-6.58
9. Donkeys	0.65	0.44	-9.40
10.Yaks	0.06	0.08	8.51
11. Total Livestock	485.00	529.70	2.23
12. Poultry	489.01	648.88	7.33

Source: GOI (2010).

In the following paragraphs, we will discuss about population of different livestock species in detail.

Cattle Population

The current population of cattle in the world is around 1427.50 million. Distribution of cattle population among different regions of world has been shown in figure 1.3.1. India stands first in cattle population (14.74% of world cattle population) in the world and other leading countries in cattle population are Brazil, United States of America, China and Ethiopia. Cattle population of world and India over a period of ten years has been shown in Figure 1.3.2.

Table 1.3.2 : Leading countries in cattle population.

Country	Population (in millions)			% of world population (2010)
	2006	2008	2010	
1. India	195.54	202.15	210.38	14.74
2. Brazil	205.89	202.31	209.54	14.68
3. United States of America	96.70	96.03	93.88	6.58
4. China	87.55	82.82	83.80	5.87
5. Ethiopia	43.12	49.00	50.88	3.56

Source: FAOSTAT (2010).

However, according to the 18th livestock census (2007), total cattle population in India was around 199.07 million which consisted of 33.06 million (16.6%) crossbred and 166.01 million (83.4%) of indigenous cattle. During 2003-07, crossbred cattle population registered an annual growth rate of 7.58% whereas it was only 0.85% for indigenous cattle population.

Among different states of the country, highest crossbred cattle population was observed in the state of Tamil Nadu followed by Maharashtra, West Bengal, Karnataka and Bihar. Indigenous cattle population was highest in the state of Madhya Pradesh followed by Uttar Pradesh, West Bengal, Maharashtra and Rajasthan. Highest cattle population was observed in the state of Madhya Pradesh (11% of total cattle population) followed by West Bengal, Uttar Pradesh, Maharashtra and Bihar.

Buffalo Population

There are 111.89 million buffaloes in the world and with a share of 58.05% in this population India ranks first. Pakistan, China, Nepal and Egypt (Table 1.3.3) are other leading countries in buffalo numbers (FAOSTAT, 2010). Distribution of buffalo population among different regions has been shown in Figure 1.3.3 and buffalo population of world and India over a period of ten years has been shown in Figure 1.3.4.

Table 1.3.3 : Leading countries in buffalo population.

Country	Population (in millions)			% of world population (2010)
	2006	2008	2010	
1. India	103.43	107.24	111.89	58.05
2. Pakistan	27.34	29.00	29.41	15.26
3. China	22.50	23.27	23.60	12.24
4. Nepal	4.20	4.50	4.84	2.51
5. Egypt	3.94	4.05	3.82	1.98

Source: FAOSTAT (2010).

However, as per 18th livestock census (2007), total buffalo population in the country was around 105.34 million. Highest buffalo population was observed in the state of Uttar Pradesh (22.6%) followed by Andhra Pradesh (12.59%), Rajasthan (10.52%), Madhya Pradesh (8.66%) and Gujarat (8.32%).

Sheep Population

Presently there are around 1078.33 millions of sheep in the world. China ranks first and India second in sheep population in the world (Table 1.3.4). Indian sheep population increased from 59.45 million in 2000 to 73.99 million in 2010.

Figure 1.3.1 : World cattle population (regional perspective, 2010).

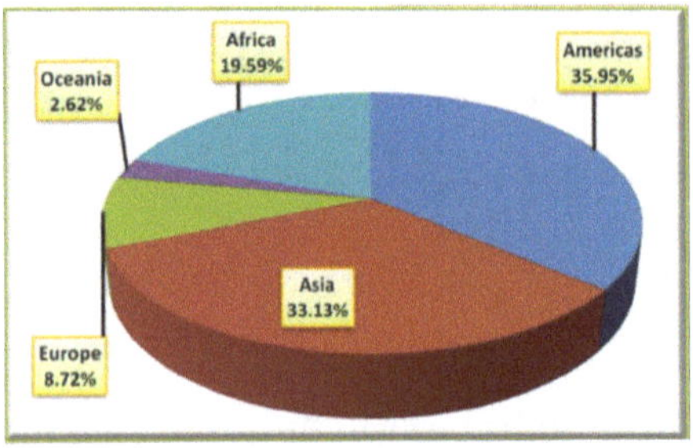

Source : FAOSTAT (2010).

Figure 1.3.2 : Cattle population of India and world (2000-2010).

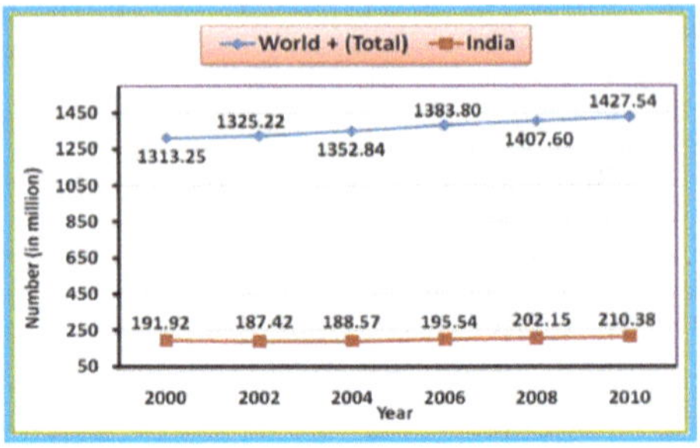

Source : FAOSTAT (2010).

Figure 1.3.3 : World buffalo population (regional perspective, 2010).

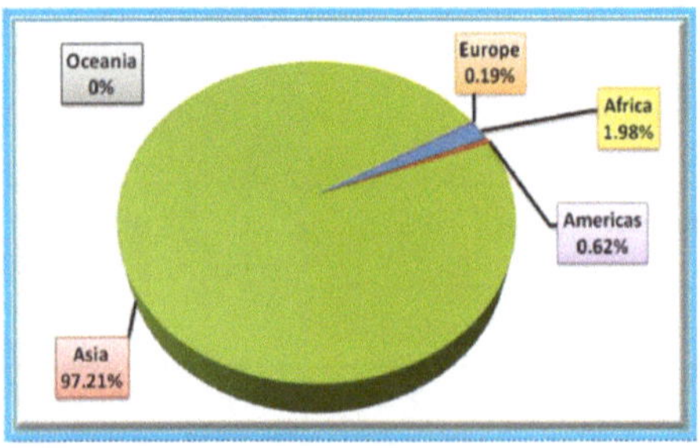

Source : FAOSTAT (2010).

Figure 1.3.4 : Buffalo population of India and world (2000-2010).

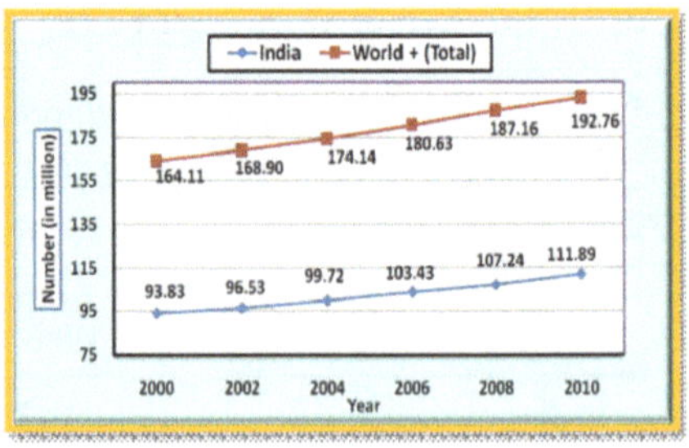

Source : FAOSTAT (2010).

Distribution of sheep population among different regions has been shown in Figure 1.3.5 and sheep population of world and India over a period of ten years has been shown in Figure 1.3.6.

As per 18th livestock census, about 71.56 million of sheep were present in India. Andhra Pradesh (35.68 %) was the leading state in sheep population followed by Rajasthan (15.63%), Karnataka (13.35%), Tamil Nadu (11.16%) and Jammu & Kashmir (5.76%).

Table 1.3.4 : Leading countries in sheep population.

Country	Population (in millions)			% of world population (2010)
	2006	2008	2010	
1. China	151.34	142.28	134.02	12.43
2. India	68.89	72.36	73.99	6.86
3. Australia	91.03	79.94	68.09	6.31
4. Sudan (former)	50.39	51.10	52.08	4.83
5. Iran (Islamic Republic of)	53.80	49.98	49.50	4.59

Source : FAOSTAT (2010).

Goat Population

At present there are 909.85 million of goats in the world and about 16.93% of this population is present in India. World's leading countries in goat population have been depicted in Table 1.3.5. Distribution of goat population among different regions has been shown in Figure 1.3.7 and goat population of world and India over a period of ten years has been shown in Figure 1.3.8.

Table 1.3.5 : Leading countries in goat population.

Country	Population (in millions)			% of world population
	2006	2008	2010	
1. India	136.29	145.00	154.00	16.93
2. China	146.86	143.60	150.71	16.56
3. Pakistan	53.79	56.74	59.86	6.58
4. Nigeria	51.21	53.80	56.52	6.21
5. Bangladesh	43.60	47.30	51.40	5.65

Source: FAOSTAT (2010).

However, according to 18th livestock census (2007), out of the total goat population (140.53 million) of the country, Rajasthan had highest number of goats (15.3%) followed by West Bengal (10.72%), Uttar Pradesh (10.52%), Maharashtra (7.39%) and Bihar (7.23%).

Figure 1.3.5 : World sheep population (regional perspective, 2010).

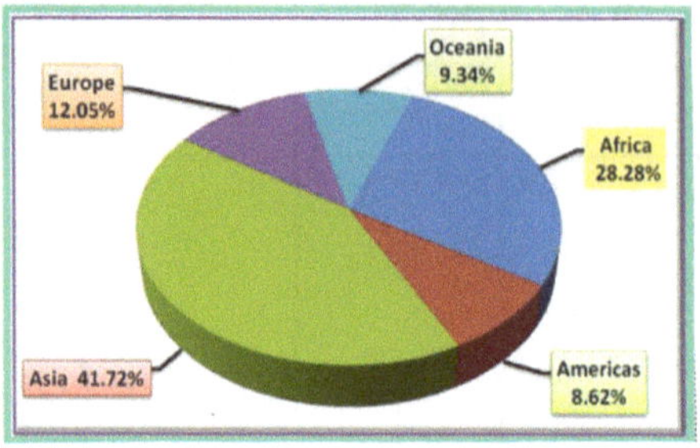

Source : FAOSTAT (2010).

Figure 1.3.6 : Sheep population of India and world (2000-2010).

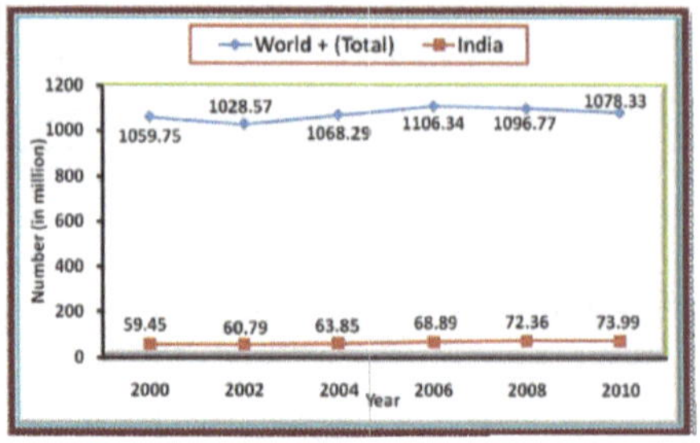

Source : FAOSTAT (2010).

Figure 1.3.7 : World goat population (regional perspective, 2010).

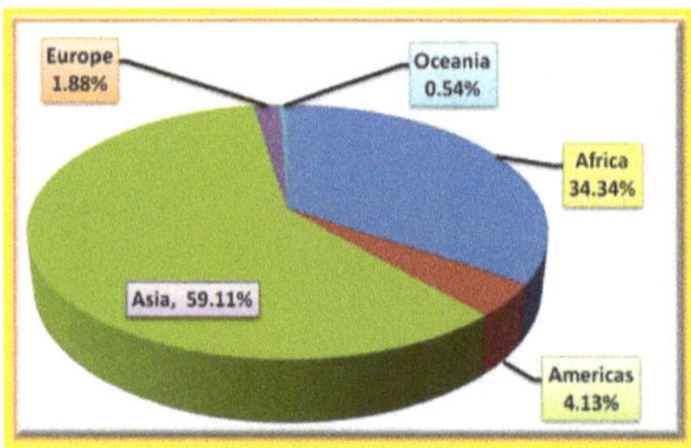

Source : FAOSTAT (2010).

Figure 1.3.8 : Goat population of India and world (2000-2010).

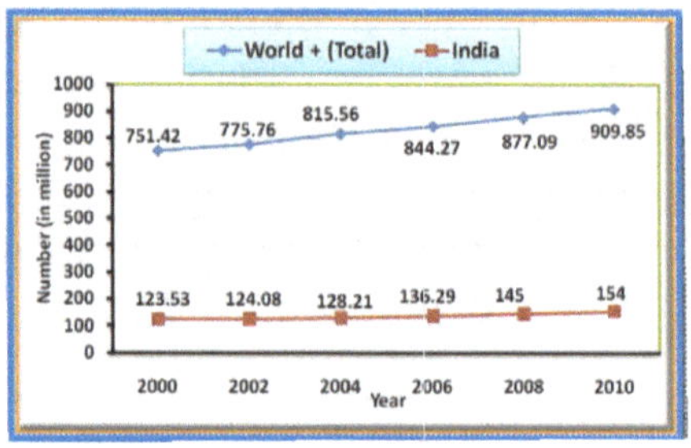

Source : FAOSTAT (2010).

Pig Population

Current pig population of the world is around 967.62 millions. China is having highest number (49.22%) of pigs in the world. USA, Brazil, Vietnam, Germany etc are the countries in which pig is present in good numbers (Table 1.3.6). Indian pig population is about 1% of total pig population of the world. Distribution of pig population among different regions has been shown in Figure 1.4.1 and pig population of world and India over a period of ten years has been shown in Figure 1.4.2.

As per 18th livestock census, India had 11.13 million of pigs. Indian pig population registered negative growth (@ -4.74 per cent per year) during 2003-2007. The crossbred pig population showed an increasing trend while indigenous pig population showed decreasing trend during this period. Among the different states of India, Assam had highest number of pigs (2 million). Uttar Pradesh, West Bengal, Jharkhand, Nagaland, Bihar and Orissa had 1.35, 0.81, 0.73, 0.69, 0.67 and 0.61million of pigs, respectively. Pig population in Uttar Pradesh showed a negative growth rate (-12.32%) whereas in Assam it registered a highest growth rate (6.71%) during 2003-2007. Annual growth rate of pig population was negative in most of the states during this period.

Table 1.3.6 : Leading countries in pig population.

Country	Population (in millions)			% of world population (2010)
	2006	2008	2010	
1. China	440.55	446.66	476.27	49.22
2. USA	61.45	68.18	64.93	6.71
3. Brazil	35.17	36.82	38.96	4.03
4. Viet Nam	26.86	26.70	27.37	2.83
5. Germany	26.52	26.69	26.51	2.74

Source : FAOSTAT (2010).

Camel Population

At present there are around 24.38 million of camels in the world. Highest number of camels are present in Somalia, Sudan, Niger and Chad. India has around 2% of total camel population of the world (FAOSTAT, 2010). In India, camels are mainly present in the state of Rajasthan (81.62%), Haryana (7.54%) and Gujarat (7.35%).

Figure 1.4.1 : World pig population (regional perspective, 2010).

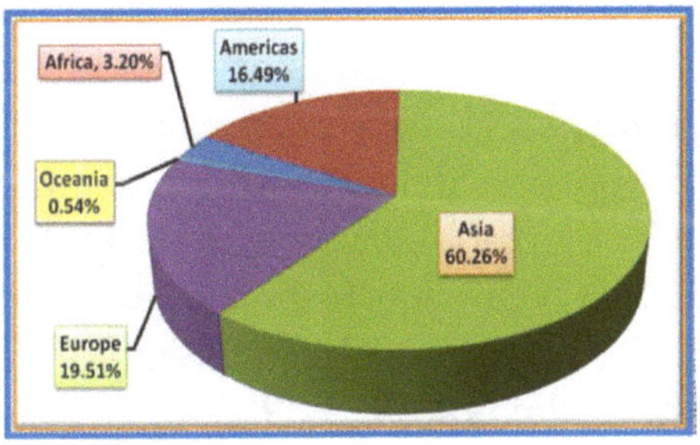

Source : FAOSTAT (2010).

Figure 1.4.2 : Pig population of India and world (2000-2010).

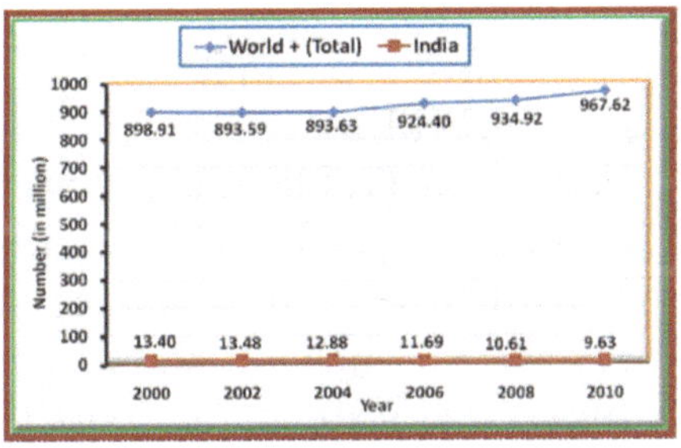

Source : FAOSTAT (2010).

Figure 1.4.3 : World chicken population (regional perspective, 2010).

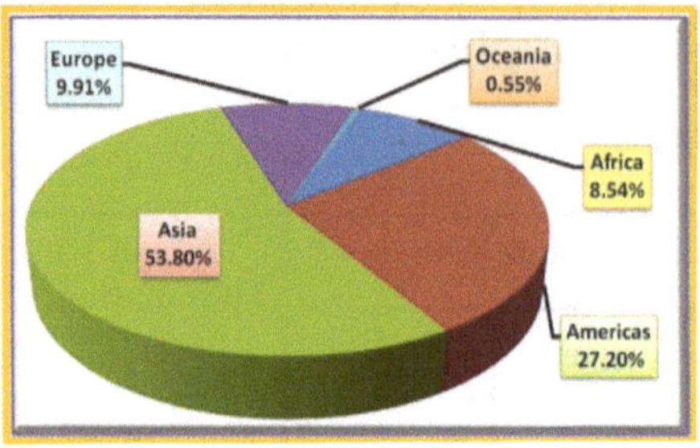

Source : FAOSTAT (2010).

Figure 1.4.4 : Chicken population of India and world (2000-2010).

Source : FAOSTAT (2010).

Other Livestock Species

As per 18th livestock census, horses & ponies were predominantly present in the states of Jammu & Kashmir (27.33%) & Uttar Pradesh (19.96%) while mules in Jammu & Kashmir (30.65%) and donkeys in Rajasthan (23.28%) & Uttar Pradesh (19.17%). Yaks were mainly (74.69%) present in Jammu & Kashmir and Arunachal Pradesh (16.86%) while Mithun in Arunachal Pradesh (82.95%) and Nagaland (12.5%).

Among different states of the country, Uttar Pradesh (11.37% of total livestock population) had highest number of livestock followed by Andhra Pradesh (11.36%), Rajasthan (10.69%), Madhya Pradesh (7.68%) and West Bengal (7%).

Poultry Population

Current population of chicken in the world is around 19606.95 million. With a share of 23.42% in total global chicken population, China ranks first (Table 1.3.7). Distribution of chicken population among different regions has been shown in figure 1.4.3 & chicken population of world and India over a period of ten years has been shown in figure 1.4.4. Total duck population of world was around 1187.68 million and India ranks sixth in duck population and has about 2.19% of global duck population (FAOSTAT, 2010).

According to the 18th livestock census, total poultry, fowl, duck and turkey population in India was around 648.83, 617.73, 27.64 and 3.45 million, respectively. Highest numbers of fowls were present in Tamil Nadu (20.53%) followed by Andhra Pradesh, West Bengal, Maharashtra and Karnataka while highest numbers of ducks were present in West Bengal (43.98%) followed by Assam, Tamil Nadu, Kerala and Andhra Pradesh.

Table 1.3.7 : Leading countries in chicken population.

Country	Population (in millions)			% of world population (2010)
	2006	2008	2010	
1. China	4451.87	4355.40	4592.72	23.42
2. USA	2050.00	2059.00	2100.00	10.71
3. Indonesia	1188.81	1253.43	1349.63	6.88
4. Brazil	1011.52	1202.02	1238.91	6.32
5. India	573.00	680.73	841.87	4.29

Source : FAOSTAT (2010).

Production Statistics

Milk Production

India is the largest milk producing country in the world (16.25 per cent of total world milk production). Leading milk producing countries of the world have been depicted in Table 1.4.1. India is largest producer of buffalo and goat milk and second largest producer of cow milk in the world. United States of America is highest producer of cow milk in the world (FAOSTAT, 2010). Contribution of different regions and species to the total world milk pool have been shown in Figure 1.5.1 & 1.5.2.

Table 1.4.1 : Leading milk producing countries of the world.

Country	Population (in millions)			% of world population (2010)
	2006	2008	2010	
1. India	99.35	108.62	116.90	16.25
2. USA	82.46	86.18	87.47	12.16
3. China	36.47	40.20	41.15	5.72
4. Pakistan	31.21	33.26	35.49	4.93
5. Brazil	26.33	28.58	30.86	4.29
World (Total)	668.82	697.80	719.22	--

Source: FAOSTAT (2010).

Species contribution

As per GOI (2012), total milk production in India (2010-11) was around 121.8 million tonnes (Figure 1.5.3). Milk production in India registered annual growth rate of 4.22% during 2000-01 to 2010-11. Buffaloes, cows and goats contributed around 62.35 (51.17%), 54.90 (45.05%) and 4.59 (3.77%) million tonnes to the total milk pool of the country, respectively. Further, out of the total milk produced by cows, indigenous cows contributed around 25.34 (46.16% of cow milk) million tonnes while contribution from crossbred/exotic cows was around 29.55 (53.83% of cow milk) million tonnes. Average yield per animal in milk for crossbred/exotic, indigenous cow, buffalo and goat was around 6.63, 2.2, 4.58 and 0.40 kg/day, respectively.

Contribution from different states

Leading states in milk production and their share in the total milk pool of the country has been shown in Figure 1.5.4. Leading state in buffalo milk production was Uttar Pradesh (23.24%) followed by Andhra Pradesh, Rajasthan, Punjab and Gujarat while Tamil Nadu (10.96%), Uttar Pradesh, Rajasthan, Maharashtra and West Bengal were leading states in total cow milk production. Tamil Nadu registered highest milk production from crossbred cows. Leading state in goat milk production was Rajasthan. Average yield per animal in milk (kg/day) for crossbred/exotic (10.95), indigenous cow (6.5), buffalo (8.59) goats (1.48) was highest in Punjab.

In the year 2010-11, per capita availability of milk was highest in Punjab (937 g/day). Other states in which per capita availability of milk was higher than national average (281g/day) were Haryana (679g/day), Gujarat (435g/day), Rajasthan (538 g/day), Uttarakhand, Jammu & Kashmir, Himachal Pradesh, Andhra Pradesh, Uttar Pradesh and Madhya Pradesh.

Problems and Prospects of Dairy Industry in India

Prospects of dairy industry in India

- Increased milk production with consequently increased availability of milk for processing.
- Increasing demand for milk as well as milk products.
- Large domestic demand.
- Low labour cost.

- Dairy sector open for investment by private and foreign players.
- Enhanced export potential for milk products of western as well as traditional types.
- Improved transportation facilities for milk and milk products.
- Availability of indigenously manufactured equipment.
- Increasing number of dairy plants in public, cooperative and private sector.
- Availability of highly trained and qualified manpower.
- Availability of diverse germplasm with ability to survive and produce under tropical climate.
- Availability of animal production technologies for quicker improvement and effective implementation.

Problems of dairy industry in India

- Low genetic potential of animals.
- Tropical climatic conditions.
- Seasonal fluctuations in milk production.
- Species-wise variation in milk quality received at milk plants.
- Dairy sector mainly dominated by informal and unorganized players.
- Lack of marketing avenues for the dairy produce.
- Introduction of foreign products in Indian market.
- Weak marketing for processed dairy products.
- Increasing chemical contaminants in the milk.
- Poor microbiological quality of milk.
- Inadequate availability of feeds and fodders.
- Excessive grazing pressure on marginal and small community lands.

Figure 1.5.1 : World milk production (regional perspective, 2010).

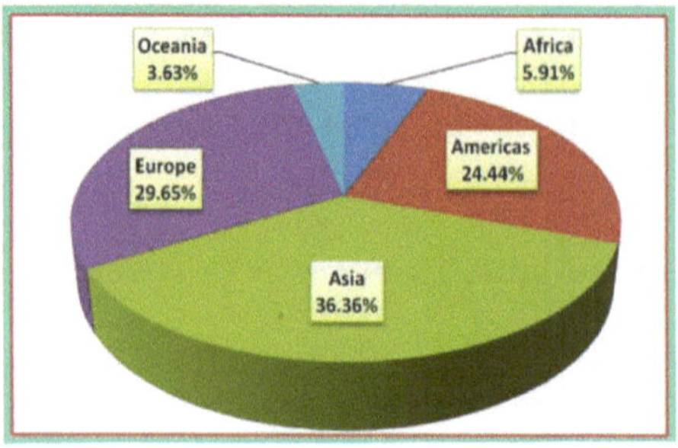

Source : FAOSTAT (2010).

Figure 1.5.2 : Contribution of different live stock species in world milk production (2010).

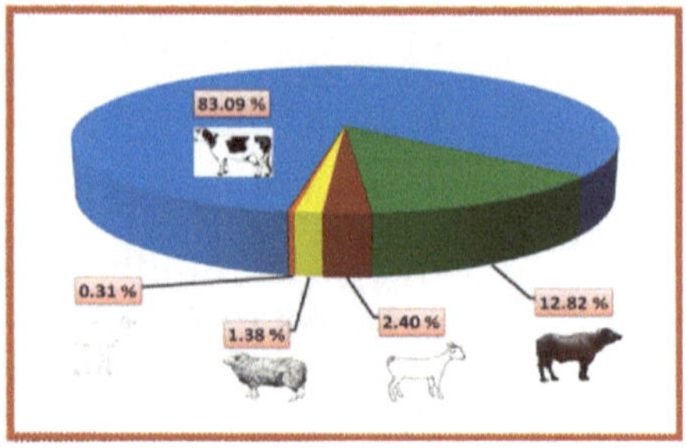

Source : FAOSTAT (2010).

Figure 1.5.3 : Milk production in India.

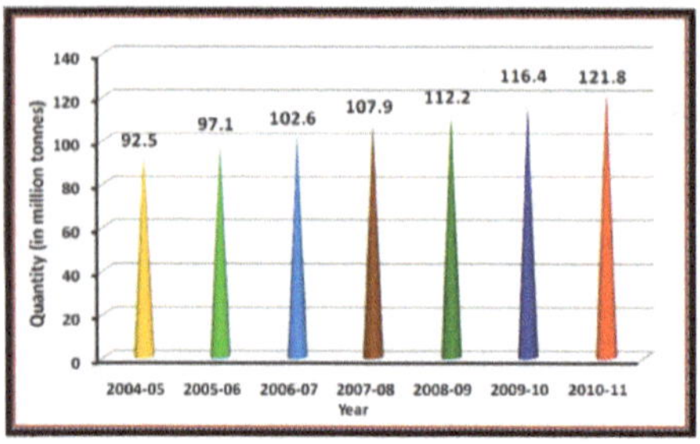

Source : GOI (2012)

Figure 1.5.4 : Leading milk producing states in India (2010-11).

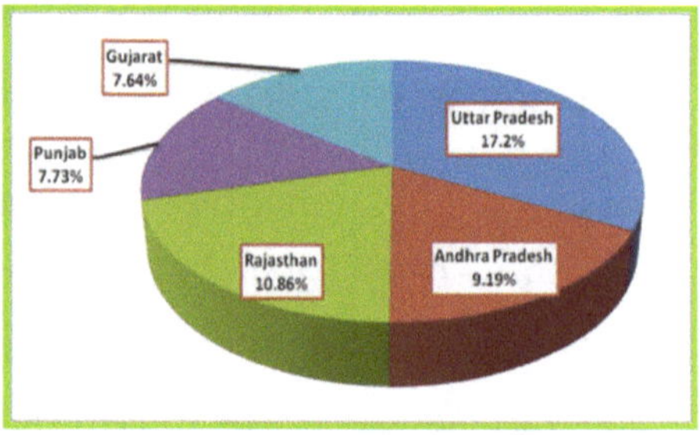

Source : GOI (2012)

Meat Production

At present global total meat production is about 296.11 million tonnes and about 41.71 % of this volume is being contributed from Asia (Figure 1.6.1). Figure 1.6.2 shows status of meat production in world and India. Pig and chicken together contributes more than 65% of total meat produced in the world (Figure 1.6.3). China ranks first in total meat (27.40%) production as shown in Table 1.4.2.

As per FAOSTAT (2010), India is leading country in buffalo meat production (42.55% of world's buffalo meat). China is leading country in chevon (25.11%), mutton (36.83%), and pork (47.25%) production while USA is largest (19.53%) chicken meat producing country in the world.

According to GOI (2012), total meat production from India was around 4.9 million tonnes (2010-11). Contribution of different species of livestock has been shown in Table 1.4.3. Poultry meat was largest contributor to the total meat pool followed by goat & buffalo meat (Figure 1.6.4). Average yield per animal for cattle, buffalo, sheep, goat, pig and poultry was around 89, 104, 13, 10, 38 and 1.2 kg, respectively.

Table 1.4.2 : Leading meat producing countries of the world.

Country	Population (in millions)			% of world population (2010)
	2006	2008	2010	
1. China	72.69	74.58	81.12	27.40
2. USA	40.64	43.00	42.17	14.24
3. Brazil	20.44	22.86	23.63	7.98
4. Germany	7.02	7.71	8.22	2.78
5. France	5.30	5.83	5.75	1.94
World + (Total)	267.58	282.86	296.11	

Source : FAOSTAT (2010).

Table 1.4.3 : Meat production in India.

Species	Meat Production (in 000 tonnes)		
	2008-09	2009-10	2010-11
1. Cattle	250	223	211
2. Buffalo	623	670	805
3. Sheep	312	341	369
4. Goat	786	814	846
5. Pig	388	391	402
6. Poultry	1884	2087	2193
Total	4280	4566	4869

Source : GOI (2012).

Figure 1.6.1 : World meat production (regional perspective, 2010)

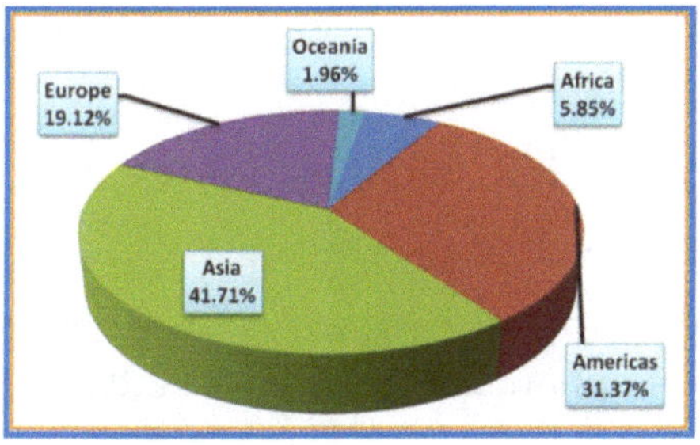

Source : FAOSTAT (2010).

Figure 1.6.2 : Meat production from India & world.

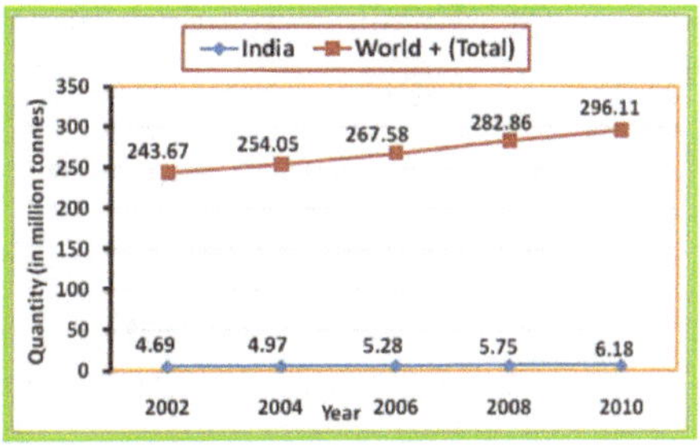

Source : FAOSTAT (2010).

Figure 1.6.3 : Contribution of different livestock species in world meat production (2010).

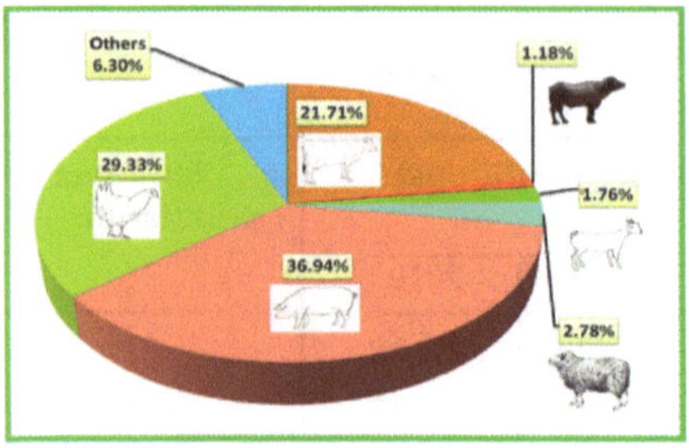

Source : FAOSTAT (2010).

Figure 1.6.4 : Contribution of different live stock species in meat production from India (2010-11).

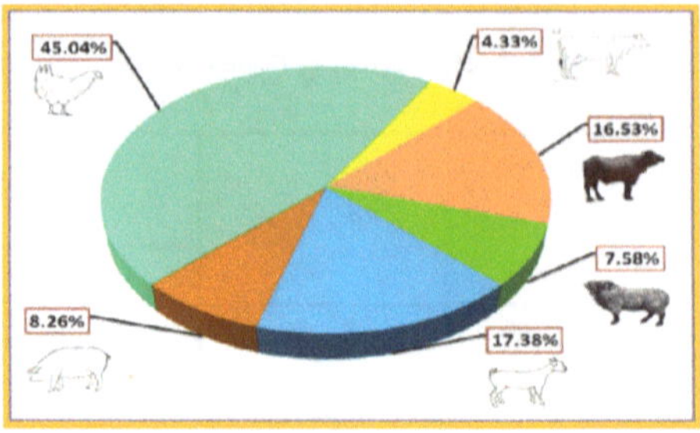

Source : GOI (2012)

Contribution from different States

In the year 2010-11, share of Uttar Pradesh, Andhra Pradesh, West Bengal, Maharashtra and Tamil Nadu in the total meat production of India was about 17.35, 15.34, 11.85, 11.56 and 9.57 %, respectively. Leading state in cattle, buffalo, sheep, goat, pig and poultry meat production was Kerala, Uttar Pradesh, Andhra Pradesh, West Bengal, Uttar Pradesh and Andhra Pradesh, respectively (GOI, 2012).

Problems and Prospects of Meat Industry in India

Prospects of meat industry in India

- Availability of large number of animals for meat purpose.
- Low production cost.
- Animals are generally raised on pasture and crop residues hence raised on green livestock production system.
- Increasing level of farm mechanization.
- Indian livestock is free from Bovine Spongiform Encephalitis, Rinderpest and CBPP.
- Indian meat remains free from radiation.
- Fat and cholesterol content of Indian buffalo meat is low.
- Indian Buffalo meat blends well with other meat.
- In India, generally hormones, antibiotics and growth promoters are not used for enhancing growth and fattening of livestock.

Problems of meat industry in India

- Lack of modern slaughter houses.
- Lack of productivity augmenting technologies.
- Poor phyto-sanitary and quality standards.
- Unorganized market.
- Lack of vertical integration between production & processing.
- Poor genetic potential.
- Lack of disease free zones
- Tropical climate.
- Socio-religious taboos.
- Low feed efficiency.
- Premature slaughtering of animals for meat.
- Discrimination against Buffalo.

Wool Production

Presently global total wool (greasy) production is about 2.05 million tonnes. Asia is the largest producer of wool (41.23% of total production). World's leading wool producing countries have been depicted in Table 1.4.4. India has produced around 2% of global wool production in the year of 2010 (FAOSTAT, 2010).

Table 1.4.4 : Global wool (greasy) production.

Region/Country	Production (in million tonnes)				
	2002	2004	2006	2008	2010
World + (Total)	2.16	2.20	2.26	2.18	2.05
A. Asia + (Total)	0.73	0.84	0.86	0.85	0.84
B. Oceania + (Total)	0.82	0.73	0.74	0.66	0.55
C. Europe + (Total)	0.25	0.26	0.26	0.27	0.27
D. Africa + (Total)	0.19	0.21	0.22	0.23	0.23
E. Americas + (Total)	0.18	0.17	0.19	0.18	0.15
a. China	0.31	0.37	0.39	0.37	0.39
b. Australia	0.59	0.51	0.52	0.46	0.38
c. New Zealand	0.23	0.22	0.22	0.20	0.17
d. Argentina	0.07	0.06	0.07	0.07	0.05
e. Russian Federation	0.04	0.05	0.05	0.05	0.05

Source : FAOSTAT (2010).

In India, highest wool production was recorded in the year of 2002-03 (50.5 million kg), thereafter it showed mixed trend and presently it is around 43 million kg. Wool production in India registered an annual growth rate of -1.18% during 2000-01 to 2010-11. Estimates of wool production from India over the period of 10 years and contribution of different states in total wool production have been shown in Figure 1.7.1& 1.7.2.

Problems and Prospects of Wool Industry in India

Prospects of wool industry in India

- Availability of large local market.
- Huge export potential.
- Availability of large pool of skilled and cheap work force.
- Huge woollen textile production capacity.
- Efficient multifiber raw material manufacturing capacity.
- Entrepreneurial skills.

Figure 1.7.1 : Wool production in India.

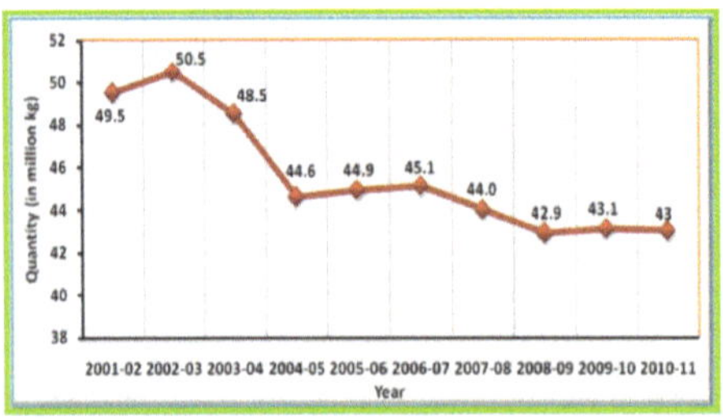

Source : GOI (2012)

Figure 1.7.2 : Leading wool producing states in India (2010-11).

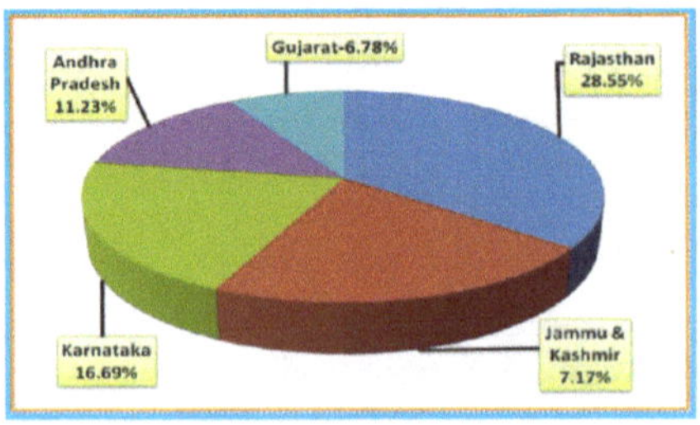

Source : GOI (2012)

Figure 1.8.1 : Egg production in India.

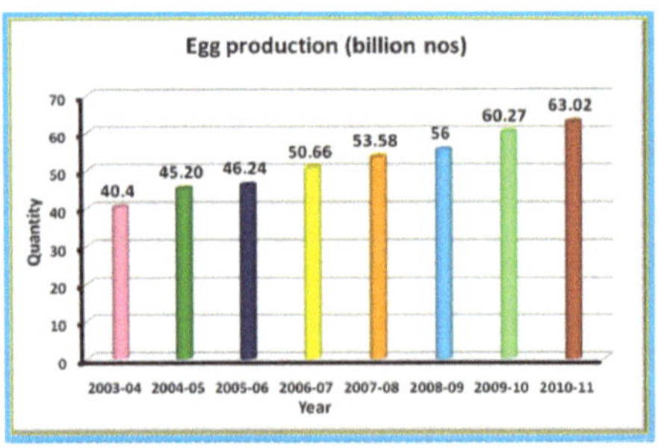

Source : GOI (2012)

Figure 1.8.2 : Leading egg producing states in India (2010-11).

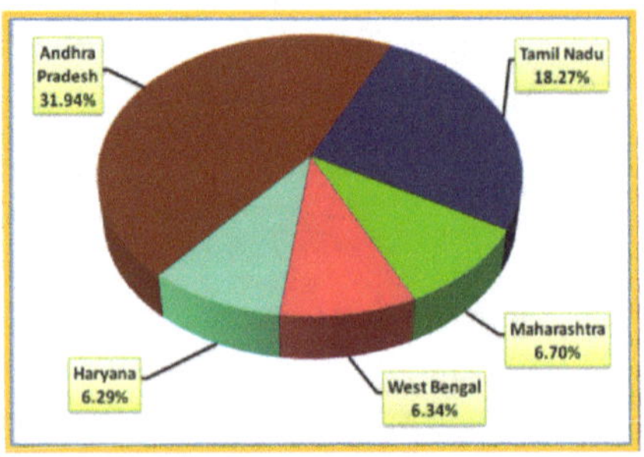

Source : GOI (2012)

Problems of wool industry in India

- Low productivity.
- Use of outdated manufacturing technology.
- Shortage of grazing lands.
- Poor supply chain management.
- Large unorganized and decentralized sector.
- High production cost.
- Low quality product mainly used to manufacture hand-knitted carpets.
- It is small in size and widely scattered.

Egg Production

At present, global egg production is about 1198.89 billion numbers. Asia has contributed more than half of this number (Table 1.4.5). With a share of about 40 % in world`s total egg production China ranks first. Other leading countries in egg production are USA (7.63%), India (5.12%), Mexico (3.97%) and Japan (3.49%).

Table 1.4.5 : Global Hen egg (in shell) production

Region/country	Number (in billion)				
	2002	2004	2006	2008	2010
World + (Total)	1007.65	1044.41	1088.22	1162.45	1198.89
A. Asia + (Total)	585.25	613.23	637.47	702.87	727.14
B. Americas + (Total)	200.65	207.03	224.89	228.75	234.92
C. Europe + (Total)	171.93	173.53	173.40	174.19	178.62
D. Africa + (Total)	46.27	47.15	48.22	52.56	53.82
E. Oceania + (Total)	3.55	3.46	4.24	4.08	4.39
a. China	393.18	410.01	418.72	465.48	476.79
b. USA	87.22	89.09	91.80	90.01	91.48
c. India	39.82	45.20	50.66	55.40	61.42
d. Mexico	38.01	40.03	45.80	46.74	47.62
e. Japan	42.16	41.35	41.61	42.57	41.90
f. Russian Federation	36.11	35.56	37.65	37.80	40.39

Source : FAOSTAT (2010).

However, as per GOI (2012), total egg production from India was estimated around 63 billion numbers (Figure 1.8.1) and per capita availability was around 53 eggs per annum (2010-11). During 2000-01 to 2010-11, egg production registered an annual growth rate of 5.58%. Out of the total egg produced, fowls contributed around 61.4 billion eggs where as ducks contributed around 1.6 billion eggs. During this period average number of egg produced per layer (fowl) per year was 202.

Contribution from different States

In the year 2010-11, Andhra Pradesh was leading state in egg production followed by Tamil Nadu, Maharashtra, West Bengal and Haryana (Figure 1.8.2). These five states produced about 70% of the total egg produced in the country. West Bengal was leading state in egg production from ducks (71% of total duck eggs). Average number of egg produced per layer (fowl) per year was highest in Punjab (280) followed by Gujarat (250), Haryana (249) and Andhra Pradesh (246). Per capita availability of eggs was highest in Andhra Pradesh (232) followed by Tamil Nadu and Haryana (GOI, 2012).

Livestock and Livestock Products Market and Marketing

In India there are over 2000 livestock markets and most of them are under the jurisdiction of state governments and union territories though local bodies like village panchayats and municipal corporations look after the direct operation and supervision work. Apart from these government livestock markets few privately owned markets also exists.

Livestock market is not well developed and facilities are mostly either inadequate or poorly maintained. There are no species specific markets. Farmers generally bring their animals to the weekly/ fortnightly/ monthly/ half yearly/annual markets/animal fairs mostly on foot and sometimes by motor vehicles. Milch cattle, buffaloes and draught animals are assembled in the same yard. Most of the trade occurs through brokers. Buyer perform animal grading based on breed, age, appearance, milk yield, lactation number and meat yield etc and prices are estimated and final price is arrived at through negotiations by direct or indirect involvement of brokers or through auction. Apart from this local trader buy the animals from villages and sell it in the market by haggling over price. Sometimes agents, middlemen visit the villages and pay advance cash to the farmers and later on animals are collected for sale.

Figure 1.9.1 : Marketing channels for livestock in India.

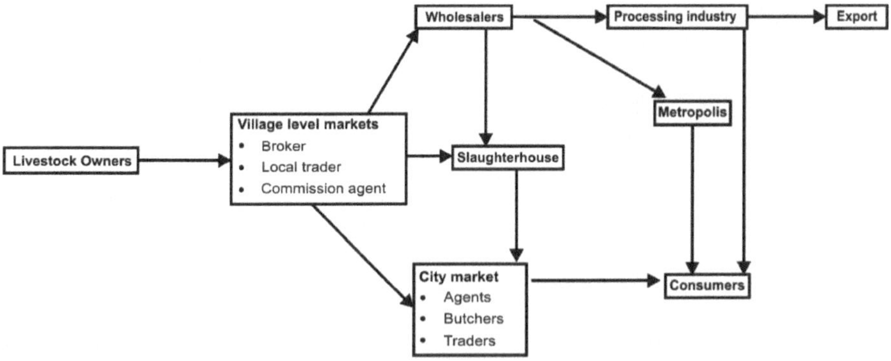

Marketing of meat involves many agencies, intermediaries and middlemen. It is the middlemen who in most of the trades earns major portion of the total price charged to the consumers. In majority of the cases, municipal bodies manage the slaughter houses and state governments mainly operates them except some of the private owned slaughter houses supplying meat to the other countries. There are about 3900 slaughter houses in the country which are licensed by local bodies. Most of the slaughter houses operate as a service abattoir. Butchers slaughter the animals for nominal fee and municipal inspectors certify it fit for human consumption.

Hides and skins are collected from villages, towns and cities and are supplied to the major terminal markets (primarily) and district level markets.

In India, milk is collected, processed and sold by various agents engaged in the dairy business from the farmers doorstep to the consumers either in the form of fresh liquid milk or processed milk/ milk products. These agents can be classified into organized or formal (dairy cooperatives, government run milk producing units and large private dairies) and unorganized or informal segment (village vendors, local halwais, curd and tea shops, contractors and small-scale private dairies etc). Though, the share of organized sector in the market is increasing steadily over the years still informal sector handles significantly higher volume of milk than the formal sector in total volume of milk going to the market. Marketing channels for milk in India has been illustrated in Figure 1.9.2.

Figure 1.9.2 : Marketing channels for milk in India.

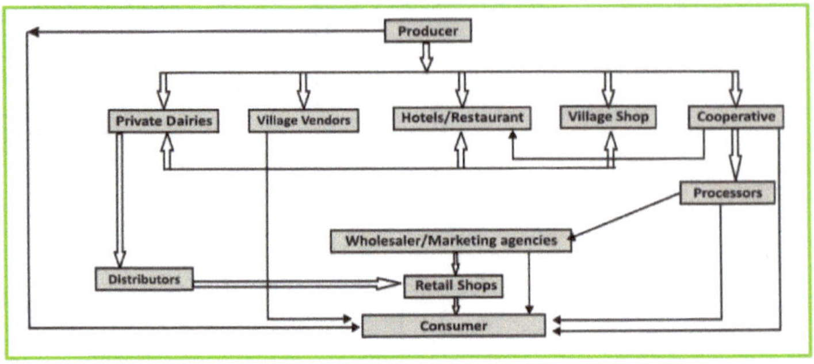

In India, there are regional differences in the quality of wool produced. Fine quality wool is produced in colder regions whereas poor quality wool is produced in Southern India. Out of the total wool produced in the country, around 85% is carpet grade, 10% is coarse grade and remaining 5% is apparel grade. They are used in manufacture of hand-knitted carpets, apparel, blankets, finished textiles, garments, knitwear, etc. The world average for wool productivity has been about 2.4 kg/sheep/year however it is only about 0.8 kg/sheep/year in India. Wool selling is a seasonal process. There is no organized market for wool in India and wool marketing channels (Figure 1.9.3) vary from state to state. Wool marketing is mainly in the hands of the private wool merchants & traders and wool passes through various intermediaries before reaching the consumers.

Figure 1.9.3 : Marketing channels for wool in India.

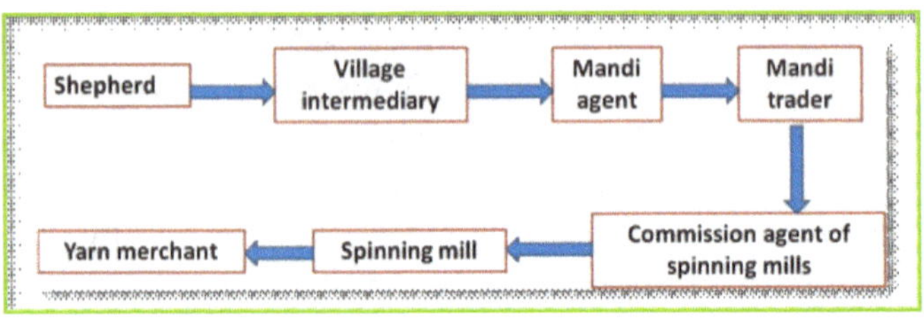

Local dealer, agents, brokers, merchants etc purchase wool from farmers by negotiations or open auction as per the fleece basis or grading of wool. Wool is brought to the market and sold through commission agents. The government buys around 10-15 per cent of marketed wool. There are ten important wool market in India which can be divided into producer market and resale market. Bikaner, Beawar and Abohar are mainly producer market whereas Ludhiana and Panipat is mainly a resale and dual purpose market, respectively.

Chapter - 2

General Principles of Livestock Management

Rana Ranjeet Singh

Body Parts of Farm Animals

It is important to know about the various terms used to indicate different body parts of the animals so that whenever a description of the animal/body part is presented, one could specifically understand it. Some of the terms used to point out different body points of common livestock species have been shown in Figure 2.1.1-2.1.5. Terms which are more frequently used for different livestock species have been depicted in Table 2.1.1.

Approach, Handling and Restraining of Livestock

Approaching Animals

Before approaching an animal always enquire about their temperament (quite or vicious) towards the stranger and watch out for warning signs of animal fierceness, especially with bulls, newly calved cows and horses. Always try to speak to the animal before touching them and don't carry a stick with you. Usually, animals are approached from behind and on their left side, touch the shoulder and neck first

Figure 2.1.1 : Body parts of a cow.

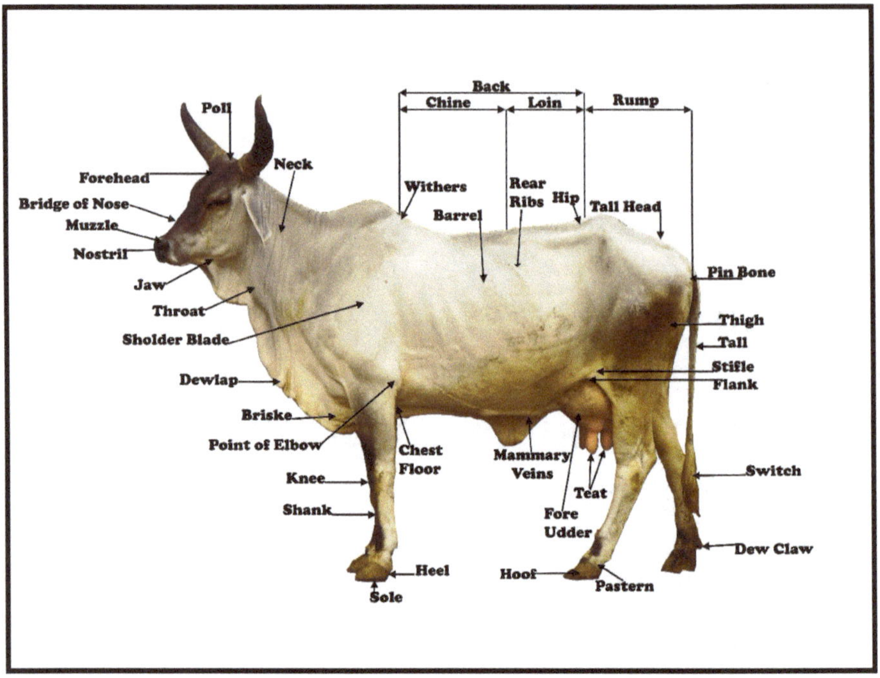

Figure 2.1.2 : Body parts of a sheep.

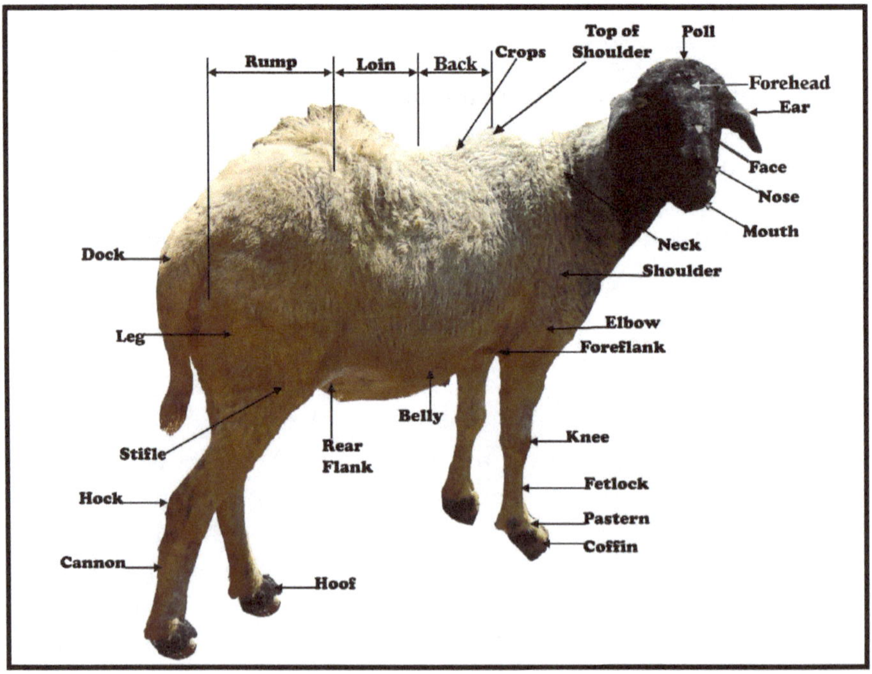

Figure 2.1.3 : Body parts of a goat.

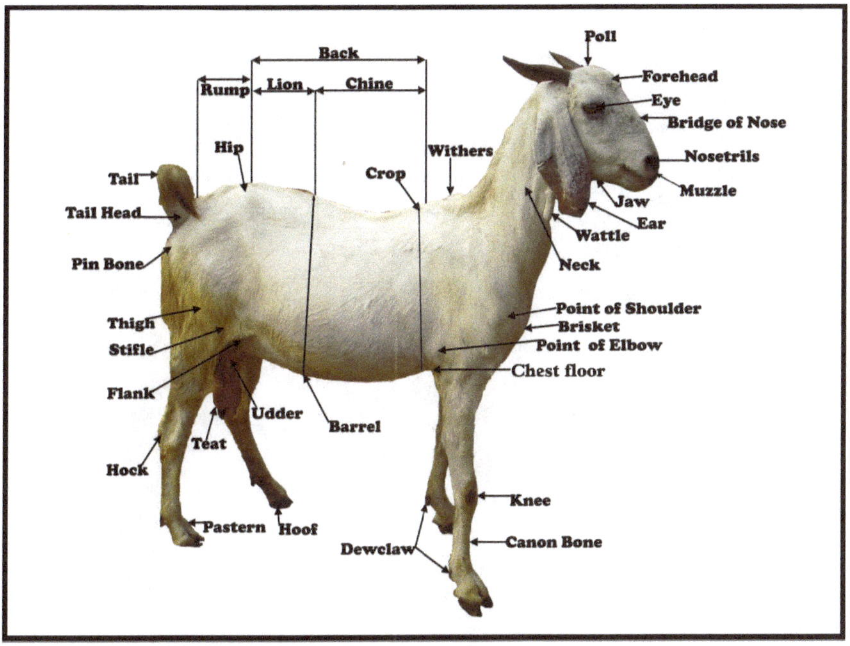

Figure 2.1.4 : Body parts of a pig.

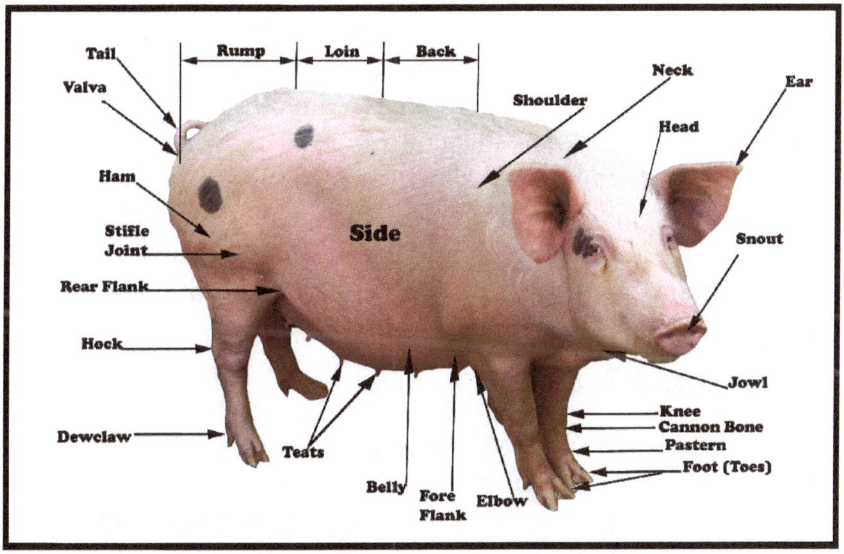

Figure 2.1.5 : Body parts of a horse.

with firm hand and if required head region may be approached. However, they can also be approached from either the left or right shoulder area but never approach an animal from the front unless it is secured in a crush. Handle strange animals with suspicion without letting the animal to know that you are afraid of them (*i.e.* desired actions should be performed slowly with firm hold) and always have the owner/person taking care of animal alongside you while handling this type of animals. Caution should always be exercised when working with the bulls. In anticipation of a dangerous situation always plan an escape route.

Handling and Restraint

Farm animals are handled routinely on day to day basis to perform various operations (vaccination, weighing, ear tagging, branding, hoof trimming, shoeing, drenching, shearing, clipping, castration, dehorning and various surgical operations etc) which requires certain degree of control over animal's movement and activity. The process of restraint should minimize the danger to the handler, animal, should not cause unnecessary pain & fright to the animal and should allow the handler to perform important management technique. A wide variety of tools

Table 2.1.1 : Common animal husbandry terms.

Details	Cattle	Buffalo	Sheep	Goat	Pig	Horse	Chicken
Species called as	Bovine	Bovine or Bubaline	Ovine	Caprine	Porcine	Equine	Gallus
Group of Animals	Herd	Herd	Flock/Drove /Band (larger)	Flock/ Band / Trip	Drove/ Stock / Herd	Pack	Flock
Adult breeding Male	Bull	Buffalo Bull	Ram/Tup	Buck/ Billy	Boar/Bran /Hogg	Stallion/ Stud (above 4 years of age)	Cock
Adult Female	Cow	She/Cow buffalo	Ewe	Doe/nanny	Sow	Mare	Hen
Young Male	Bull calf (young ox from birth to 6 or 9 months of age)	Buffalo bull calf	Ram lamb/ Tup lamb/Pur lamb (birth to weaning)	Buckling/ Male kid	Boarling	Colt foal (up to one year of age)	Cockerel
Young Female	Heifer calf (young female from birth to 6 or 9 months of age)	Buffalo heifer calf	Ewe lamb/ Gimmer lamb (birth to weaning)	Goatling/ Doeling	Gilt	Filly foal (up to one year of age)	Pullet
Growing Female	Heifer (A year-old female up to the 1st calving)	Buffalo heifer	Hogget/ Sheeder ewe/ Ewe teg (weaning to shearing)	Doeling	Gilt/Yelt (up to her first litter)	Filly (under 4 years of age)	-----

Contd.

New-born	Calf	Buffalo calf	Lamb	Kid	Piglet/ Pigling	Foal	Chick
Castrated male	Bullock/Steer	Buffalo bullock	Wether/ Wedder	Castrated	Stag steg, or seg (castrated late in life) and Hog/Barrow (castrated before reaching maturity)	Gelding/ Geld	Capon
Act of parturition	Calving	Calving	Lambing	Kidding	Farrowing	Foaling	Oviposition.
Act of mating	Serving	Serving	Tupping	Serving	Coupling	Covering	– – -
House / Dwelling	Shed/Yard/ Barn	Shed/Yard/ Barn	Pen/ Yard	Pen/Yard	Sty/ Piggery	Stable	Pen/ Cage
Meat	Beef/Veal	Buffalo beef/ Carabeef	Mutton	Chevon	Pork	Meat	Meat

are available which can be used to restraint an animal. An animal handler must be familiar with these wide varieties of tools for handling the animal safely, humanely and effectively. These tools have been classified broadly into different categories as- a. psychological, b. sensory diminishments, c. chemical sedation or immobilisation, d. use of confining chutes and alleys and barriers and e. use of equipments/tools and physical force etc .

a) **Psychological restraint:** Each domestic animal species displays its own behavioural pattern, level of apprehension and other traits (leader follower relationship in sheep, pulling back of rope/snare when applied on upper jaw of pig, voice commands of handler etc). Hence, these psychological attributes of the animal should be understood properly and implemented correctly into the restraint practices which will result in easier and efficient restraint procedures with minimal stress and consequently better welfare to the animal.

b) **Sensory diminishments:** Reducing or eliminating an animal's visual communication with its environment (blindfolding the domestic horse and pig) make it possible to introduce it into a new environment, such as a trailer or a new stall without inducing fright. Similarly use of cotton ear plug, touching animals in the proper direction in the appropriate areas of the body may act as important tool to restrain them.

c) **Chemical sedation or immobilisation:** Chemical restraint is used to quietened very vicious and highly nervous animals before casting. It minimizes struggling which will lead to easy application of ropes and appliances. Different agents are used for chemical restraint (xylazine, ketamine, detomodine, diazepam, zolazepam, tiletamine etc alone or in combination) but none of them are satisfactory in all cases.

d) **Use of confining chutes, alleys and stock etc:** All these devices act as a tool of restrain by confining the animal to a smaller area which results in easy handling of animals. However, the suitable degree of confinement may vary noticeably, depending on the species and the situation.

    i) **Chute:** It provides immobilization via a head catch and squeezable sides. The side panels can be dropped to examine feet and legs, whereas the side bars can be dropped to examine the dorsum of the animal. Many models and brands are available commercially in the market farmer can choose one as per their requirements.

ii) **Stocks/Trevis/Crush:** It is made up of hard seasoned hard wood or metal tubing. Different types of crush are used. Usually, it consists of four posts well driven into the ground and cemented there, with two or more crosspieces at each side and end. Front end may have two or three removable cross bars whereas rear end has one such bar. There may be cross beams too.

iii) **Alley way:** It is the narrow passage area that restricts cattle from turning around when they approach the chute. To prevent the backing up in the alley way poles or back gates can be used.

e) **Use of equipments/tools and physical force:** Various types of equipments/tools/physical force is used to restrain the animal. We will discuss some of them (species wise) in the following section.

Figure 2.2.1 : Devices used to confine the animal to a smaller area.

Handling and Controlling of Cattle and Buffalo

Cattle & buffaloes can be driven simply by arm gesture and other similar techniques. As they are herd oriented hence just move few animals of a herd others will follow them. You can merely move the animal by entering into their flight zone (refers to how close you can get to an animal before it moves away). Cattle have wide angle vision and a blind spot directly behind them so to move them forward the handler must enter the flight zone from behind at 45^0 to the point of balance at shoulder and in front of the point of balance to move them backwards. Handling of cattle should be gentle because they have very good memory and learn quickly to differentiate between good and bad handler. Following restraint tools/techniques can be used to control them.

a) **Rope:** Rope is one of the simplest and most frequently used tools for restraining animals. Ropes can be made of cotton, hemp, coir, nylon, polyester, polypropylene and polyethylene fibres etc. Type of rope chosen by farm manager will depend on type of work, species, age of the animal and strength required. Hard and rough rope may cause abrasion to the animals hence should be avoided. Cotton rope are soft, flexible, causes least rope burn and intermediate in cost. Two sets of ropes should be kept ready at farm. i) for casting and securing animals-it should be around eight meter length and about 2.5 to 3.0 cm thickness ii) for securing small animals, horn and legs- it should be about two meter in length and around 1 to 1.5 cm in thickness.

b) **Halters:** Halter is one of the commonest and least callous of all restrain methods used for controlling large animals. Halters are made of leather but simple ones can be made from rope too. Rope halters are used to lead, tie, secure or steady the head of the animals. For most of the domestic animals excellent rope halters are available commercially. A temporary rope halter can be prepared by making a small loop at one end of the rope and second loop about ten centimetres below the first loop. Then, free end of rope is passed through first and second loop sequentially to complete the halter. Temporary rope halters can be modified to fit any species and size of the rope should be varied as per the strength of the animal. Further, a lasso can be improvised into a halter by throwing a lasso around animal's neck then about 25 centimetre fold of free end of lasso is passed through the lasso loop twice behind the lower jaw to make a nose piece.

Alternatively, a rope halter can be made by taking about three to four meters long rope and laying it on the floor. Now, at about one meter from the left hand end, strands are untwisted sufficiently to enable the right hand part to be pass through it so that loop of about 2.5 cm diameter can be made. Then rope is laid down in the similar manner as it was done earlier. Now, another untwisting is done close to the loop (on right hand side) and left end is passed through it. Now it is pulled tightly and splice is tied with a few turns of string. To form the poll-piece, strands of the right hand end is untwisted at about 25 to 30 cm from the loop already made and left hand end is passed through it for a short distance. An overhand knot is made at the left hand end to fix and fastened on either side with string. Now long right hand end is passed through spliced loop first made, that will form the nose. To suit different categories of animals, the knot in poll-piece can be adjusted to fit the head. The adjustable portion should pass under the chin and not across the nasal bridge. The lead portion of the rope should be on left side of the animal.

Figure 2.2.2 : Halters.

c) **Nose tong:** Animals are restrained by grasping the nasal septum between thumb and finger through the nostrils, forming a nose tong but to restrain a large animal in this manner is difficult hence metallic nose tong is used. To reduce damage to nasal cartilage tongs having a smooth surface and space between the balls are used. The nose tongs should be placed in the nostrils with a rotating motion i.e. first, one side of the tong is inserted then rotated across the nostril and inserted in opposite side then tong is quickly closed. To keep the tong in place, pressure must be maintained either by holding the tong by an assistant or the rope may be tied above and to the side of the fixed object.

d) **Bull nose rings:** Bull rings are used for restraining bulls. It is generally made of two semicircular pieces hinged together and may be made of non rusting materials like copper, aluminium, or some other alloy. According to the size and age of the animal small or large ring can be used. The point and socket type of ring can be push through without prior punching while for ordinary type, a hole is first punched in the nasal septum about 1 cm back from the nostrils with the help of a special bull puncher. When hole is punched too far back then too little is available to accommodate a rope or a chain. The free ends of the two pieces fit together or in a form of point and socket and secured either by flush spring or screw with counter sunk head resulting in a smooth joint. Bulls are generally rung at the age of 10 to 14 months. Some bulls need a new ring every year due to greater amount of wear and tear.

e) **Muzzles:** Muzzles are used to stop animals from biting people or other animals, eating their bedding etc. Calves are muzzled to prevent them from sucking their mothers or other mates. Muzzles are made up of leather strap, wire netting, string, rope, bamboo splits etc. Leather box muzzle is better but it is expensive.

f) **Cradle:** It is a device commonly used to immobilise the neck of horse, cattle and buffalo. Ten or twelve pieces of wood having pointed ends are strung on the two pieces of rope and applied to the neck and tied along top line of the neck. This device is used to prevent an animal from flexing his/her neck so that it may not suck/lick/bite its limbs having wounds or sucking their own milk.

Figure 2.2.3.1 : Common livestock restrain tools.

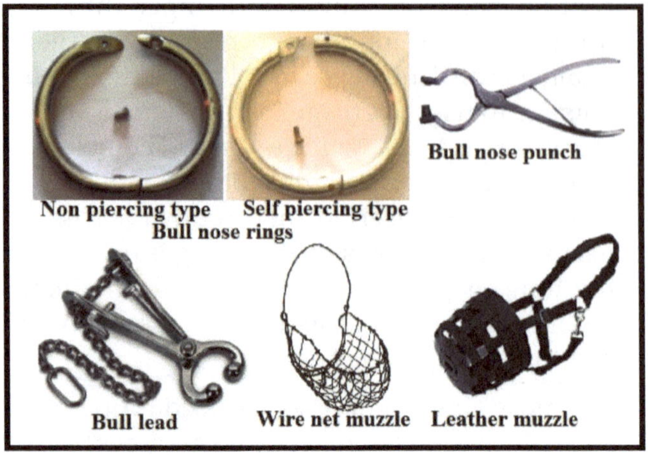

g) **Gags:** Gags are used for keeping the two jaws apart for examination of mouth, teeth or for passing the tube in cattle, buffaloes, horse, sheep and goats. There are many types of gags in use. These are mostly made of metal though wooden gags are also available. The mouth piece gags should be covered by rubber or leather to prevent damage to the gums.

Figure 2.2.3.2 : Common livestock restrain tools.

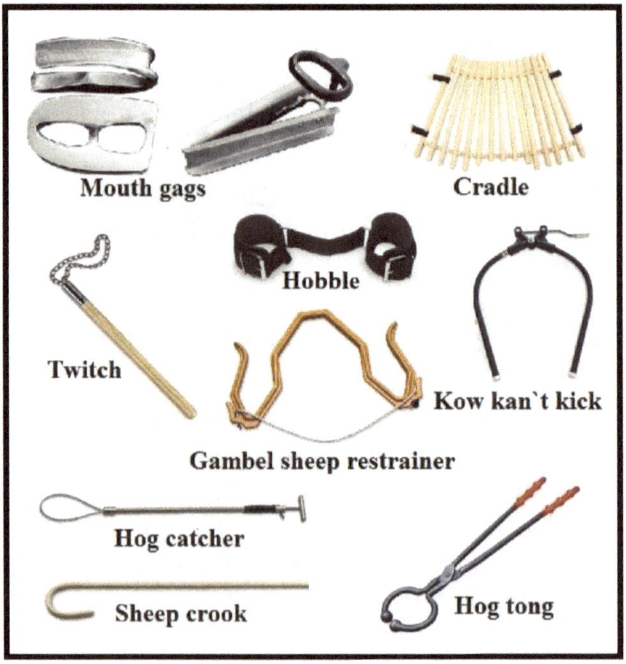

h) **Tail restraining:** In this method, either animal is tied to a post or placed in a stanchion. Now, handler stands directly behind the animal and lifts the tail with one hand and uses another hand to grip the tail at its base. Then base of tail is grabbed with both hands as close as possible and tail is pushed upward straight over the back. In this manner pressure exerted on the base of tail will make animal calm down and will divert the attention from manipulation. Care must be taken otherwise improper application of technique may result in fracture of coccygeal vertebra.

i) **Preventing cows from kicking:** Many contrivances are used to prevent cow from kicking. Some of them are chain or rope hobble, milk man' rope, a short length of rope tied around the flank of the animal and "can't kick" clamp.

Figure 2.2.4 : Livestock restraint techniques/tools.

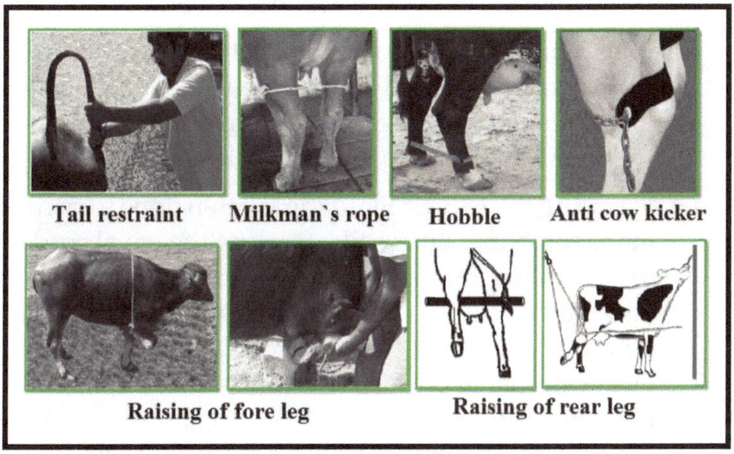

j) **Leg raising techniques:** There are various physical methods for raising feet of the animal.

   i) Foreleg of an animal may be lifted manually by standing in front of fore leg facing towards posterior end of the animal. Then hands are griped at the fetlock joint of fore leg and thumbs are placed under the dewclaws. Now, leg is pulled in upward direction, flex the leg at elbow and shoulder and rest her leg on handler's knee or thigh. Pushing the animal's weight onto the other leg will make raising the foot easier. Use of a nose lead or rope tightened around the flank may reduce the struggling by as it will divert the attention of the animal.

ii) Another technique by which the front leg of an animal can be lifted manually is by attaching a rope around the pasterns of the leg to be lifted and passing the free end of the rope over the withers to the other side. Operator lean against the animal on the side in which leg to be handled and person standing on opposite side pulls the rope and in this way the leg is lifted.

iii) Hind limbs can also be lifted manually by standing beside the animal facing the rear of the animal and handler position his thigh below the stifle joint and grips the hind limb at fetlock joint below dewclaws and pull the leg in upward direction and at the same time flexes the fetlock.

iv) The rear leg can also be raised up by fastening a rope just above the hock and passing the rope over a beam above and slightly behind the cow. An assistant pulls the rope whereas operator grabs the foot and rests on his knee.

v) In another method a loop is placed above the pasterns and the rope is passed over a beam, rope is pulled which raises the leg. When desired height is reached free end of rope is tied back right above the hock.

k) **Flanking the calves:** Flanking is often done to perform various management practices like treating the navel cord with iodine or removing extra teats, ear tattooing, castration and ear tagging etc. In this method, the calf is approached, cornered and caught either by hand or rope. Now, stand on the left side of the animal and grasp the ventral neck area with left hand, turn the head towards the body and almost immediately grasp the right flank with right hand. Bend your leg at knee and put under the animal. When calf jumps/struggles, use this momentum to lift animal off its feet thus weight is transferred from feet of calf onto your leg. Bend your knees to push the left side underneath the calf, and quickly press animal down onto its left side. Young calves can be held down by placing one knee on the calf's neck and one on the calf's side, leaving your hands free. Calves weighing up to 90 kg may be satisfactorily laid on their side by this method.

Figure 2.2.5 : Flanking the calf.

Casting Cattle & Buffaloes

Casting refers to placing a cow in lateral or sternal recumbency by exerting pressure on its muscle and nerves by a series of carefully placed and tightened ropes. In short, it is the process of throwing of animals on ground. When near total immobilization is required to perform various operations like surgery, dehorning, branding, trimming of feet etc. animal are generally thrown on the ground.

The best place to cast an animal is a well grown grass field, a sandy area or bedding prepared from wheat straw etc. There should be ample working space, surface should be reasonably level. There should be no stones, bricks, metal pieces, wooden pieces on the ground and no objects nearby (pole, wall, tree etc) against which the animal may hurt itself. Casting should be avoided during extreme weather conditions. Before casting adequate number of men should be arranged for assistance. Cattle can be thrown on either side but normally they are thrown on their left side. Pregnant animals should not be thrown except in certain unavoidable circumstances (correction of uterine tortion). To avoid danger of injury to digestive organs, animals are fasted for twelve hours before casting. Narcotic drugs are used to calm down vicious animals so that they can be easily roped for casting.

i) **Reuff's or half-hitch method:** Reuff's method is most commonly used and efficient method of casting cattle. First, halter the animal and lead it to the lay down area then secure the lead rope to a post or ring using a quick release knot. A running noose is made at one end of 8-10 meter long rope and fixed around the base of horns. First half hitch is made round the neck, second just in the back of the shoulders and third over the loin and under the flanks in front of udder or scrotum. The first half hitch may be dispensed in with if a sufficiently long rope is not available and in polled cattle, the noose can be fixed to the neck. The man standing around head region should steady and turn the head in opposite direction it has to be thrown and loose end of rope is pulled backwards and towards the side it has to be thrown by two men standing behind the animal. Pulling of rope exerts pressure on the nerves and blood vessels, causing the animal to sink on the ground. When the animal sinks to the ground one man must keep the head and neck down while other must tie the animal's legs together or with a fixed object. After completion of the management practice the process is reversed.

Figure 2.2.6 : Reuff's method of casting.

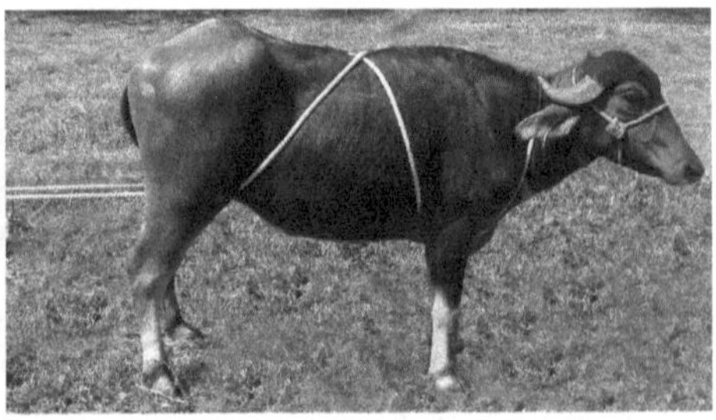

Figure 2.2.7 : Crisscross method of casting.

ii) **Crisscross method:** This is an alternative technique where two people are required to apply the rope. First, halter is applied then animal is lead into the lay down area and lead rope is tied to a post or ring using a quick release knot. Approximately a 12 m long rope is divided into two equal halves and each half is coiled to the centre. The centre of the rope is placed over the neck and each half of the rope is passed crossways to the opposite side between the front legs. Then each half of the rope is moved in backward and upward direction and the ropes are then crossed over the top of the thorax to the other side and both the persons exchange free ends of the rope and the rope further passed between the hind legs through the space under the thighs. All slack is removed from the ropes and either one or both persons can exert the pressure necessary from the rear to cast the animal on the ground. The advantage of using this method is that since no ropes cross the abdomen, there is no danger of injury to the milk vein or the penis.

Handling and Control of Sheep & Goats

It is easier to handle sheep and goats than the other farm animals. However, it is better to handle them in group because particularly sheep have very strong flocking instinct. Always try to handle them in group because individual animal in large pen may panic and try to escape. All the animals of a pen/group should be rounded up into a corner and the desired animal is identified, approached quietly with stretched hand, after reaching the animal one hand should be placed under the chin and lifted immediately. Now, second hand can be put either behind the neck or around its dock to prevent backward movement. Alternatively they can be held by grasping its hind limb above hock joint. Horned animals can be held by their horns. Neck chain, halter, Linton's gag, Gambel humane sheep restrainers, shepherd's crook and portable sheep stocks etc can be used to restrain these animals.

Turning up a sheep: The sheep should be held by a person standing on its left side positioning his left hand under the chin/breast and right hand around the dock. Right knee of the operator should be against the left flank of the animal. Now he should move the right hand from the dock to the right flank region. Now, head is bent over its right shoulder with left hand and simultaneously pressure is applied on the flank region towards its legs with right hand and animal is turned. These actions will cause the animal to lose its balance and will fall against operator's legs. Now jaw is released and both forelegs are grasped and pulled upward to set the animal straight and simultaneously allowed to lounge back against legs of operator, position is adjusted and finally sheep is set up on its rump at approximately 60 degrees to the vertical. An adult goat may be thrown to lateral recumbency by flanking (as used in case of calf).

To restrain a young lamb/kid, right fore and hind legs are grasped in right hand and left fore and hind legs in left hand by keeping head upright and body supported on some fixed object or against operator's body. For drenching purpose, lamb is placed in standing position with its rib cage or neck in between operator's legs standing just behind lamb and for tail docking purpose position of lamb is reversed.

Figure 2.2.8 : Turning up/setting up a sheep.

Figure 2.2.9 : Holding of Sheep & Goat by one leg.

Figure 2.2.10 : Holding of young animal for castration/docking.

Handling and Control of Pigs

Pigs are very unpredictable animal which often become aggressive without exhibiting much warning signs hence one must take care while approaching them. They are intelligent animal with little herd instinct and resist handling efforts. Generally they exhibit tendency to move in opposite direction to the actually required by the handler. Further, due to their short neck and lack of a definite separation at the head and neck application of halters or ropes has limited utility in their restrain.

A pig board, pig net, snout rope (a running noose slipped over the upper jaw caudal to tusk/canines and pulled tightly), pig catcher, a bucket or blindfold over the head of the animal, long-handled neck tongs and headgate can be used to control pigs. Sometimes nose ring is applied to control as well as to prevent them from rooting.

Size of the animal to be handled will depend on strength of the person and nature of job to be performed. Light weight young pigs can be restrained by hands in many ways. Irrespective of methods used animals to be restrained is placed in a small pen or alley first so that it becomes easier to catch the animal. Sometimes a hog hurdle or portable partition is used to corner the animals so that grasping becomes easier. Some of the methods which can be used to restrain a light weight pigs are mentioned below

One of the rear legs is grasped firmly by one or both hands and entire rear end of the body is lifted off the ground so that pig cannot put much effort to escape. Another hand is used to grasp the front leg on the same side of the pig. Now using hold on front and rear limb, pig is lifted completely from the floor and gently placed on the ground on its side and enough pressure is applied on the side of the pig by the knee to have complete control over the animal.

One or both hands are used to grasp either of the pig's hind limb. Now, position of hands and body is changed in such a way that handler can grasp pig's left foreleg in his left hand and right foreleg in his right hand. Then the position of pig set with its head up, back towards handler's legs and body of pig pressed between legs.

One of the rear legs is grasped with one or two hands. Now the grip is adjusted quickly in such a way that left hand should hold pig's right rear leg and right hand its left rear leg. Then pig is lifted off the ground by both of its rear legs to about waist height. This will place pig's back in front of handlers legs with its nose toward the ground. To have more control body of pig can be squeezed between two legs of handler.

Casting Pigs

Many methods are in practice to place a pig in lateral recumbency. However, successful execution of all these techniques requires firm control of animals with snare. Hence, animals are snared first and then any one of the following techniques can be used to throw the animal on the ground.

i) A loop is made at the end of a long rope, this loop is placed around the neck of the pig and a bowline knot is positioned slightly right or left of the pig's neck then the rope is brought back and a half hitch is made just behind the front legs. The rope is further brought back and second half hitch is placed just in front of rear legs. Now the pig can be thrown upon its side by pulling the free end of the rope from behind the animal and direction of the throw can be decided by rope and snare. After completion of the management job first half hitches are loosen then the noose and finally snare is removed.

ii) One end of a short rope is tied to the foreleg and another end to the hind leg of same side (side in which animal is supposed to be thrown). Now, the rope beneath the body of the animal is passed towards the opposite side in upward direction over the back. Handler positions himself on the same side in which the ropes are fastened to the animal and pulls the rope over the top of the body. This pulling drag the legs out from under the body and pig will be thrown on its side. If animal is required to be immobilized for longer time then legs should be secured otherwise handler can place the knee to have control over the animal.

iii) A short rope is used to tie both forelegs above the fetlocks. Another short is used to tie both hind legs in the similar fashion. Now a longer rope is tied to the rope joining the two fore legs and run it backwards between the hind legs. Another long rope is tied to the rope which is tied to both hind legs and run forward between the forelegs. Now pulling of ropes in opposite direction will throw the animal on the ground.

Handling and Control of Horse

While approaching a horse speak calmly, avoid loud noises and sudden movement. When standing near a horse, place your hand gently on the animal. Position of ear, dilated nostrils and eyes reflect temperament of the animal. Never stand directly behind in the blind spot of a horse. When tying a horse always use a quick-release knot and lead ropes should be tied short and high enough to a solid post. Various restraint tools/techniques which can be used to control horses are

a) **Halter & lead/chain shank:** Halters and lead shank are more frequently used to handle horses. Permanent halters are made up of leather or nylon rope however a temporary halter can also be made from rope. The lead shank may be a simple rope or chain tied/attached to the halter ring usually with the help of a snap. Further, chain shank can be used as a bit or bridle.

b) **Twitch**: It is a simple but most important manual tool of horse restraint. It works on the principle that when pressure is applied on sensitive body parts, endorphins are released which calm down the animals hence they show an increased tolerance for discomfort. Twitches can be by applied by grasping of a large quantity of loose skin in the neck area (just in front of the shoulder) and squeezing it as hard as you can, with some twisting action (for young horses or foals) or by seizing the horse's nose with hand and twisting or pinching the upper lip or by using a hardwood (about 2 feet in length) handle with a short piece of chain or rope attached at one end.

c) **Ear hold** (earing down): It involves grasping the base of the ear in the web between the thumb and the forefinger tightly against the base of ear and closing of hand squeezes the ear and tends to lift the ear from the head. In the process any evasive movements of the horse will be a form of self-punishment.

Other contrivances like neck cradle, muzzle, bibs, bridle, side stick, breeding hobbles and stabilizer etc are also used to restraint horses.

d) **Leg raising techniques:** Legs of horses are lifted to examine, manipulate and to restraint the animal for other purposes. There are various methods for raising feet. Some of them are mentioned below.

    i) To lift a near foot, stand in front of the shoulder facing the tail of the animal. Passing the hand on to the forelimb down to knee make the horse automatically to lifts its leg. If the horse resents, try to screw pastern, if it continues to resent then try to lift the leg while touching the operators left elbow to the back of the horse's knee and further try to pick up forcibly.

    ii) Following steps can be followed to lift the hind left limb of a horse-face the back of the horse, position the left hand on the point of the hip and right hand over the rump then move it down slowly over the thigh, hock and finally on the mid tarsal area. Pull the leg forward and upward and lift the left up to the height of hock of opposite side. Step towards the tail up in under the lifted leg. Finally, support the lifted leg of horse by your legs and body. Use both free hands to examine and manipulate the limb.

e) **Foal restraint:** There are many ways in which a foal can be restrained. First corner the foal and grab the neck of the foal. Now adjust yourself in such a way that one of your hands is placed around the chest and another around the hip below the tail head. If a foal has to be thrown on the ground then you can do it by i) from the restraint position mentioned above just pick up the foal from its position and lay it over upon its side. ii) from the restrain position, use one hand to turn the head and neck of the foal towards back of the opposite side and with another arm put pressure on the rump. These actions will make the foal to sink to the ground. iii) for heavier foal, restrain the foal as mentioned above and simply by pulling the tail between the hind legs and maintaining pressure will throw the animal on the ground and cast position can be maintained by pulling the tail up in front of the stifle and putting pressure on the neck.

Casting Horses

Horse can be casted by using general anaesthesia or by tying without sedation or and tying with sedation. A procedure of casting using combination of tying and sedation may be followed as mentioned below.

One person should control the head of the animal with the help of halter and lead. Make a doubled bowline knot at the centre of a sufficiently long and strong rope. Place hobble and ring in each of the hind limb. Position bowline knot over the withers and one working end of rope should extend towards right hind limb and another towards the left hind limb of the animal and run into hobble of respective side from inside to outside. After coming out of the hobble these working ends of rope should run forward, upward and inserted into the neck loop of respective side. Now, sedate the animal with proper chemical agent. To cast the animal on right or offside, the nearside working end of rope is run backward around the hip of the horse so that handler can apply a pull at a 90^0 angle to the body of animal. Offside leg is lifted and held up around 6-8 inches from the ground by pulling the working end at offside. When horse shows signs of sedation, should be thrown on to its offside by pull on left hip applied by near side handler, pulling of head to the left by person holding the lead rope while offside handler keeps the offside hind limb up. To cast the animal on its left side, position of the handlers and ropes should be adjusted accordingly. Once the animal is casted all four legs should be flexed and secured properly.

Methods of Animal Identification

Animals need to be identified for various purposes like maintenance of proper records on the farm, to accomplish feeding, breeding, milking, healthcare practices properly, for recording of parentage, registration of offspring, registration of pure bred animals, for disease and residue trace back, insurance and to prove ownership when animals are theft or lost. An ideal method of identification should be i) permanent ii) visible and readable from a distance iii) widely acceptable and iv) cheaper. There are several methods of animal identification in use, some of them are mentioned in following paragraphs.

Table 2.2 : Common methods of animal identification.

Species	Methods of identification practiced
Cattle & buffalo	Tattooing, ear-tagging, branding, ear notching, ankle band, brisket tag, neck chain, tail tag, Photograph or sketch, dermatoglyphics (muzzle and palatine printing) and RFID etc.
Sheep	Ear tags, tattooing, ear notches, neck chain, temporary methods (use of stencil, grease stick, spray paint) etc.
Goat	Tattooing, ear tags, ear notches, neck chain, freeze branding etc.
Swine	Ear notching, ear tags, tattooing, paint branding etc.
Horse	Signalment (natural marks & colour), Lip tattoo, branding, hoof branding, ocular scans, trichoglyphs (hair whorls), photograph or sketch, RFID etc.
Poultry	Wing or leg band

a) **Ear-Tagging:** It is one of the most popular methods of animal identification however always used in combination with a permanent method of identification as there are chances that tags can pull out of the ear. Ear tags can be applied at any age to any of the two ears depending upon the handling facilities. Commercially these tags are available in many styles, colours and sizes. Tags are made up of plastic (one or two piece) or metal, which may be either pre-numbered (number or alphanumeric codes) or blank. Now-a-days tags with RFID and printed bar codes are also available.

Figure 2.3.1 : Different types of ear tags.

Plastic tags are better than metal tags as it can be read from longer distance and stays in the ear for longer period of time. These tags are of two types, self piercing and non-piercing type. Non piercing metal and plastic tags require a hole to be punched first in the upper edge of the ear close to the head leaving enough space between edge of the ear and tag for the growth of the ear. One end of the tag is placed through hole and ear tags are clamped together with pincers. In two piece ear tag, male part is placed over the pin of applicator and female portion is placed into a clip and by closing the applicator they are fixed to the ear. Larger plastic tags are used for cattle and buffalo while small tags are used for sheep, goats, pig and calves. To prevent infection and facilitate healing antiseptic should be applied to the wound.

b) **Tattooing:** This is a permanent method of animal identification in which several small holes (in the form of letters and/ numbers) are punched inside of the ear and then filled with tattoo ink or paste which after healing leaves a permanent tattoo mark. A tattooing set consists of a tattooing forceps, a series of letters and numbers made in the form of dies having sharp, needle like projections and tattooing ink or paste (insoluble carbon black or green vegetable pigments that are inert to tissue). Generally, upper third part of inside of the ear, midway between base and tip of the ear is selected as tattoo site, avoiding blood vessels. This area is cleaned with alcohol. Desired letters or numbers or both are placed in tattooing forceps and checked on paper before applying it on the ear. Tattooing ink is smeared on the area, on needles, instrument is positioned on the area and then handles of forceps are clamped together quickly and completely and released. Tattoo ink or paste is rubbed into all holes. If performed properly with clean instrument there is little chance of infection. Major disadvantage with this method is that close inspection of ear is required to read the tattoo mark and in dark pigmented ear, marks are not easily readable. In dark pigmented cattle and buffalo it can be performed on under surface of the tail head. This method is most suited for baby calves and pigs. Tattooing is avoided in case of lamb/kid until they are at least six months of age.

Figure 2.3.2 : Different types of tags.

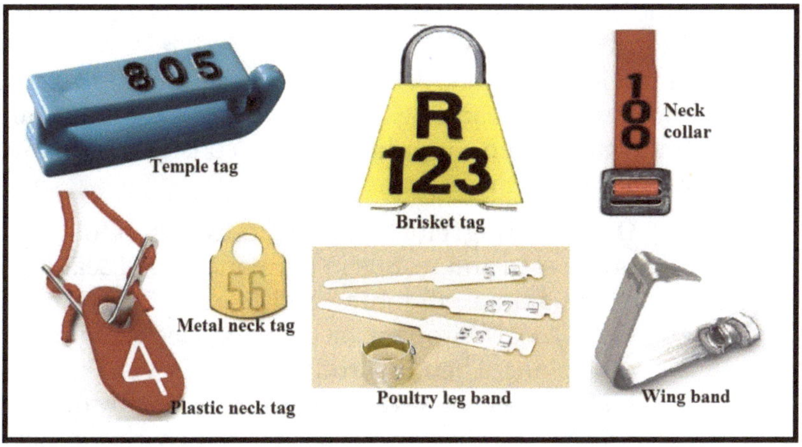

Figure 2.3.3 : Branding rods.

Figure 2.3.4 : Tattooing set.

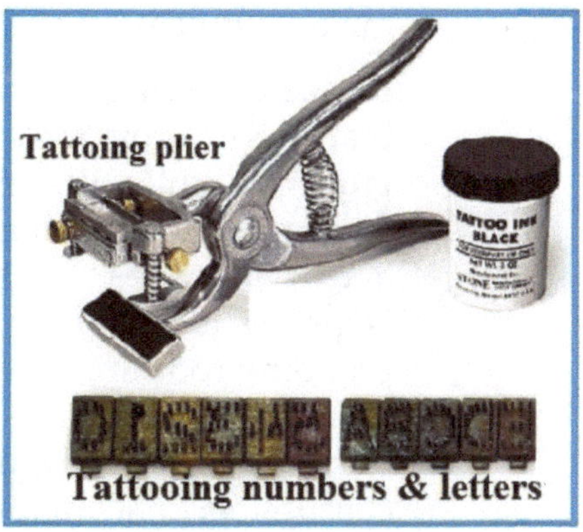

c) **Branding:** Branding is a permanent method of identification and is suitable for marking cattle, buffalo, horses and camel. In this method numbers, letters, designs or a combination of them are singed on the skin with the help of a hot iron or chemicals (corrosive acids, caustic soda paste, caustic potash and refrigerants). Branding iron rod consist of numbers 1 to 9 and 0 while alphabet A to Z. Letters and numbers made up of copper or copper alloy are better as it retains heat better and hairs don't stick to it. Similar type of brands can be used in hot or freeze branding only difference is that brands used in freeze branding method has a shallow groove on the contact surface. Both of the methods give satisfactory results when performed properly. Branding can be done on different areas of the body but commonly done on thigh, hip and shoulder of the animal but to avoid hide depreciation, can be done on cheek and hind limb. Animals marked by this method can be identified quickly from a distance. Animals should be well restrained by casting and tying their legs. A bright sunny day should be chosen for the branding. Branding in cattle is done when calves are about one year of age. Branding causes considerable pain to the animal and reduces value of hide.

i) **Hot branding:** Hot branding destroys hair follicles located under skin and results in permanent bald scar on the skin of the animal. Branding rods can be heated by burning wood/coal/ propane or by using electricity till it is gray ash colour. Then this rod is pressed over skin uniformly for 3 to 5 seconds. Take out the rods one by one and same process is repeated till all desired characters have been branded. The space between the numbers/letters/designs should be at least 2.5 cm. A small amount of mustard oil mixed with zinc oxide is smeared over the burns.

ii) **Freeze branding:** In freeze branding, application of cold branders destroys melanocytes and the hairs from this area grow out white. Branding rods are cooled in refrigerant like liquid nitrogen or a mixture of isopropyl alcohol (99%) and dry ice. Area to be branded is clipped closely and then isopropyl alcohol (95%) is rubbed on the area. After that, branding rods are applied for at least 30 seconds. Freeze branding is less damaging for hide and less painful than hot branding. Disadvantages are: it is more expensive, time consuming, final brand takes up to 4 months and technique is less suited to light-coloured animal.

d) **Ear notching:** In this method V shaped notches are made at specified places along the borders of the ear with the help of a pairs of sharp scissors or pincers. Too small notches will close up as the animal grows and too large notches will deform the ear. Ear notching is most commonly used method of identification in baby pigs though it can also be used in other animals (cattle, sheep & goat). Ear notching should be done as early as possible after birth to have an early identification for the piglets and less bleeding. Further, in older animals this process may stimulate ear biting. There are several systems of ear notching in use. Purebred associations specify which system they require for their records. Before notching, determination of right and left ear is made by viewing the pig from the rear. Right ear shows the litter number while left ear shows individual pig number. A standard ear nothing system for pigs has been described below.

The right ear has five locations for ear notching, their numerical values are 1, 3, 81, 9 and 27 (Figure 2.3.5). Except for the 81 notch, there may be one or two notches at each of the other four locations and to determine the litter number of a pig, the numerical values assigned to each notch are added.

The left ear has three locations for ear notching and their numerical values are 1, 3 and 9 (Figure 2.3.5). As with litter number, the number values are added to find the individual pig number. Some of the examples are given in Figure 2.3.5.1.

e) **Electronic livestock identification:** Electronic animal identification is presently based on the use of radio frequency waves in the low frequency band. Radio frequency identification device (RFID) has four components i) Transponder (transmitter and responder) is a miniaturized electronic radio-frequency device consisting of microchip and an antenna, enclosed in a water-proof protector. Transponder with passive technology (without batteries) is used. ii) Reader or transceiver (transmitter and receiver), it receives the information telegram. Now-a-days, readers are handheld wand like devices or arch like portals through which animal walks. iii) Data accumulator, stores the collected data and iv) Software, it manipulate the data and converts it into usable form. There are three main types of transponders for animal identification devices-i) Injectable transponders- it can be used in all animal species. The transponders are covered by a bio-

Figure 2.3.5 : Universal Swine notching system.

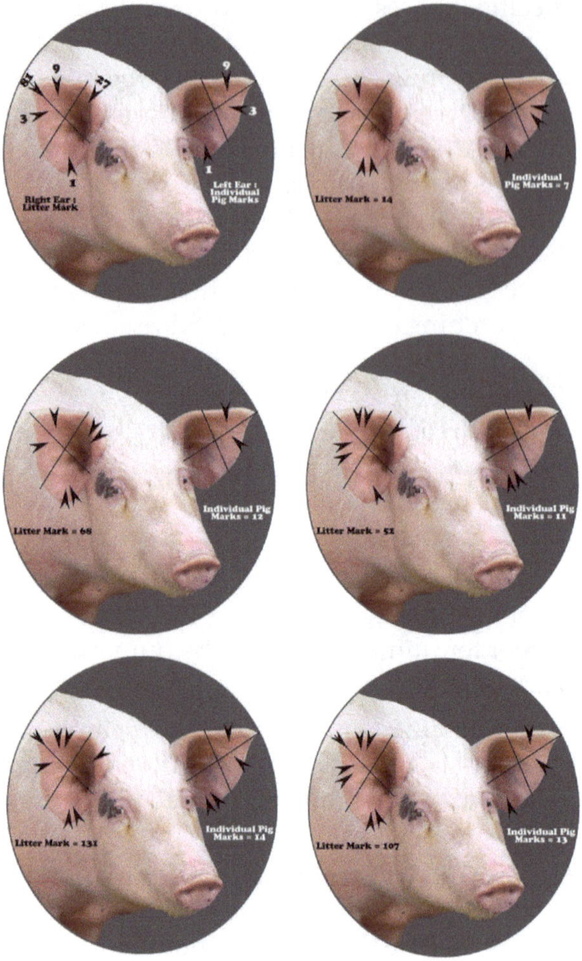

Figure 2.3.5.1 : Ear notcher

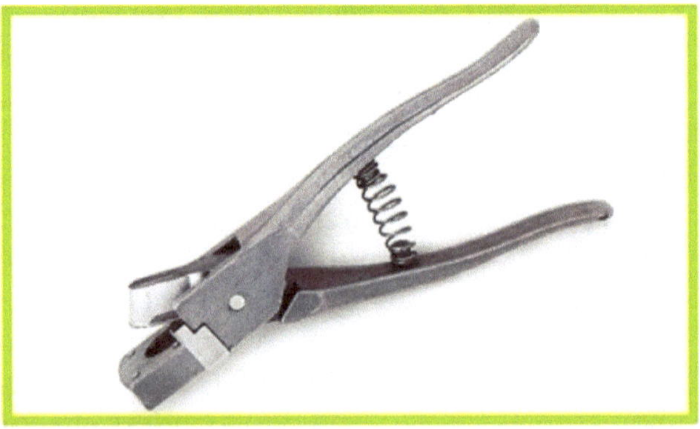

compatible glass capsule and are injected subcutaneously through a needle at different body sites. ii) Electronic ear-tags- it can be used for almost all livestock species. Transponders are included in a plastic round button-tag and used as the female part of a conventional plastic ear-tag. iii) Bolus transponders- it is meant for ruminants only. Transponders are placed into a high specific gravity capsule (bolus) and orally applied to ruminants. They are retained in the fore-stomachs, mainly in the reticulum.

Determination of Age of Animals by Various Methods

Knowledge about the age of the livestock is necessarily required for a variety of purposes like for selection and fixing the price of animals, determining the dose of the drugs, to ascertain proper breeding age, to evaluate body weight gain strategy etc. Methods which are used to ascertain the age of an animal are- record, dentition, horn ring, size and characteristic of hoof and size and general condition of the animal etc. If properly maintained, determination of age by farm record is the most accurate method. Other methods are generally used only when record is not available or sometimes to crosscheck the age in doubtful cases.

a) **Determination of age of animals by general body condition:** A young animal can be compared with old animal as young animals are small in size with active disposition, soft and tight skin and soft hair coat where as old animals are large in size with normal outlook, rough hair coat and loose skin.

b) **Determination of age of animals by hooves:** Hooves of the younger animals are small in size, smooth and evenly levelled at sole while in older animals it is larger in size, rough and may not be evenly levelled at sole. Both of the above mentioned methods can only classify the animals into young, yearling, adult and old category.

c) **Determination of age of animals by horn rings:** Horns of the animals can be used to estimate the age of animals with some level of accuracy. When a calf is about 10 days old two small, hard rounded buttons emerge from the skin. At about 12 to 15 months of age horn has its permanent natural shine through the surface. In the second year, first horn ring appears and second ring appears at the age of three years but later when animal grows older these two horn rings are so close that they appear as one. Appearance of horn ring continues as one per year. Hence, age of an animal with two and three horn ring will be four (3+1) and five (3+2) years, respectively.

d) **Determination of age of animals by dentition:** The approximate age of an animal can be find out up to certain extent by observing the time of appearance and the degree of wear and tear of temporary and permanent teeth. Among many factors which influence dentition, main factor is breed and plan of nutrition (for groups within the breeds). Additionally, there are individual variations between groups. There is a distinct difference between the ages at which the teeth are erupted in cattle of *Bos taurus* and the *Bos indicus*.

Structure and classification of teeth: Teeth are the hardest structure in the body. They remain implanted into the alveolar sockets of the upper and lower jaw bones. A tooth has three parts -body or crown (appears above the gums and is covered with very hard material known as enamel), root or fang (the implanted narrow part which is covered with yellowish material cementum) and neck (the part, which joins crown and fang). The bulk of the teeth are formed by dentine and within dentine there is a central canal known as pulp cavity. Teeth are classified as per their placements in the oral cavity. The incisors (I) are most anterior in position followed by canines (C), pre molars (P) and finally molars (M). Premolars and molars are known as cheek teeth as they are placed laterally. Incisors help in cutting and picking up food while canine and cheek teeth help in tearing and mastication of the food. There are two sets of teeth in each species deciduous or milk or temporary teeth, which are either found at the time of birth or erupt within few weeks after birth and are replaced by permanent set at certain age. These sets of teeth are characterized by snow white colour, smaller in size, distinct neck and short fang whereas permanent teeth appears at a certain age and are generally brown in colour, larger in size, no distinct neck and have long and stout fang. Except molars all the teeth are replaced by the permanent teeth.

Table 2.3.1 : Dental formulas of different livestock species.

Species	Dental Formula	
	Temporary teeth	**Permanent teeth**
Cattle, Buffalo, Sheep & Goat	2(I 0/4 C 0/0 P 3/3 M 0/0) =20	2(I 0/4 C0/0 P3/3 M3/3) =32
Pig	2(I 3/3 C 1/1 P 3/3 M 0/0) =28	2(I 3/3 C 1/1 P 4/4 M 3/3) = 44
Horse	2(I 3/3 C 0/0 P 3/3 M 0/0) =24	Male: 2(I 3/3 C 1/1 P 3 or 4/3 M 3/3=40 or 42 Female: 2(I 3/3 C 0/0 P 3 or 4/3 M 3/3=36 or 40

i) **Estimation of age of Cattle & Buffalo by dentition:** The front teeth of cattle can be easily examined hence attention is generally confined to them, the stage of development of the molar teeth being noticed only when additional confirmatory evidence is necessary. On each side of the lower jaw there are four incisors or front teeth and six molar teeth in the adult bovine. Incisors are absent from the upper jaws of buffalo, cattle, sheep and goat and their place being taken by the 'dental-pad', which is a hard mass of dense fibrous tissue covered by tough mucosa, against which the eight lower incisor teeth bite. The teeth are loosely embedded in the jaw so that a slight amount of movement is normally possible. The incisors are named centrals, first intermediates or medials, second intermediates or laterals and corners. However, it is more convenient to enumerate them from the central pair as first pair, second pair, etc. At birth calves are having two or more milk teeth and within one to two months all the eight temporary incisors appear whereas in case of buffalo calves first and second pair of deciduous incisors is through the gums and others are hardly visible through gums at this stage. Average age (in months) of eruption of permanent teeth in cattle and buffaloes has been shown in Table 2.3.2. In adult cattle sometimes it becomes difficult to distinguish old temporary and permanent teeth. In that case if all six pairs of cheek teeth are present the incisors are permanent. After eruption of all permanent teeth age is estimated by wear and general appearance of the teeth. At six years of age central incisors are supposed to be worn level while second, third and fourth pairs wear gradually. This wear increases with age become extensive after ten years of age. Cattle and buffalo after eight years of age are generally designated as aged animals without trying to estimate the exact age.

Table 2.3.2 : Average age (months) of eruption of permanent teeth in Cattle & Buffaloes.

Type of teeth	Average age of cattle & buffalo at eruption of their permanent teeth (months)	
	Cattle	Buffalo
Incisors		
First pair	24-30	30
Second pair	36	42
Third pair	48	54
Fourth pair	54-60	70-72
Molars		
First pair	24	
Second pair	24	
Third pair	36	
Fourth pair	6	
Fifth pair	18	
Sixth pair	24	

ii) **Estimation of age of Sheep & Goat by dentition:** Mature sheep and goats have a total of 8 incisors (4 pairs-lower jaw only), 12 premolars (lower and upper jaw, 3 pairs each), and 12 molars (lower and upper jaw, 3 pairs each) making a total of 32 teeth. The two central incisors are called pinchers, the adjoining ones- first intermediates, the third pair- second intermediates and the outer ones- corners. At birth lamb may have 0-2 pair of incisor teeth and by the age of 6-10 months all temporaries are present. Until sheep and goats are four years of age, the front teeth of the lower jaw furnish a fairly reliable guide as to their development. With more advanced age, the teeth merely wear down and spread apart and the degree of wearing or spreading is an indication of age. The average age of eruption of incisor teeth (months) in sheep & goat has been summarised in Table 2.3.3.

Table 2.3.3 : Average age (months) of eruption of Incisor teeth in sheep & goat.

Type of teeth	Average age of Sheep & Goat at eruption of their permanent teeth (months)	
	Sheep	**Goat**
First pair	14-20	15 (±2 months)
Second pair	21-25	24
Third pair	26-32	36
Fourth pair	More than 32	48

Starke & Pretorius (1955) & Battaglia (2007).

In animals, beyond 4 years of age teeth look longer and there is more space between the teeth, provided no teeth have been lost and animal is known as spreader. At about 7 to 8 years of age, sheep may lose some of the tooth and known as broken mouth and when loses their entire incisor teeth they are known as gummers. Sheep grazing in sandy area may have worn teeth (after 4 years of age) and known as short mouth animals.

iii) **Estimation of age of Pigs by dentition:** Dental formula of pig has been given in Table 2.3.1. The pattern of eruption of incisors is not as per their position in the jaw. First corner incisor erupts followed by centrals and intermediaries. At about 5 months of age all temporary incisors will be present. At the age of six month, one year, and one and half year permanent corner, centrals and intermediaries will replace corresponding temporary teeth, respectively.

iv) **Estimation of age of horse by dentition:** Dental formula of horse has been given in Table 2.3.1 and Table 2.3.4 shows average age (months) of eruption of different types of teeth. Young horses have a total 24 temporary or deciduous or milk teeth i.e. 12 incisors and 12 premolars. Mature male horses have 40-42 permanent teeth and mares have 36-40 depending on the number of canine teeth present. Canine teeth erupt at about 4-5 years of age in male horses while it appears only 20-25% of the time in mares and are usually smaller than canine present in males. All horses don't posses first premolar or wolf-tooth but when present it only appears in the upper jaw.

Table 2.3.4 : The average age (months) of eruption of teeth in horse.

Type of Tooth	Approximate age at eruption	
	Deciduous	Permanent
1st Incisor	Birth or 1st week	2.5 years
2nd Incisor	4-6 weeks	3.5 years
3rd Incisor	6-10 months	4.5 years
Canine or bridle	---	4 to 5 years
1st Premolar (wolf tooth)	---	5 to 6 months
2nd Premolar	Birth to 2nd week	2.5 years
3rd Premolar	Birth to 2nd week	3 years
4th Premolar	Birth to 2nd week	4 years
1st Molar	---	9-12 months
2nd Molar	---	2 years
3rd Molar	---	3.5 to 4 years

Source: Parker (2008).

At about 5 years of age when all permanent teeth have erupted, the horse is said to have a full mouth. Further, some changes in the teeth like disappearance of cups (deep indentures in the centre of the surface of young permanent teeth), angle of contact and shape of the surface of the teeth can he used to estimate the age of horses as mentioned below. However, age of horses can be determined accurately by the teeth up to age 10, thereafter it involves some guesswork.

- Usually, these cups become smooth in the lower centrals, intermediates, corners, upper centrals, intermediates and corners at around 6, 7, 8, 9, 10, and 11 years of age, respectively.
- At 11 or 12 years of age when cups are absent from all lower and upper incisors the horse is said to be smooth-mouthed.
- The angle of contact which remains usually 160 to 180 degrees in young animal changes to less than a right angle in aged animals.
- At the age of 7 years a notch or hook appears on the caudal corner of the upper incisor.
- At about 8 years of age when all cups are absent in lower incisors, dental star appears in the lower central incisors.
- When hook or notch disappears at the age of 9 years and at this time distal end of Galvayne's groove may be visible at the margin of the gum on the upper corner incisors.

- This groove is half-way down labial surface of incisor at about 15 years, extends entire length of labial surface of incisor at about 20 years, disappears from upper 1/2 of tooth at around 25 years and disappears completely from the tooth at about 30 years of age.

Dehorning

Dehorning is a process by which horns of an animal is removed by mechanical/chemical/ electrical dehorner. Dehorning and disbudding are adopted in horned cattle breeds. Dehorning should be performed as early as possible in the animal's life as in older animal it is painful and there are chances of substantial bleeding. Disbudding should be preferably done when the calf is in between 3 to 10 day after birth because up to this age horn buttons doesn't get attached to the skull. The objective of dehorning is reduce space required, chances of bruise, injury to the herd mates as well as to the operator and handling of animals becomes easy. Dehorning should be done in spring or post-monsoon season but if the process is required to be carried out in fly season, fly repellent should be used to the wound. Among the several methods available for dehorning, age is an important factor which determines the method to be used. Some of these are discussed below.

1. **Chemical method:** This method is used to dehorn very young calves (1-2 weeks old). The process of dehorning is accomplished by applying caustic potash on the horn button. First buds are located with the help of fingers and hairs around it are clipped and area is cleaned with a mild antiseptic. Vaseline is applied on the periphery of the clipped area to protect the adjoining tissue. The horn button is vigorously rubbed with a caustic potash stick till a slight bleeding appears. To avoid burning of fingers caustic potash stick is wrapped in cotton wool or holder. The chemical along with destroyed horn tissue should be cleaned by cotton or soft and clean cloth to avoid further damage to the area. Calves dehorned by this method should be protected from rain or splashing water around the operated area.

2. **Bell shaped dehorner:** Use of bell shaped hot iron or electric dehorner is an excellent method of removing horn of a calf less than 4 weeks of age. In electric dehorner a temperature of about 1000^0F is maintained and application of this heated rod for 10 seconds is sufficient to destroy the horn buttons.

3. **Clipper and saws:** Specialized clippers and saws are used to dehorn the older cattle. Local anesthesia should be administered around the base of the horn. The tool is positioned to have deep cut to remove a ring of skin with the horn. To arrest the bleeding either horn artery is tied with cotton or silk thread or some other technique is used. An antiseptic solution is applied to the wound to prevent the infection.

4. **Rubber bands:** In this method, a groove is made around the base of horn and by using elastrator a specially designed rubber band is slipped into the groove. This rubber band shuts off the blood circulation and horn gradually comes off.

Figure 2.4.1 : Dehorning devices.

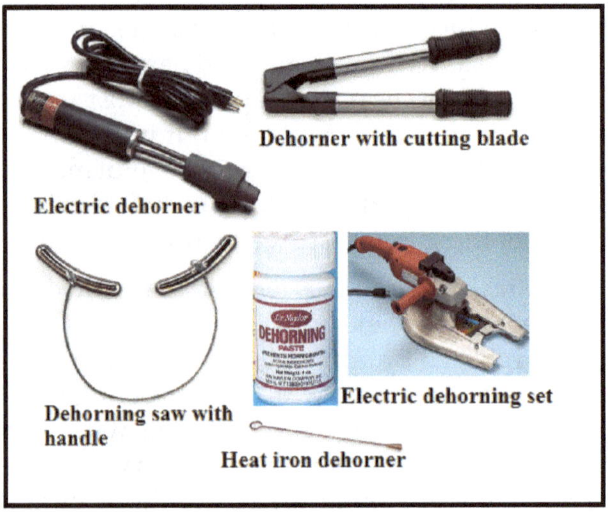

Castration

Castration is removal or destruction of all or part of the male reproductive system. It is done to check undesirable breeding, to decrease aggressiveness, to enhance on-farm safety of animals & attendants, to enhance body weight gain, to produce desirable type of meat and in buffaloes it results in fineness of neck which makes easy fitting of yoke. Castration can be performed either by surgical or non surgical methods.

a) **Bovines:** Calves may be castrated at any time up to about one year of age but it should be done preferably around 8-10 weeks of age as this practice is very painful in older animals.

1. **Surgical method:** There are many methods of surgical castration and all of these methods can accomplish successful removal of the testicles with the help of scalpel or sharp knife. Castrated animals should be rested for few days and if required wound is dressed once or twice. Animals should be castrated in cold season and to avoid the fly menace should be avoided in rainy season. Haemorrhage and infection is major problems associated with castration particularly with surgical methods.
2. **Nonsurgical method:** Nonsurgical method of castration is performed either by using emasculatome (Burdizzo®) or elastrator and rubber ring. Animals can be castrated throughout the year by below mentioned methods.
   i. **Emasculatome method:** The emasculatome method of castration is a bloodless castration in which testicles are made non-functional by crushing the cord and blood vessels associated with spermatic cord without disturbing the central septum of scrotum. In this method animal is restrained properly then cord is positioned towards the outer edge of the scrotum with the help of fingers. Now, emasculatome is placed at about 2 inches above the top of the testicle and then jaws of the instrument is clamped and held in position for approximately 30 seconds. This process is repeated on the same cord but about ½ inches below the first clamping site. Likewise process is repeated on second testicle.

Figure 2.5.1 : Castration devices.

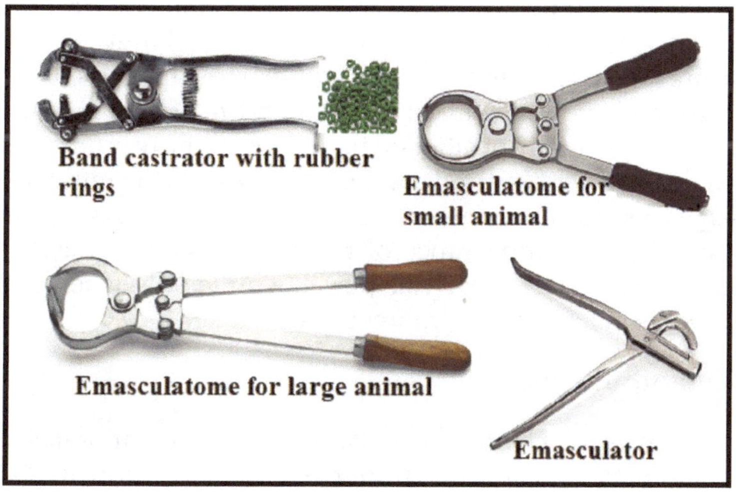

ii. **Elastrator & rubber ring method:** In this method ring is placed on the scrotum between animal's body and testicles with help of an elastrator. The rubber rings stop blood circulation to the testicles and lower scrotum, that atrophies and slough off in about 3- 4 weeks. This method works best on animals of less than 1-2 months of age. Care must be taken that both the testicles remains below the rubber band and when scrotum slough off animal become prone to tetanus. Hence, tetanus antitoxin should be given to the animal.

b) **Sheep & Goat:** Male lambs are usually castrated between 2 to 4 weeks of age but if a male lamb has to be marketed before five months of age then they are not castrated to attain faster growth, higher feed efficiency and leaner meat. Castration can be done with knife/emasculatome (smaller) /elastrator. Male kids can also be castrated at the same age and by same method as in male lambs.

c) **Pigs:** Apart from other benefits castration of pig results in removal of boar odour or tainted odour from the pork. Pigs are mainly castrated by surgical methods (scalpel, knife, razor blade and side cutter method etc) when they are about 1- 14 days old.

d) **Horse:** Colts are mostly castrated between 12 to 14 months of age which may improve an unruly horse's temperament.

Removal of Supernumerary (Extra) Teats

A normal udder of a cow should have four squarely placed teats but many calves are born with more than usual four teats, these extra teats are known as supernumerary or extra teats. These teat/teats can develop and grow like a normal teat and may be blind or leaky. An extra teat is generally positioned higher up on the rear udder or between normal and larger front and rear teats. These teats can interfere in normal milking process, may increase susceptibility for mastitis and may overall detract from the appearance of the animal. Hence, these extra teats should be removed as early as possible preferably when calf is below six months of age by surgical methods.

Trimming the Feet

Animals which are mostly kept in confinement or lacks adequate exercise often grow long hooves. This condition makes animal unattractive, bring changes in normal posture and gait of the animals and increases susceptibility to other foot troubles like foot rot etc. Hence, hooves should be periodically trimmed. Hooves of young animals can be trimmed with a long handled chisel while standing on a hard ground or flat timber floor. Large animals need to be well restrained for trimming sole and cleft. Hoof trimmer or wood chisel (hoof knife) and hammer can be used to shorten the toe. A shoeing knife can be used to trim the sole and dead hoof, and then a rasp may be used to smooth down the surface.

Figure 2.6.1 : Hoof care tools.

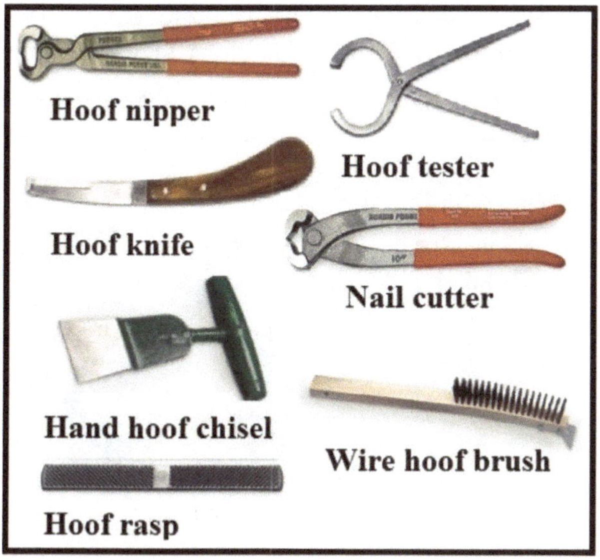

Exercising

Exercise is necessary to keep animal thrifty and active. Animals which are mostly kept in confinement or lacks adequate exercise may have weak leg muscles, overgrown hooves and puts strain on legs. Exercise for the animals kept under loose housing system is not a problem but if they are fie stall then they must be turned into an open area or to pasture at least once a day and preferably twice a day. To maintain good condition of breeding bulls they must be given exercise regularly. Providing a loafing area of about 120 square meters or parading for an hour daily or use of bull exerciser will serve the purpose. Confinement of horse has poor effect on health and may result in development of various vices. A stabled horse must be given a minimum of 30 minutes exercise daily.

Grooming

Grooming involves brushing, combing and rubbing of hair coat of the animal. Grooming is done to remove dirt, loose & dead hairs, skin secretions, lice, other skin parasites from hair coat and to massage & stimulate skin, hair follicle, oil glands and muscles. It also provides an opportunity to make the animal stand and submit to your wishes and also provide information regarding general condition of the animal. Apart from horses other farm animals not groomed regularly and elaborately. Grooming is essential for all show animals. Equipments which are generally used to groom an animal are body brush, wisps made of paddy straw or hay, rubber or metal curry comb, dandy brush, mane and tail comb, towel or burlap cloth, sweat scraper and hoof pick.

Figure 2.7.1 : Grooming, shearing and clipping tools.

Horses that are stabled should be groomed comprehensively once daily. Work horses should be groomed both before and after the work. The handler should stand well away from the animal, facing the tail and should hold curry comb in the right hand and brush in the left hand and begin on the left side of horse at the neck and then proceed to withers and shoulders, fore-leg up to knee, back of the animal, side, belly, croup and hind-leg down to the hock. Brushing should be done in sweeps and moved in the direction of flow of hair but against the fall of hair to remove dried mud, dung etc adhered to the hair. Sometimes instead of using curry comb and brush simultaneously, first curry comb is used then brush. Curry combs should be used gently and brush vigorously. Body parts like area between fore-legs, hind legs, flank etc should be brushed carefully. After completing the grooming on the left side of the body, right side should be groomed in similar fashion. Body brush can first be used in brushing the head then the mane in downward direction. Tail can be combed and brushed with either body brush or with dandy brush. The effectiveness of grooming can be assessed by passing the finger tips against the direction of hair and if it is not properly done, the fingers will become soiled with dirt.

Usually dandy brush and curry comb are the main tools used for grooming of dairy animals. A buffalo/cow generally licks herself practically all over and removes a fair amount of dirt in the process. Grooming of dairy animals comprises brushing of rump, sides, thighs and washing of udders with warm water containing suitable antiseptic solution and wiping of udder with sterilized warm damp cloth. Wiping is done to remove the redundant water. Surface of the udder should be damp as this prevents dry dust, etc. from falling into the milk. The process of grooming dairy animals is similar to that employed for horses. These processes should be done just before the milking and done in loose house and then animal is moved into the clean and dust free milking shed. Working bullocks should be groomed before and after the work and normally legs, shoulder and hips are groomed with wisps made of paddy straw or hay to relieve from tiredness and fatigue.

Clipping

Clipping is the process of removal of excess hairs and it is done to improve their general appearance, to make milking process more hygienic and to increase its acceptability to a show judge, buyers. Dairy cow may be clipped twice a year. Clipping should be done against the lay of hair, slowly and gently. Udder, flank, underline and end of tail switch are clipped in a dairy cow. Entire head except eyebrows, eyelashes and hair on muzzle, ears, neck, tail above two inches, udder and inside of thigh are clipped for a cow going to participate in a show or judging competitions. Excessive hairs from coronet, legs, bridle path, ear, muzzle and jaw is clipped in case of horses and should be repeated every 8-10 weeks. Thinning and shortening of manes and tail is also practiced in horses time to time. A milking goat should be clipped around tail head up to three inches of rump, flank area up to five inches from front udder, entire udder including sides and rear portion and entire rear legs.

Bedding

Bedding helps in keeping animals clean and comfortable. It soaks up urine, make manure handling easy, and prevent losses of ammonia and potash by leaching. Provision of bedding to the animals is more important in temperate countries than tropical countries like India. However, certain parts of northern India where mercury level goes significantly lower in winter together with cold wind which necessitates provision of bedding to the animals, more importantly to the young animals. Ideal characteristics of bedding materials are it should be absorbent, dust free, readily available, easily disposed off and cheap. Generally wheat straw, chopped grass, crop residue and sugarcane bagasse is used as bedding material and thickness of bedding should be 10-15 centimetres.

Methods of Controlling External Parasites of Livestock

To control external parasites a variety of drugs are available in the market which can be applied topically by various methods like by dipping, spraying, 'pour-on', 'spot-on', dusting, collar and tag. Usually, drugs which are used to control external parasites of livestock are liquid concentrates which is diluted with water to produce an emulsion for application.

a) **Dipping:** To eradicate external parasites from the hair, wool, fur and skin animals are immersed from time to time into a chemical solution. To ensure a reasonable level of residual protection, sheep should be dipped preferably within 3 weeks after the shearing. Dipping is generally practiced for sheep but sometimes also used for cattle. Following precaution must be undertaken while implementing the dipping process:

- Dipping of advance pregnant, sick and diseased animals with wounds, cuts scratches and other skin injuries should be avoided.
- Extremes of weather should be avoided and dipping is not recommended when it is raining.
- Preferably dipping should be started in the morning and must be completed by the mid-day. Therefore, last animal must get ample time for drying before evening.
- The animals should get adequate rest before dipping. In other words, tired animals should not be dipped.

- Lambs should be dipped separately from ewes with attention being paid to pairing-up ewes and lambs after dipping.
- It is essential to dip the head of sheep while dipping but animal's head should not be restrained in such a manner that it may swallow dipping solution.
- When dipping for the treatment of blowfly larvae, lice, mange, mites and ticks, the body of the sheep should be kept immersed at least for thirty seconds.
- For treatment of sheep body should be kept immersed at least one minute.
- The animals are offered free access of drinking water before dipping otherwise they may drink dip solution.
- Animals should be detained in a draining pen for 15 minutes after dipping to drain out the solution from their body.
- Once the procedure is complete dipping solution must be discarded properly so that it should not pollute the water supplies of the farm.

There are two methods by which dipping can be practiced i) hand bath and ii) swim bath. In the first method animal is lifted by a person and placed in a tank made up of galvanized iron (1.2×1.0×0.5 m) for two minutes and then placed on a draining board to just drain off surplus dip back into the tank. This method is practiced on a farm having small flock. Second method is used on a farm with large flock size and followed in following manner-the animals are first gathered in a collecting pen then driven through a narrow race (which is too narrow for them to turn around). In this race they pass through a foot bath which removes mud from the feet before they enter the dipping tank. Then the animals are moved forward into the dipping tank and are completely dipped in the solution. The animals walk out of the tank along the race of opposite side which has a sloping concrete floor to allow all solution dripping off the animals to run back into the tank. The width of this race allows only one animal to pass at a time and may have gates to allow animals to be held and marked for particular treatment.

b) **Spraying:** Under certain conditions and in some countries spraying is practiced instead of dipping for the treatment of ectoparasites. Spraying can be as good as dipping and can be used to treat pregnant animals which cannot be dipped. It requires less amount of drug and water, less risk is involved, can be employed when dipping is not possible. Various types of spray machines are available in the market which can be used as per the need of the farm or spray race can be constructed. The animals should be tethered or tied up securely between posts or held in a crush or race. Spraying can be done as: first hooves, under the tail and down to the scrotum or udder and the inside of the back legs should be thoroughly wetted then spray the belly, sides and back, and spray the outside and insides of the front legs and up under the brisket and lastly spray the head and neck making sure that inside the ears is also sprayed. Care must be taken to avoid spraying the eyes, nose, and mouth.

Other methods like dusting where drugs are applied in powder form to the animal body. In 'pour-on' method specified amount of solution may be applied along the dorsal midline for fly infestation, to the top of the head and around the base of the horns for head fly infestation, or to the site of the lesion for blowfly strike. In 'spot-on' method all the specified amount of solution is applied to a single site on the dorsal midline behind the shoulders or to the base of the poll of the head. Tapes and tags are used for fly control in extensive husbandry systems on animals.

Determination of Body Weight using Different Body Measurements

Body weight of livestock is essentially required for a variety of purposes like to know about amount of feed needed, efficiency of feeding programme, for estimating body weight gain or loss, to know about potential health problems, to determine time of breeding in heifers, to provide proper amount of medication and for determining show-ring classification. On an organized farm weighbridges are normally used to know the body weight of the animal but the use of weighbridge is cumbersome, costly and more importantly it is barely available under field condition. Therefore, various workers developed many formulae for the estimation of body weight of animals using their body measurements. Some of them are mentioned below.

1. For Cattle & Buffalo (Sastry, 1999)

a) Shaeffer's formula

$$\text{Live weight(lbs)} = \frac{\text{Length} \times \text{Girth}^2}{300}$$

L = body length as measured from the point of the shoulders to the pin bones in inches

G = heart girth in inches. Shaeffer's formula is used to calculate body weight of cattle, buffalo, sheep and goats.

b) Mullick's formula for buffaloes

$$X = 25.256\,(Y) - 960.232$$

X = Estimate of body weight in pounds

Y = Heart girth in inches

c) Bhandari and co-workers formula for buffaloes

i) For dry and non-pregnant buffaloes :

Body weight (lbs) : –2123.73 + 26.356 (girth in inches) + 21.50 (length in inches)

ii) For dry and pregnant buffaloes :

Body weight (lbs) : –1934.48 + 20.52 (girth in inches) + 25.90 (length in inches)

iii) For lactating buffaloes :

Body weight (lbs) : –2387.60 + 27.12 (girth in inches) + 24.55 (length in inches)

2. Formula to Estimate Body Weight of Goat & Sheep (Khan *et al.*, 2004)

$$W = \frac{(G+L)^2}{X}$$

Where;

W = Live weight of goat in lbs, L= Length in inches

X = 17.0 when girth ranges from 15-19 inches

X = 13.5 when girth ranges from 20-25 inches

X = 12.0 when girth is more than 25 inches

3. a) Formula to estimate body weight of a horse (Parker, 2008)

$$\text{Live weight(lbs)} = \frac{\text{Length} \times \text{Girth}^2}{330}$$

W = Weight in pounds

L = body length as measured from the point of the shoulder to the point of the buttocks in a straight line (in inches).

G = heart girth is measured just behind the elbow (in inches).

b). Formula to estimate body weight of foals (from 1 to 6 weeks of age)

$$\text{Live weight} = \frac{\text{Hearth girth (inches)} - 25.1}{0.07}$$

Transport of Livestock by Rail, Road, Air and on Foot

Transport of livestock involves movement of animals accomplished by one or more means of transport including related operations like loading, transfer, rest and finally unloading at the place of destination. The movement of animals from one place to another is necessary for various purposes like for their sale, purchase, participation in various shows, fairs, slaughter and treatment of illness etc. Care must be taken so that during the transport animals are subjected to minimum discomfort and reach the destination without deterioration of their health and minimum economic loss.

General Conditions for Transport of Livestock

1. Healthy animal should be transported & a qualified veterinarian should certify it.
2. Advance pregnant and diseased or sick animals should not be transported, transport of such animals is allowed only in the case of emergencies.
3. Young animals should be separated from adult animals & advance pregnant animals should be separated from other animals.
4. During transport all the animals should be given humane treatment.

5. Adequate feed and water should be provided during longer journey but feeding large amount of concentrate and succulent feeds should be avoided.
6. The transport vehicles should be clean, the floor, walls shall be undamaged & free from nails & other sharp edges.
7. Bedding materials such as straw or hay should be placed on the floor to avoid injury & it should not be less than 5 cm thick.
8. In case of journey for more than 12 hrs, an attendant should be present at all the time & should ensure the proper condition is maintained during transport.
9. The animals should be transported with maximum comfort and safety.
10. The animals should be transported by most economic means viz. truck, railway wagon, ship etc.
11. Animals should be protected from adverse and extreme climatic conditions.
12. Avoid overcrowding and slippery floors in the transport vehicle.
13. Apart from the above general condition some species of livestock require specific condition for transport & it also depends upon the mode of transport.

Effects of transport: Poor transportation process may have severe detrimental effect on the welfare of livestock and can lead to significant loss of quality and production. When transport facilities or conditions are improper, animal will be in stress, may suffer from bruise, injuries, dehydration, suffocation, exhaustion, heat stroke, bloat, poisoning, predation and finally animal may die due to any of the above mentioned reasons. Even if animal doesn't die the quality of meat obtained from these animals is of inferior type.

Means of transport: Animals are generally transported by trekking (on hoof), road motor vehicle, rail, ship and aircraft.

1. **Trekking/on hoof/feet:** Moving animal on the hoof is suitable only where road and rail infrastructure does not exist, or when distances from farm to destination is short. The animals are moved during early hours of the day and during the journey they are given rest at intervals under sheds. They are provided feed and water there. The animals are driven on soft ground than on hard road. The animals may be conveniently tied into

small groups with their neck ropes in the movement of cattle, sheep and goat. This method is slow and full with risks to the welfare and value of the animals. Maximum distance covered per day, maximum number of hour animals are moved per day, period of rest and suitable environmental condition for trekking for different species of animals have been given in the Table 2.4.1.

Preferable temperature range for adult animals is about 12^0 to 30^0C and for young animals around 15^0 to 25^0C. At every 2 hours water and at every 4 hours feed should be provided to adult animals while for young animals it should be at every 1 and 3 hours, respectively.

Table 2.4.1 : Conditions of transport of livestock on feet.

Species (Animal)	Maximum distance covered/day & speed	Maximum no. of walking/day of hours
Cattle (Cows)	30 km/ day, 4 km/hr	8
Buffaloes	25 km/ day, 3 km/hr	8
Calves	16 km/ day, 2.5 km/hr	6
Horses, Ponies, Mules & Donkeys	45 km/ day, 6 km/hr	8
Young ones (Foal)	25 km/ day, 4 km/hr	6
Goats and Sheep	30 km/ day, 4 km/hr	8
Kids and Lambs	16 km/ day, 2.5km/hr	6
Pigs	15 km/ day, 2 km/hr	8
Piglets	10 km/ day, 1.5 km/hr	6

Source: GoI, 2001.

2. **Road transport**: This is by far the handiest, first choice and most user friendly method of transport of animals. In this method animals are transported in trucks or tractor trolley to the longer distances by road with the advantage of direct loading and unloading at point of origin and destination. Vehicle should move at uniform speed and better road should be preferred. Animals should be given rest and provided feed and water during long journey. The animals should be protected from extreme weather conditions. The truck platform should be bedded with straw.

    - **Cattle:** Specially fitted goods vehicles with a special type of tail board and padding around the sides should be used. Ordinary goods vehicles shall be provided with anti-slipping material such as coir matting or wooden board on the floor. Suitable rope and platform should be used for loading and unloading of cattle from vehicles. No goods vehicle shall carry more than six cattle. Each goods vehicle shall be provided with one attendant. To prevent cattle being frightened or injured, they should preferably, face the engine.

- **Sheep & Goat:** Goods vehicles (capacity of 5 or 4 ½ tons), which are generally used for transporting animals, shall carry not more than forty sheep or goats. In the case of large goods vehicles, partitions shall be provided at every two or three metres across the width to prevent the crowding and trampling of sheep and goats. In the case of ewes, goats or lambs or kids under six weeks of age, separate panels shall be provided.
- **Horse:** Bamboo poles of at least 8 cm diameter between each animal and two stout batons at the back shall be provided to prevent the animal from falling. To prevent horses from being frightened or injured their heads should face left away from the passing traffic. Each vehicle shall not carry more than four to six equines. These vehicles shall be driven at a speed not more than 35 kilometres per hour.
- **Pigs:** Goods vehicles, which are generally used for transportation of animals, shall carry not more than twenty pigs. In the case of transport of pigs under six weeks of age, separate panels shall be provided.

3. **By rail**: Rail transport is useful for short-haul journeys where loading ramps are available at platform and communication is direct to destination. The train transport is preferred for longer distance as it saves tax expenses, maintenance, handling charges and checks the body weight loss during transit. The railway wagons used for transportation of animals must be roofed, provide adequate ventilation and floor must be non-slippery. Along with bedding material a sand layer (10-15cm thickness) can be provided which soaks up urine, prevent slipping of animals and make lying down comfortable. The animals should be provided with adequate rest, feed and water at regular interval.

   - **Cattle:** The average space provided per cattle in railway wagon or vehicle shall not be less than two square metres.

In case of railway wagon the dropped door of the wagon may be used as a ramp when loading or unloading is done at the platform. Cattle shall be loaded after they are properly fed. Cattle-in-milk shall be milked at least twice a day and the calves shall be given sufficient quantity of milk to drink. Two experienced attendants should accompany the wagon. The side doors should be left open for ventilation. Partitions of suitable size should be made of wooden poles to keep the animals on their feet while train is moving. Cattle shall be loaded parallel to the rails, facing each other and wagon should be attached in the middle of the train. An ordinary goods wagon shall carry not more than ten adult cattle or fifteen calves on broad gauge, not more than six adult cattle or ten calves on metre gauge, or not more than four cattle or six calves on narrow gauge wagons.

- **Sheep and goats:** Generally, sheep and goats are transported separately but if their number is small then they can be separated by means of special partition. Male animals shouldn't be mixed with females in the same compartment. At least five centimetres thick bedding materials (straw) shall be provided. No animals should be tied except there is a danger of their jumping out. The leg of animals shouldn't be tied down. Broad, meter and narrow gauge wagons shouldn't accommodate more than 70-100, 50-60 and 25 sheep or goat, respectively.
- **Horse:** Horses shall be transported by passenger or mixed trains only. Ordinary goods wagon when used for transportation shall carry not more than eight to ten horses or ten mules or ten donkeys on broad gauge and not more than six horses or eight donkeys on meter-gauge. Two breast bars shall be provided on each side of the wagon, one at a height of 50 to 80 cm and the other at 110 cm.
- **Pigs:** Adequate ventilation shall be provided in every wagon and the upper door of one side of wagon shall be kept open. Not more than 35-50 pigs in broad gauge and 25-30 pigs in meter gauge wagons should be transported.

4. **By ship**: This mode of transport is only used when other means of transport are not possible. The animals should not be allowed to see the seawater, which can be avoided by provision

of big walls. As the Journals time is more, adequate provision of feed, water and attendant should be made. The animals should be tied with ropes or chains during transportation or should be restrained with any other proper means.

5. **By air**: This means of transport is only used for very important and specific animals for a longer distance journey to cover the distance in shorter time and with maximum comfort. Troublesome animal should be given tranquillizer before loading especially when being transported by aircraft. The animals are restrained in special type of crate boxes, which are well bedded and covered with canvas.

Identification of Sick Animals

The symptoms shown by an animal will help to diagnose whether it is healthy or diseased. Some of the important symptoms of ill health in animals are described below:

i. A healthy animal is alert, aware of its surroundings and holds its head up watching what is happening around it while animals not showing interest in its surrounding, not willing to move, standing with head down and a tendency to separate from the other animals are warning signs of ill health .

ii. The healthy animal will walk easily and steadily with all of its feet taking its weight. Irregular movement indicates some kind of suffering.

iii. The eyes of a healthy animal should be bright and alert with no discharge at the corners.

iv. Healthy animals eat greedily. Loss of appetite and stoppage of rumination in case of ruminants are early signs of diseases.

v. In short-haired animals (cattle & goat) the hair or coat of the healthy animal will be smooth and shiny. Horses should not sweat when resting. Raised hair coat, falling of brittle and lustreless hair is undesirable. The coat should not have patches.

vi. Muzzle and nostrils of healthy animals will be moist and free from any discharges. Healthy animals frequently lick their noses with their tongues. The muzzle will be dry in animals having high temperature.

vii. Dung of healthy cows should be semi solid in consistency and free from gas bubbles, blood clots and mucous. Dung of sheep and goats is in the form of aggregates of small globules. If the animal has difficulty in defecating (constipation) this is also a sign of ill health.

viii. The urine should be clear and the animals show no signs of pain or difficulty in urinating. Difficulty in urination, dark or bloody colour urine and any abnormal odour (sweet odour in ketosis) are signs of ill health.

ix. In lactating animal, sudden change in quality and quantity of milk yield is one of the early symptoms in several diseases. Any sign of blood or other matter in the milk, swelling of udder and signs of pain when udder is touched indicate infection in the udder.

x. Presence of abnormal discharges from vulva and around tail region indicates some health problem.

xi. Change in normal voice of animal i.e. groaning, grunting and grinding of teeth indicates some kind of illness.

xii. Inflammatory conditions and oedema of any body part, changes in colour of visible mucous membrane and signs of haemorrhage over them are other signs of illness.

xiii. **Temperature:** Variation in body temperature can be measured by inserting a clinical thermometer into the anus of an animal for half a minute. A fair amount of idea may be perceived by placing palm over skin and base of the ear. Young animals, female in late pregnancy and excited animals frequently show higher temperatures while on chilly days, emaciated and malnourished animals and cows suffering from milk fever may have sub-normal body temperature. The normal body temperatures of different livestock are given in Table 2.5.2.

xiv. **Pulse:** Pulse refers to the expansion and elongation of the arterial wall caused by column of arterial blood due to contraction of left ventricle of the heart. Variation in pulse rate reflects the rate at which the heart pumps blood through the body. While recording pulse rate it is necessary to keep animal at rest with least external disturbances. Generally, pulse rate found to be higher in young and pregnant animals. The normal pulse rate in different livestock species are given in Table 2.5.2. Pulse can be recorded at below mentioned places in different species of animals.

xv. **Respiration rate:** Respiration rate can be measured by counting movement of flank/movements of ribs and sternum/ the hot stream of expired air blowing against back of palm kept near the nostril of an animal/placing stethoscope over thorax. Together with respiration rate rhythm, depth and type of respiration should also be noticed for identifying a sick animal. Respiration rate may change under the influence of various disease conditions. The normal respiration rate in different livestock species are given in Table 2.5.2.

Table 2.5.1 : Site of recording pulse rate in different livestock species.

Species	Sites
Cattle & Buffalo	• Middle coccygeal artery (around 10 cm below the level of anus). • Facial artery (external). • Transverse facial artery. • Femoral artery (calf).
Sheep & Goat	• Femoral artery (inside the hind limb rather more than half-way between stifle and hip joints).
Pig	• Femoral artery (small pig). • Middle coccygeal artery.
Horse/Donkey/Mule	• External maxillary artery (medial aspect). • Transverse facial artery. • Great metatarsal artery.

Table 2.5.2 : Normal temperature, pulse rate and respiration rate in farm animals.

Species	Normal temperature(^{0}C)	Pulse rate/ minute	Respiration rate/ minute
Cattle	37.8-38.9	40-50	20-25
Buffaloes	37.2-39.0	40-45	16-20
Horses	37.2-38.2	36-45	8-18
Goats	38.8-40.5	70-80	12-25
Sheep	38.3-40.0	70-80	12-25
Pigs	38.6-40.9	60-80	10-20
Camels	34.4-37.0	28-32	5-7
Fowls	40.6-41.7	300	15-30

Source: ICAR (2002).

Methods of Drug Administration

Administration of medicines through the proper route for prevention, diagnosis and treatment of diseases is the important part in health management of animals. Drugs can be administered parenterally (means apart from the intestines), orally and locally (applied directly over the site where action of the drug is required). Choice of route of drug administration mainly depends on nature of drug and some animal factors. Some of these routes of drug administration are :

1. **Oral administration**: This route of drug administration is the most common, convenient, less likely to cause adverse reaction and don't require sterile condition. However, effect of drugs administered through this route is comparatively slow. Oral administration delivers medicine directly to the animal's gastrointestinal (GI) tract and through this route liquid (solutions, suspensions and emulsions), solid (tablets, capsules, boluses and powders) and semi solid (electuary) medical preparations can be given. Tablets, capsules and boluses can be given to an animal by hand or pilling/balling gun. For administration of bolus, the head of the animal is raised, mouth is opened, and tongue is grasped with left hand and from right hand bolus is pushed into the mouth back to the pharyngeal region. The powder can be smeared on the tongue of animals after mixing in feed or in drinking water. The electuary is smeared on the back of the tongue, gums, molar teeth and lips of animals. Oral liquid drugs may be given by dropper, syringe, and drench (given by forcing the animal to drink using long neck bottle or stomach tube). Drenching is more frequently used particularly to administer anthelmintic drugs. While practicing drenching care must be taken that animal shouldn't chew the bottle or tube, drug should be poured slowly and if animal coughs instantaneously during drenching, immediately stop the procedure because it indicates wrong technique i.e. that the drug has entered into the respiratory tract.

   - **Drenching by bottle**: After restraining animal properly, mouth is opened by left hand leaving the tongue mobile. The neck of bottle is then inserted in between the cheek teeth and cheek and the drug is allowed to dribble slowly. The sheep and goat can be drenched by narrow necked bottle or with small tube (2-3" in length) fixed with funnel at one end.

- **Drenching by stomach tube**: It is the more convenient and hygienic method than previous method. The drenching by stomach tube has many advantages over drenching by bottle like: there is no risk of choking, no wastage of drug, drugs having unpleasant taste and odour can also be given and can be performed with least resistance and struggling. The stomach tube measures around 9-10 feet in length with 1/2" to 5/8" of diameter. At about 3/4", there is an opening on its one end. Two marks, one at about 10-14 inches and another at 4 feet are made on the tube to indicate the position of tube in the stomach when the first mark is in nostril then the solid end of tube is in pharynx and when the second mark is in nostril then the solid end is in stomach.

2. **Parenteral/Injectable Route:** It refers to administration of drugs directly into the body with the help of syringe and needle. This is the most efficient and reliable route for administration of drugs into the body as drugs administered by this route act rapidly, utilized more efficiently, generally act longer. Further, drugs which cause gastric irritation, destroyed in or not absorbed from digestive tract can be given through this route. This method has certain disadvantages also as it is expensive, requires sterile technique, somewhat painful, causes local irritation, chances of systemic allergic reaction and requires expert hands for administration. Various routes for administration of drugs through injections have been mentioned below.

   i) **Intra-muscular injection** (I/M): It is the most common parenteral route of drug administration. I/M injections come in aqueous solutions, aqueous suspensions, oily suspensions, and injectable pellets. The drugs are injected by inserting the needle deep into skeletal muscles particularly gluteal muscles and sometimes muscles of neck, back or hindquarters. Large amount of drug at one site may result in necrosis hence, small amount (<12-15 cc) of drugs must be deposited in one site. While performing I/M injections care must be taken to ensure that the syringe is not injecting the drug directly into a blood vessel. To verify just pull back the plunger of the syringe and observe if blood become visible in the syringe, if so then pull out the needle from injection site. Choose a new site and do it again till blood doesn't come in syringe.

ii) **Intra-venous injection** (I/V): The drugs are injected directly into the lumen of blood vessels. In large animals, sheep and goat, injection site is jugular vein present in jugular furrow in the middle third of neck whereas it is cephalic vein or tarsal vein in case of dog & cat and ear vein at the base of ear in pig. Advantages of administering drugs through this route is that it is the fastest method of reaching drugs into the bloodstream, response is quick, large volume of fluid can be given and drugs irritant to muscles can also be given. Intravenous injection should be given with the needle pointing in the direction of blood flow. This method has certain disadvantages like risk of systemic allergic reaction, air embolism in blood vessels, phlebitis and insoluble drugs cannot be given by this route.

iii) **Subcutaneous injection (S/C)**: The injection is given underneath the dermis of skin but on the top of muscle layer. In large animals the site of injection is middle third of neck behind the shoulder and in small animals, axillary space (under the front leg), flank and abdomen. Fold of skin is held in between the finger and thumb of left hand and needle is pierced swiftly into the skin, then the drug is injected slowly. The area is massaged to facilitate the dispersion of the drug. Response of drug is faster than oral but slower than intramuscular administration. Irritating or hyperosmotic solutions shouldn't be given through this route as it will damage the tissue.

iv) **Intra-dermal injection (I/D or I/C):** The drug is injected into the layers of the skin (between dermis and epidermis). Very small quantity (1-2 cc) of drug can be injected at a site, which leaves a bead like elevation at the injection site. This route is preferred when slow absorption of drug is required and mostly preferred for local treatment and sensitivity test of drugs and sera.

Other parenteral methods of drug administration are -Intra-cardiac injections (IC), Intraperitoneal (IP), Intramammary (IM), Epidural, Subdural, Intrathecal, Intra-articular and Intramedullary which is given within the heart, peritoneal cavity, within the teat and udder sinuses, above the dura mater of the meninges, under the dura mater of the meninges, or into the subarachnoid space of the meninges, within the joint and within the medullary cavity of bone, respectively.

3. **Local Routes:** Locally administered drugs are those which are applied directly on the site where the action of the drug is desired. Local administration of drugs includes topical application (skin, eye or ear), and other routes like rectal, vaginal and urogenital.

   i) **Topical application :** In topical medication, drugs are directly applied over external surface of skin or mucous membranes. Topical medications come in ointment, cream, gel, paste, lotion, powder, and aerosol dose forms. Topical applications are used for localized treatment of conditions such as localized skin infections (lesions), abrasions and localized skin allergies. Medicines are directly applied on affected body part of skin by hand or by spray in liquid form over whole of the body. Topical treatment also includes eye and ear medications. Eye preparations are available as liquid drops and ophthalmic ointments. Ear preparations are available as drops, ointments and creams.

   ii) **Urogenital tract**: Various preparations viz. pessaries, ointments, vaginal and uterine douche, and many liquid medicaments are administered through this route. Vaginal drug delivery systems like controlled internal drug release (CIDR) devices, progesterone-releasing intravaginal devices (PRID) and vaginal sponges are used for estrus synchronization in sheep, goats and cattle. Pessaries are inserted into the vagina to treat vaginal infections and the ointments are applied in vagina with help of vaginal speculum. Non-irritant antiseptics like potassium permanganate, lugol's iodine etc. in different concentrations are used as vaginal and uterine douche to treat metritis, vaginitis, bovine sterility etc. To remove urine, to wash urethra and urinary bladder catheterization is done with help of a catheter using antiseptic solutions in urogenital affections like cystitis, urethritis, etc.

   iii) **Rectal**: Medicated fluids, buffer water or soap water (enema) are given through this route. Due to availability of smaller surface area rate of absorption of drug is slower than oral route. Solid preparations called suppository are inserted into the rectum in small animals. Local irritation is a possible side effect of drug administered through this route.

iv) **Inhalation:** The medicated vapours of volatile drugs, produced by heat or water bath are used for inhalation. The inhalation route is also used to treat local respiratory conditions and includes bronchodilators, mucolytic enzymes, antibiotics and glucocorticoids.

Animal Behaviour Terminology & Common Vices of Animals, their Prevention and Control

Animal behaviour is a composite process involving the interaction of inherited abilities and learned experiences to which animal is subjected individually or in groups. Primary function of behaviour is to enable an animal to adjust to some changes in condition (external or internal) and thus enhances their likelihood of survival. Etholgy is the study of animal behaviour, its causes and its biological function whereas an ethogram is a detailed description of the behaviour features of a particular species. Behaviour is essentially inherited through genes and modified by the environment. Behaviour is controlled by the nervous system and executed by muscles, bones, etc. within the body. None of the behaviour is independent of genetic information and at the same time not independent of all environmental factors. Behaviour could have many causes like genetic constitution of the species (determines behaviour pattern), presence and absence of the primary stimulus (produces behaviour). For a behaviour to occur it is required that there must be some change in external or internal environment. Finally, learning by which an animal organises their behaviour and act according to what they have learned by their earlier experiences. Knowledge of behaviour can be used as a tool in disease diagnosis, handling, recommendations for proper animal rearing methodologies, identification of abnormal behaviour and evaluation of welfare status of the animals. Some of the terms which are related with animal behaviour have been discussed below.

- **Instinct:** Instinct deals with that component of behaviour which is inherently present at the time of birth. With this animals remain capable of producing a large gamut of adaptive responses without prior learning or experience. Young ones of mammals have the instinct to nurse though they have to learn first to locate the teat. Chicks soon after hatching start pecking to get the food.
- **Learning:** It is defined as the acquisition of a new response or a qualitative change of an existing response, or an inhibition or facilitation of an existing response by a new stimulus.

- **Conditioning:** There are two types of conditioning i) classical conditioning and ii) operant conditioning. In former type, an animal learns to exhibit an existing response to a new stimulus e.g. Pavlov's study, where unconditioned response (salivation) normally shown to the unconditional stimulus (bell) and milk let-down by dairy cows in response to the characteristic sounds of a milking parlour. In later type, an animal learns to respond in a particular way to a stimulus as a result of reinforcement (positive or reward and negative or punishment) when the proper response is made. Cattle come to manger when they see a feed truck coming is an example of reward or positive reinforcement while avoidance of an electric fence by the cattle is an example of negative reinforcement or punishment.
- **Habituation:** It refers to the gradual quantitative change in response which may lead to a complete loss of response as a result of repeated stimulation. For example when an animal first see or hear low flying aircraft/sound of heavy machinery they become frightened and may try to escape but after they are frequently exposed to it they become habituated. In contrast, sensitisation is the increase in response to a repeated stimulus. Presence of a predator might bring out a greater response second time than animal first see it.
- **Imprinting**: Imprinting includes processes whereby helpless young ones develop social preferences, usually to their own species. It varies from species to species e.g. odour and licking of fluid by dam develops bond in cattle and sheep. Imprinting is a rapid type of learning which occurs during a definite period of early life, whose effect is irreversible and remains powerful without repeated reinforcement. Imprinted animal responds to the species general characteristics of stimulus object rather than the individual characteristics of the object. Critical period or sensitive period is the age at which stimulation given in early life noticeably influence behaviour in later part of life.
- **Systems of animal behaviour:** Farm animals exhibit a number of systems/patterns of behaviour and each of these patterns has a definite adaptive function. Organization of behavioural systems differs from species to species. Major behavioural systems exhibited by farm animals are

    i) **Ingestive behaviour:** Ingestive behaviour of farm animals includes consumption pattern of food both solid and liquid and it differs distinctly among the various livestock species viz. chicken-pecking, duck by nibbling or scoop, cattle & sheep by rumination, pig by rooting etc. Ruminants while eating their food don't chew their feed thoroughly at the beginning, they mix it with saliva and swallow it and once certain amount of feed has been consumed they regurgitate the feed for chewing. Cattle usually graze and ruminate for 4-9 hours per day and may regurgitate and chew around 300-400 boluses per day while sheep and goat graze for 9-11 hours , ruminate for 7-10 hours and chew around 400-600 boluses per day. A goat often chooses to browse whereas sheep prefers to graze closer to the ground. Cattle cannot graze plant material closer than one cm. Pig pushes its nose into the ground in forward and upward thereby moving the dirt and exposes roots, tubers, grubs and earthworms. Horse utilises both its upper and lower jaw to cut the grass, chew it thoroughly and don't ruminate.

ii) **Eliminative behaviour:** Pattern of defecation and urination are examples of eliminative behaviour. Cattle, sheep, goat and pig void their faeces while standing or walking and usually urinate while standing. Cattle, buffalo, sheep, goat, pig void their urine indiscriminately. Cat burry their urine and faeces while dogs deposit them on some spots called as scent post.

iii) **Sexual behaviour:** Each animal species has unique pattern of sexual behaviour and is important in successful implementation of breeding programme. Males and females of some species produce chemical substance known as pheromones which attracts members of opposite sex. In a herd of cow, bull identifies the cow in heat by visual means and follows her, smell and licks her genitalia and rests his chin on the rump of cow and when in full heat allows the bull to mount. The buck snorts when detects a goat in heat. Boar and stallion cannot detect heat by olfactory or visual means. A sow in heat will stand still and flicks her ear when approached by a boar whereas a mare in heat stands, crouch a bit, urinates and vulva shows winking when approached by a stallion. Bulls, rams and stallion smells vagina and urine of female and exhibits a characteristic behavioural sign by lifting its head and curling its upper lip known as Flehmen reaction.

iv) **Care-giving behaviour (Epimeletic):** Most of the care giving behaviours is related with the mother hence known as maternal behaviour and is very important for the survival of the young animals. Licking or grooming of new born by the mother, nest building, brooding, fighting tendency of mother in response to bawl or squeal of young in the event of a threat, reduction in milk production after certain time in order to make the calf to go in search of forage consequently early development of rumen etc are examples of this behaviour.

v) **Care-soliciting behaviour (Et-epimeletic):** Generally young animals of most of the domesticated livestock species cannot care themselves completely hence solicit for help from others in the form of call or signal for help. Calves bawl, lamb bleat, pig squeal and chicks chirp to draw attention of their mother when they are disturbed, distressed or hungry. Usually loudness of the call gives an idea about amount of distress implicated. Adult also when separated from group or hungry solicit help from farmer.

vi) **Agonistic behaviour:** This includes behaviour activities of fight, flight and other associated reaction when an animal is in contact of another animal or human beings. Generally fighting is more evident among males of all domestic mammals and birds. Mixing unfamiliar males results in fighting behaviour which passes through stages like offense, defence, escape and passivity, results in establishment of social dominance. Fighting ends when defeated animal escapes and assumes a passive posture. Fighting is less intense in cow, sow and mare than their males.

vii) **Allelomimetic behaviour:** This behaviour engages two animals performing the same thing at same time with some degree of mutual stimulation. This behaviour induces a phenomenon known as "social facilitation", means any increment of activity due to the presence of another animal. It is well known that animals eat more, less distressed, healthier and productive when in group than alone. Animals which have strong herd instinct become stressed and agitated when kept alone.

viii) **Investigative behaviour:** When placed in a new environment, all domestic animals have tendency to explore and investigate it by utilizing their sense organs (eyes, nose etc). This ability varies significantly from species to species.

ix) **Shelter seeking behaviour:** Animals look for an environment in which they feel comfortable. In doing so they attain shelter from trees, certain part of barn and bodies of other animals. In hot weather condition cattle and sheep look for shady area for rest and rumination while pig prefers wet area. In cold weather animals huddle together and particularly in pig it is so pronounced that some animals dies due to crushing.

x) **Behavioural disorders:** Animals manifest maladaptive or abnormal behaviour in a situation when they are unable to adapt to their environment. In this situation animal will either give proper reaction but expressed in the direction of another object/ animal as a substitute to the original stimulus (known as displaced behaviour) e.g. when an animal is threatened by a stick, hook their herd mate instead of the handler or animal reacts improperly (known as displacement of behaviour) e.g. when a cock involved in fighting all of a sudden stop and start pecking the ground as it is feeding. Hence, abnormal behaviour can be defined as behaviour that differs in pattern, frequency or context from that which is shown by most members of a species in a condition which allows a full range of behaviour (Fraser & Broom, 2007). Occasionally, word vice is used to denote abnormal behaviour. A common type of maladaptive or abnormal behaviour is stereotypy. Stereotypy is a repeated, relatively invariate sequence of movements having no obvious purpose. It may adversely affect the health and performance of the animal as well as of other member of the group. Some vices are dangerous while others are injurious to the owner, animal or both and even sometimes make the animal almost useless. Some of the abnormal behaviour of livestock species has been discussed in Table 2.6.1.

Table 2.6.1 : Common vices of animals, their prevention and control.

Name of vices	Common Species	Description, prevention and control strategies
1. Weaving	Horse	It involves swinging of head, neck and anterior part of the body from side to side in such a manner that weight rest alternately on each fore-limb. It occurs due to long period of idleness and results in loss of weight. It can be controlled by tying the animals with crossreins to restrict sideways movement of the head. Further, animal should be turn out to pasture or adequate exercise should be given.
2. Wind sucking (aerophagia)	Horse & cattle	It occurs in many forms. Initially animal may jerk head upward, opens mouth takes air in and raises floor of mouth as the neck is flexed. Characteristic sound is produced when air is expelled. It can be controlled by using wind sucker strap, creation of fistulae on each side of the mouth between the buccal cavity & outer cheek, use of flute bit, use of metal gullet strap or plate and by changing the environment of the animal.
3. Pawing and stall kicking	Horse	It is performed repeatedly by lowering the head, pulling ears back, arching the back, and kicking with hind leg against wall or door behind. It can be controlled by use of hanging mat or barriers behind, turning the animal to pasture or by providing adequate exercise.
4. Crib-biting	Horse	Animal grasps the edge of the manger or some other object with incisor teeth and press down, raises the floor of mouth, soft palate is forced open and produces a distinct grunt sound. It can be controlled by tying a strap round the throat (restricts arching of neck), use of hollow, cylindrical perforated bit (checks making mouth air tight), use of a thick rubber or wooden bit (prevents closing of jaw) tying the horse with two ropes each on either side of neck to avoid lateral movement and surgery.
5. Hyperphagia	Horse, dog, cattle	It is the habit of rapid eating and can be controlled by any means which make it difficult to the animal to consume rapidly.
6. Tail rubbing	Horse	The animal moves its hindquarter from side to side to a fixed object with tail pushed into the perineal region in absence of any parasitic infestation to the region. It can be controlled by keeping tail region clean and using a tail bandage/guard/board.

Contd.

7. Licking or crib-whetting	Horse	Animal applies the tongue repeatedly to an area of its own body or to an object in the surrounding. It can be controlled by providing adequate space.
8. Pacing or route tracing	Horse, zoo animals and poultry	Animal follows a path repeatedly that returns to its point of origin with minor changes. It can be controlled by providing adequate space & food, changing the environment of the animal.
9. Head shaking	Horse	It involves repetitive bobbling up and down of the head. It can be controlled by changing the environment of the animal and by applying a mesh or veil over the nose.
10. Coprophagia	Rabbits, Horse, Pigs	It is habit of eating faeces. It can be controlled by providing regular exercise and providing adequate space.
11. Kicking	Cattle & Horse	It involves striking with foot over others. It occurs due to lack of work, irritation, impatience, fear, mischievousness, recurrent oestrus and faulty milking practice. It can be controlled by applying anti-kicking straps, holding the tail vertically high, application of 'kicking board' behind the hind legs, tying of a bag of sand to hind legs, tying a wooden log with short chain above the hock or to the pastern and providing adequate exercise to the animal.
12. Tongue rolling	Cattle	Animal extrudes its tongue from the mouth and moves by curling and uncurling outside or inside the mouth with no solid material present in it. It can be controlled by applying wind sucker strap, inserting a metal ring through the frenulum of the tongue, providing mineral salts and adequate space for movement.
13. Inter-sucking (young)	Cattle & Pig	It is a habit of sucking and licking of own body, objects in the pen and body parts of herd mates. It occurs due to early weaning, bucket feeding and poor welfare conditions. It can be controlled by housing animals individually, use of automatic nurser, tying of calves for one hour after bucket feeding, using muzzle and providing adequate roughage.
14. Inter-sucking or milk-sucking by adult animals	Cattle	It involves sucking of milk by another cow or bull. Rarely some milking cow shows habit of sucking of milk from their own udder. It can be controlled by application of pointed prongs to the nose and face region of sucking animal, tying of an electrical device (gives

Contd.

		electrical shock to the wearer) on the forehead of sucking animal.
15. Tonic immobility	Cattle	It is habit of lying or freezing in extended recumbency with low level of reaction to a stimulus which in general causes animal to change its posture. It can be controlled by transferring animal to a new environment and providing adequate space.
16. Head rubbing	Cattle & Pig	Head is repeatedly moved back and forwards against a solid object. It occurs due to long period of confinement to stall and may result in physical injury. It can be controlled by adequate space to the animal.
17. Bar-biting, tether-biting	Swine	Animal opens and closes its mouth around a bar or tether and engages its tongue and teeth with surface and performs chewing movements. It occurs when animals are housed in narrow space and don't allow animal to turn around. It can be controlled by providing straw or sawdust as litter material.
18. Anal massage	Swine	It involves rubbing of nose on anus of the other animal and as a result ingestion of faeces. It can be controlled by providing adequate space, opportunity for rooting and straw and sawdust.
19. Tail-biting	Swine	Animal take tail of another animal and chew on it resulting into wound and haemorrhage. It can be controlled by tail-docking (distal half), removal of tail biter from the group, reduction in stock density and providing opportunity of rooting.
20. Sham chewing	Swine	It involves jaw movement similar to those revealed when chewing food is shown at a time when animal has no food in its mouth. It can be controlled by providing sufficient quantity of high fibre feed and access to manipulable material such as straw.
21. Maternal cannibalism	Pig and sheep	It involves killing (crushing or biting and eating) of own young ones by a parturient animal. It is associated with hyper-excitability particularly in sows farrowing for the first time or crowded indoor condition. It can be controlled by providing adequate space, providing well balanced rations and use of anti crush devices.

Chapter - 3

Breeds of Livestock

Prem Prakash Dubey, A Kumar and S Kumar

Animal Genetic Resources

Agricultural biodiversity is the product of thousand years of activity during which humans have sought to meet their needs in a wide range of climatic and ecological conditions. Well adapted livestock have been an essential element of agricultural production systems. For livestock keepers, animal genetic diversity is a resource to be drawn upon to select stocks and develop new breeds. More broadly, genetically diverse livestock populations provide society with a greater range of options to meet future demands, so the wise management of the world's agricultural biodiversity is becoming an ever greater challenge for the international community. The livestock sector in particular is undergoing dramatic changes as large-scale production expands in response to surging demand for meat, milk and eggs (FAO, 2008).

Animal genetic resources (AnGR) include all animal species, breeds and strains that are of economic, scientific and cultural interest to humankind in terms of food and agricultural production for the present or the future. Domesticated animals are considered to be those species that are bred in captivity and modified from their wild ancestors to make them more useful to humans, who control their reproduction (breeding), care (shelter, protection against predators) and food supply (Diamond, 2002; Mignon-Grasteau, 2005). Only about 40 animal species have been domesticated since the Pleistocene. Common species include cattle, sheep, goat, chicken, duck, pig, horse, buffalo, rabbit, camel, donkey, elephants, various poultry species, reindeer, etc. All these AnGR are vital to the economic development of the majority of countries in the world playing an important role in the subsistence of many communities. Although only a subset of the diversity present in the ancestral species survived in the domestic counterparts, domestic livestock diversity has been continuously evolving. Reshuffling of genes at each generation, mutation, and cross-breeding or admixture of different gene pools has offered new opportunities for natural and human selection. Globally, domestic AnGR supply some 30% of total human requirements for food and agricultural production (FAO, 2007). They are particularly vital to subsistence and economic development in developing countries. In rural areas, livestock are an important source of food and cash, hence are crucial for the purchase of consumer goods and procurement of farm inputs. Other functions of livestock include production of such non-food items as leather, skins, wool, transportation and fuel (from dung) in some communities. They also facilitate the use of marginal lands of little or no value for crop agriculture (Anderson, 2003). Existing AnGR thus represent a massive past investment which, if managed appropriately, can provide insurance against an enigmatic global future. Although no compelling quantitative data is available, it has been estimated that about 50% of the total livestock genetic variation is between species and the remaining 50% is accounted by variation among breeds within species. However, different species tend to perform particular functions, often in specific environments that have limited overlap with other species and livestock species are unlikely to become extinct. Thus the focus on conservation of AnGR has to be within species. Moreover, the variation between breeds is likely to be much higher when a global perspective is taken and when more extreme traits, such as adaptation to harsh environments and disease resistance are considered. However, a more important consideration is the rapidity with which AnGR can be exploited to deliver new levels of production and adaptation, including disease resistance. Within breeds, the amount of genetic change that can be made per unit time is a function of genetic and environmental variance, whereas the rate of change between breeds is a function of range rather than variance. Allowing for this, it is clear that for the majority of traits and production systems the most valuable resource, in terms of providing rapid adaptation to the huge diversity of existing production systems, and for providing flexibility to respond to changing systems and environments, is the variation between breeds (Rege & Gibson, 2003). Thus genetic erosion within livestock species, including their wild ancestors, is of particular concern because of its implications for the sustainability of locally adapted agricultural practices and the consequent impact on food supply and security. Within species, an understanding of the evolutionary history of different breeds in a country or region and quantitative data on the genetic relationships among the breeds, can provide critically important inputs into the decision-making process. Current inferences of evolutionary history of breeds are based on archaeological, anthropological and ethnographic data, but increasingly this is being supplemented or replaced by results of molecular genetic studies.

Risk Status Classification

Risk status classification of breed endangerment is available from many sources. At present, the most widely reported indicators pertinent to livestock biodiversity are found in the list provided by FAO through the "Domestic Animals Diversity Information System" (DAD-IS). FAO (1992) has developed a framework for classifying breeds on the basis of level of 'threat' into various categories based on considerations of population size, fertility, length of reproductive cycle and the exposure of the population to `risk-causing factors'.

DAD-IS Monitors Breeds Worldwide and Classifies them into the following Risk Categories

- **Extinct:** A breed is categorized as extinct when there are no breeding males or breeding females remaining. Nevertheless, genetic material might have been cryo-conserved which would allow recreation of the breed. In reality, extinction may be realized well before the loss of the last animal or genetic material.
- **Critical:** A breed is categorized as critical if the total number of breeding females is less than or equal to 100 or the total number of breeding males is less than or equal to five; or the overall population size is less than or equal to 120 and decreasing and the percentage of females being bred to males of the same breed is below 80 percent, and it is not classified as extinct.
- **Critical-maintained:** Those critical populations for which active conservation programmes are in place or populations are maintained by commercial companies or research institutions.

- **Endangered:** A breed is categorized as endangered if the total number of breeding females is greater than 100 and less than or equal to 1000 or the total number of breeding males is less than or equal to 20 and greater than five; or the overall population size is greater than 80 and less than 100 and increasing and the percentage of females being bred to males of the same breed is above 80 percent; or the overall population size is greater than 1000 and less than or equal to 1200 and decreasing and the percentage of females being bred to males of the same breed is below 80 percent, and it is not assigned to any of above categories.
- **Endangered-maintained**: Those endangered populations for which active conservation programmes are in place or populations are maintained by commercial companies or research institutions.
- **Breed at risk**: A breed that has been classiûed as critical, critical-maintained, endangered, or endangered-maintained.

India has vast and varied biodiversity in farm animals distributed over its large geographical area. Various breeds of different species of livestock have been developed through natural selection in different agro-ecological zones. The animal genetic resources include all types of species and breeds that exist. Mason, I. L. 1988 had published a book "A World Dictionary of Livestock Breeds Types and Variety" it is most comprehensive book on Domestic Animal Genetic Resources.

Species, Breed and Strain

The kingdom is the highest taxonomic category. The lowest taxonomic category of the kingdom is the species. The species is the basic unit in taxonomy as well as in evolution. A species is group of individuals which are similar in morphological characteristics, reproductively isolated and genetically distinct from other group individuals. The morphological and physiological character formed the basis of dividing animal kingdom into species and breeds within a species. A species is further divided into breeds or sub species. Different breeds have both qualitative and quantitative characters which differ from one breed to other. These qualitative characters are morphological such as colour and types of hairs, shape and size of horns, presence or absence of some body parts like zebu breeds of cattle are humped whereas European breeds are humpless for which they are regarded as sub species of same genus, *Bos*. The quantitative character can be considered as differences in type and size of body, milk production and fat percent in milk etc. Each species of farm animals contains a number of breeds. A breed is genetic entity developed over a long period of time as a result of planned mating and selection. Therefore, a breed has certain well defined physical conformation different from other breeds in the vicinity and has distinct local names. There is further division of a breed into strains based on isolation from each other due to geographical conditions or due to different aims of breeding the animals. This is called as strain breeding. The individual of a breed or strain are more closely related to each other than others due to inbreeding. This group is called as a line. A line is called as inbred line when the inbreeding coefficient is reached at least 0.375 as a result of two generation of full sib mating.

Table 3.1 : Number of breeds of various livestock species in the world and India.

Sl. No.	Species	Number of breeds		
		World	India (FAO)	India (ICAR)
1.	Cattle	787	61	34
2.	Buffalo	72	19	17
3.	Sheep	920	59	44
4.	Goat	351	29	23
5.	Pigs	358	3	3
6.	Horses	384	9	6
7.	Camels	56	–	8
8.	Ass	77	3	3
9.	Chicken	606	–	18

Source: FAO (1995) & ICAR (2002).

Cattle Breeds of India

The cattle of India vary in size and body confirmation perhaps due to their origin, adaptation to different ago-climatic conditions and the use which they have been put to. The cattle breeds of India have been divided into milch breeds which produce good quantity of milk but do not produce good quality bullocks, dual purpose breeds which are fairly good milkers and produce bullock of good draft powers, and the draft breeds providing little amount of milk and excellent draft powers. Indian cattle breeds under different categories of utility along with their home tract and characteristics have been given below:

Table 3.2.1 : Indian Cattle breeds with their distinctive features.

Breed	Home tract/distribution	Distinguishing features
A. Milch Purpose Breeds		
1. Sahiwal or Lola or Montogmery or Lambi-Bar.	Montogmery/Sahiwal district of Pakistan. Punjab,Haryana,UP,Bihar, MP.	Most popular breed of the subcontinent. Coat colour is red, dunn or pale red. Deep body, loose skin, pendulous sheath, voluminous dewlap, long whip like tail almost reaching to the ground. Muzzle and eye lashes are of light colour and has white ring along the eyes. Milk yield ranges from 1400-2500 kg per lactation.
2. Red Sindhi or Scindhi or Red Karachi.	Karachi and Hyderabad districts of Pakistan.	Smaller in size with compact body and round dropping quarters, dark colour muzzle, thick horn, heavy hump & pendulous udder. Coat colour red (shades from dark to light red). White stripes are found occasionally on dewlap and forehead. Presence of dark coloured muzzle. Milk yield ranges from 1250-1800 kg per lactation.
3. Gir	Gujarat (Junagadh, Kathiawar and Baroda), Maharashtra and Rajasthan.	Coat colour is white with dark red or dark brown patches all along the body. Complete red coloured coat mottled with yellowish red to black patches also found. Presence of protruding broad & long forehead and pendulous frontward ears resembling a tiny curled leaf with a notch at tip and straight and levelled back. Milk yield ranges from 1200-1800 kg per lactation.
4. Tharparkar or Thari or Grey Sindhi or White Sindhi.	Rajasthan (Ajmer, Jodhpur & Jaisalmer).	Coat colour is white or grey. Medium size animal with compact body, short straight and strong limbs, broad poll, convex forehead, long back. Presence of white or light grey line along the spine in young animals. Males are good draught animals while females are good milk producers (1800-2600 kg).

Contd.

B. Dual Purpose Breeds		
5. Rathi or Rath	Rajasthan (Alwar & Bikaner), Haryana & Punjab.	Coat colour is generally brown with white patches all over the body however animals with complete brown or black with white patches are frequently found. Medium size animals with medium size ear and short horn. Males are good draught animals while females are good milk producers (up to 3120 kg).
6. Hariana	Haryana (Rohtak, Hisar, Karnal,Gurgaon), Western UP & Rajasthan.	Most popular dual purpose breed in the Indo-Gangetic plains. Popular coat colour is white or grey. Compact body, long and narrow face, flat forehead & bony prominence in the centre of the poll, stumpy horn, short tail with black switch reaching just below hocks. Males are good work animals and cows are average yielders (600-800 kg).
7. Kankrej	Gujarat (Kutch & Ahmedabad).	Coat colour is silver grey, iron grey or black. It is heaviest Indian cattle breed, powerful body, forehead dished, strong curved horns accompanied by skin up to some length, well developed hump and well known for *Swai Chal*. Males are powerful work animals and cows are average milkers (1400 kg).
8. Deoni	AP, Karnataka & Maharashtra.	Looks like Gir cattle but have less pronounced forehead and no notch near the tip of the ear. Coat colour is black & white or red & white with irregular spots. Horns outward and backward curved, heavy dewlap & pendulous sheath. Males are heavy work animals and cows are average milkers (900 kg).
9. Ongole	AP (Ongole, Nellore & Guntur districts).	Common coat colour is white although males have dark grey colour at extremities. Large, heavy and muscular body, broad forehead, horns thick at the base and firm without cracks, hump

Contd.

		well developed.Males are heavy work animals and cows are average milkers (1000 kg).
10. Gaolao	M.P. (Chhindwada) & Maharashtra (Wardha and Nagpur districts).	Coat colour is white. Slightly convex face, coffin shaped skull, eyes almond shaped, voluminous dewlap. Looks like Ongole breed. Males are good work animals and cows are moderate milkers.
11. Mewati or Kosi	U.P. (Alwar, Mathura and Bharatpur districts).	Coat colour is white with grey patches on the body. In general they are sturdy, powerful and docile animals. Males are good work animals and cows are moderate milkers.
12. Krishna Valley	Karnataka (Ghataprapha, Malaprabha, Belgaum, Dharwad), Maharashtra (Satara, Kilhapur) & A P.	Coat colour greyish-white, brown and black & white, usually darker shades are present on the fore and hindquarters in males. Krishna Valley is a medium-sized animal with a short body having a moderate hump. Forehead is wide and concave, ears are short, erect, horizontal and pointed, tip of the horns is generally blunt. Males are massive & powerful and cows are fairly good milkers.
13. Dangi	Maharashtra (Ahmednagar & Nasik) and Gujarat (Valsad).	Coat colour is generally broken red and white or black & white. Medium size body, oily skin, small head, projecting forehead, large muzzle, small ear and horn. Males are good draught animal for heavy rainfall area and cows are not good milkers.
C. Draught Purpose Breeds		
14. Nagauri	Rajasthan (Nagore, Jodhpur & Barmer).	Coat colour is white and grey. Face is long and narrow but the forehead is flat, eyelids are rather heavy and overhanging, eyes, hooves, dewlap is small. Males are well known for trotting (fast road work).
15. Malvi	M.P. (Gwalior), & Rajasthan.	Coat colour varies from grey to white. They have short, deep and compact body, head is short and broad with dished forehead, short

Contd.

		and pointed ears, strong pointed horns.
16. Nimari	MP (Nimar & Indore) & Maharashtra.	Coat colour is generally red with large glint of white on various body parts. Long body, straight back, flat forehead, horns emerging from outer angles of poll in outward direction and carried upwards with a gentle curve to turn at points, hooves can stand rough wear.
17. Kherigarh	U.P. (Lakhimpur Khiri).	Coat colour is white or grey. Small, narrow face with large hump in males. Males are good for light draught work.
18. Kenkatha or Kangwariya	UP (Bundelkhand, Banda), MP.	Coat colour is grey (light on barrel and dark on other parts). Small, sturdy and powerful animals.
19. Ponwar	UP(Pilibhit, Lakhimpur).	Coat colour is black and white, white markings on dewlap, forehead and limbs. Body is small, compact. Males are good draught animal and cows are very poor milkers.
20. Bachaur	Bihar (Sitamarhi, Darbhanga, Samastipur and Muzaffarpur districts).	Coat colour is grey. Face is short, forehead broad and flat or slightly convex and short neck, prominent eyes and medium sized horns. Males are good draught animal and cows are not good milkers.
21. Khillari	Maharashtra (Sholapur & Sitapur districts).	Coat colour is grey-white. Short, deep and compact body, forehead long, convex, distinct groove in the centre, long and pointed horns. Males are fast & powerful draught animal and females are average yielders.
22. Amritmahal	Karnataka (Mysore).	Coat colour varies from white to nearly black (shades of grey). Head is long with a bulging forehead, furrowed in the middle, horns emerge in upward and backward direction and with pointed tips and well known for draught ability.
23. Hallikar	Karnataka (Mysore, Mandya, Bangalore, Kolar, Tumkur).	Coat colour is grey. Medium size compact animal. Horns are long and straight up to half length then

Contd.

		a forward bend, well known for draught ability and trotting.
24. Bargur	Tamil Nadu (Coimbatore).	White red or white patches on the body. Brown horns, eyes, muzzle and hooves Dewlap fine but well mashed.
25. Kangayam	Tamil Nadu (Coimbatore).	Coat colour is grey or white. Small compact body, broad forehead, black muzzle, short and erect ears, bright and elliptical eyes, stout and thick horns. Males have black colour on head, hump, neck and quarters. Males are good draught animal.
26. Siri	Sikkim and West Bengal (Darjeeling).	Coat colour is mainly black and white with extensive solid black colour. Head square cut, wide and flat forehead, short ears and moderate dewlaps. Hump is placed slightly forward tan the other breeds. Males are good draught animal and cows are fair milkers.
27. Red Kandhari	Maharashtra (Latur, Parbahni, Nanded, Hingoli and Beed districts).	Body colour is red to almost dark brown. They are medium in size, strong, compact and are reared mainly for draught purpose.
28. Punganur	Andhra Pradesh (Chittoor district).	White, grey or light brown to dark brown in colour.
29. Umblechery	Tamil Nadu (Thanjavur, Thiruvarur and Nagappattinam districts).	Grey with White marking on the face & legs. Legs below hocks have white marks.
30. Vechur	Kerala.	Mostly red, black and white in colour with long head and narrow face. Smallest cattle breed.
The following Breeds of cattle have also been documented as new breeds (NBAGR, 2009)		
31. Motu Cattle	Motu area of Malkangiri district in Orissa.	Coat colour is mainly brown (reddish), sometimes also grey. In few animals it is also white in colour. Muzzle is black, eyelids are black and tail is mainly black but sometimes grey in colour.
32. Ghumusari	Ghumusar area of Ganjamand Phulbani districts of Orissa.	Coat colour is white though in some animals it is also grey. Muzzle is black but occasionally

Contd.

		it is grey, eyelids are black and tail is mainly black but sometimes grey in colour.
33. Brinjharpuri	Binjharpur area of Jajpurdistrict in Orissa.	Coat colour is white though in some animals it is grey, black or brown in colour. Muzzle is black, eyelids are black and tail is mainly black but sometimes grey in colour.
34. Khariar	Khariar area of Nuapada district in Orissa.	Closely related to Motu breed. Coat colour is predominantly brown however sometimes also grey. Muzzle is black but occasionally it is grey, eyelids are black in colour.

Important Exotic Breeds of Cattle : Some of the important breeds of exotic cattle have been discussed in Table 3.2.2.

Table 3.2.2 : Exotic breeds of cattle with their unique features.

Breed	Home Tract	Distinctive features
1. Holstein Friesian	Netherlands (North Holland and West Fries land).	Long and narrow head, large chest, large capacious udder with prominent milk vein, average production of 6000-7000 kg with low fat % (3-3.5%). AFC about 30 months and CI 13-14 months.
2. Brown Swiss	Switzerland.	Large wedge shape body, broad dished forehead, thick and loose skin. Medium milker with average production of 5000 kg milk per lactation, AFC and CI equal to HF cattle.
3. Jersey	Jersey (English Channel).	Compact wedge shape body, light dished head, protruding eyes and top straight line. AFC and CI almost equal to HF cattle. Average herd average of 4500 kg per lactation with 4.5% fat.
4. Ayrshire	Ayrshire country of Scotland.	Outward and upward horn, symmetrical udders. AFC and CI almost equal to HF cattle.
5. Guernsey	Guernsey in the channel island.	Herd average 4000 kg milk per lactation with 4.5-5 % fat (golden yellowish colour due to carotene pigment).
6. Red Dane	Denmark.	Compact body, outward and forward horns, udder well attached. Herd average 5000 kg milk per lactation with 4.3 % fat.

Contd.

7. Aberdeen Angus	North East Scotland.	Beef breed, barrel shape body, rectangular small short head, polled, prominent breast bone.
8. Short horn	North East England.	Dual purpose breed, wax colour short and strong horns.
9. Hereford	Hereford (England).	Beef breed used for draught purpose also. Compact body, short legs. Head, brisket, underbelly, udder or scrotum tail switch, lower parts of legs are white.

Crossbred Cattle : Crossbreeding of zebu cattle with exotic bulls started in India as early as 1875 by European missionaries. Local cows were bred with Shorthorn bulls to evolve Taylor breed of cattle near Patna, Bihar. At institutional level, crossbreeding programme started at Imperial dairy research institute, Bangalore involving Ayshire x Hariana (Year 1910), Ayrshire x Sahiwal (Year 1913), Ayrhsire x Red Sindhi (Year 1917) and Holstein Friesian x Red Sindhi (Year 1938). Allahabad Agricultural Institute, Naini also started crossbreeding as early as 1924. Crossbreeding experiments have resulted into development of some the strains of crossbred cattle with varying success rate (Table 3.2.3).

Table 3.2.3 : Crossbred cattle strains evolved in India.

Breed	Crosses made with place
1. Taylor Patna	Shorthorn and Jersey with local cows around.
2. Jersind	Jersey x Red Sindhi at Allahabad.
3. Sunandini	Brown Swiss x non descript at Kerala.
4. Frieswal	3/8 to 5/8 level of exotic inheritance of Friesian and rest Sahiwal at M.D.F.
5. Karan Fries	Friesian x Tharparkar at NDRI, Karnal.
6. Karan Swiss	Brown Swiss x (Sahiwal & Red Sindhi) at NDRI, Karnal.
7. Jerthar	Jersey bull x Tharparkar cows at Bangalore.

Buffalo Breeds of India

Asia has 96% and India has 50% of the total world buffalo population. China, Pakistan, Thailand, Philippines, Indonesia, Nepal, Vietnam, Burma, Bangladesh and Sri Lanka are other leading countries buffalo population. About 60% of the total buffalo in India are non descript and rest 40% population is descript classified into 19 breeds according to FAO and 17 breeds according to ICAR.

Table 3.3.1 : Buffalo breeds of India with their distinctive features.

Breed	Home Tract	Distinctive features
1. Murrah	Rohtak and Hisar district of Hariana, Nabha,Patiala and Ludhiana district of Punjab.	Tightly curled horns, a long tail with a white switch reaching to the fetlock, colour is jet black with white markings on the tail, face and extremities, Milk yield 1800 kg., fat 6.9-8.3 %.
2. Nili-Ravi	Ferozpur district of Punjab.	Horns coiled, long neck, tail is long almost touching the ground, pink markings on the udder and brisket, white markings on the forehead, face muzzle, legs switch and around bright eyes. Milk yield 1750 kg., fat 5-8 %.
3. Surti	Surat, Kaira, Baroda district of Gujarat.	Horns are sickle-shaped, unique straight back. Two white collars, one round jaw and other at brisket. Milk yield 1300 kg., fat 7.5-8.3%.
4. Mehsana (Murrah x Surti)	Mehsana & Sabarkanta district of Gujarat.	White markings on the face, legs or tip of the tail. Horns vary from long sickle type to curved knot, dewlap absent. Milk yield 1950kg, fat 5.2-9.5%.
5. Jaffarabadi	Junagarh,Jamnagar & Kutch district Gujarat.	Prominent forehead, heavy horns emerged out compressing head giving study eye appearance. Lactation yield 2300 kg, fat 6.5-8.5%.
6. Bhadawari	Agra, Etawah (UP) , Gwalior (MP).	Medium size wedge-shaped body, udder not well developed but milk veins fairly prominent, two white lines "cherron" at lower side of neck. Dewlap absent, short hair of black or brown colour. Milk yield 950 kg., fat 8-13%.
7. Nagpuri/ Berari/ Ellichpuri	Nagpur, Akola, Amravati (MS), MP and AP.	White patches on face, legs and tail. Breed is known as lighter type than those of north. Naval flap absent, black colour with long horns. Milk yield 1050kg, fat7-8.5 %
8. Toda	Nilgiri district of Tamil Nadu.	Hump and dewlap are absent. Along the crest of the neck, hump area and back, there is a thick

Contd.

		growth of hair like a mane which impairs a bison appearance, crescent shape horns. fat 7.5-8%.
9. Tarai	Bareilly, Rampur, Sahajanpur district of UP.	Slightly convex head with prominent nasal bone, long & flat horns with coils. Sometimes there is a white blaze on the forehead.
10. Pandharpuri/ Marathwadi	Marathwada region of Maharashtra.	White patches on forehead, legs and tail. Long twisted thin horns curving backward (sword like). Milk yield1500 kg., fat 7 %.
11. Manda	Ganjam district of Orissa and AP.	Grey but yellowish tufts of hair may present on knee and fetlock, sharp eyes with broad red margin around the lids, horns broad and semi-circular extending backward and inward.
12. Jerangi	AP. and Orissa.	Dwarf breeds of buffalo and its height does not exceed four feet, conical small horns and run backward.
13. Kalahandi / Paddakiwadi	Orissa and AP.	Broad horns and half curved running backward at the tip, prominent eyes and large with narrow red margin around the lids, due to light colour it tolerates heat more than dark coloured buffaloes.
14. Sambalpuri	Orissa and MP.	Narrow barrel and prominent forehead.

The following Breeds of buffalo have also been documented as new breeds (NBAGR, 2009)

15.Banni (Kutchi, Kundi)	Gujarat.	Mainly black, sometimes copper colour, muzzle black in majority of animals and brown in copper colour animals, eyelids black and brown also, tail switch white or black, hoof generally black and brown.
16. Chilika	Orissa.	Brownish black or black, muzzle black, eyelids black.

Important Sheep (*Ovis aries*) Breeds of India

Sheep is an important economic livestock species, contributes greatly to the agrarian Indian economy. They play an important role in the livelihood of large percentage of small and marginal farmers engaged in sheep rearing specially in arid/semi arid and mountainous areas. Sheep are mostly reared for meat and wool. Milk from sheep is of limited importance in few areas of Jammu and Kashmir, Rajasthan and Gujarat. Indian Sheep is not regarded as dairy sheep like their counterparts in Middle East and Eastern European Countries. There are 44 descript breeds of sheep. In spite of large number of breeds around 75% of the animals are non descript. There is a large intermixture amongst the breeds in region where two are more breeds exists. Most of breeds of sheep are very well adapted to harsh climate, long migration, tropical disease, poor nutrition, and shortage of drinking water and poor quality water. Very little concerted efforts for developing these breeds through artificial selection have been made. Some important sheep breeds are listed below with their unique characteristics:

Table 3.4.1 : Sheep breeds of India with their unique characteristics.

Breed	Home Tract	Utility	Identification
1. Gurej	Gurej (Kashmir).	Carpet wool	Biggest among the Jammu and Kashmir breeds. Annual greasy fleece weight varies from 0.5 to 1 kg per animal. A small proportion of animals have small, pointed horns. Tail is thin and short.
2. Poonchi	Poonch & Rajori district of J&K.	Carpet wool	Annual yield is 1.6 kg, clipped twice or thrice a year. Animal are similar to Gaddi but are smaller in size.
3. Gaddi	Kulu, Kangra & Chamba district of HP.	Carpet wool	Animals are medium sized, usually white, although tan, brown and black and their mixture are also found. Males are horned, 10 to 15% females are also horned. Clipped thrice in year, produces a total of only 1 to 1.5 kg fine lustrous wool.
4.Rampur Bushair	Rampur,Bushair, Shimla,Kinnour, Nahan,Bilaspur, Lahaul & Spiti district of HP.	Carpet wool	Nose is Roman type, males are horned while females are polled. Legs, belly and face are devoid of wool.

Contd.

5. Muzaffarnagri	Muzaffarnagar, Bulandshar, Shaharanpur, Merrut district of UP.	Meat & carpet wool	Extremely long tail reaches up to fetlock. Face light slightly convex, ears occasionally black, both sexes polled. Ears Long and dropping.
6. Hisardale	Hisar District of Haryana and hilly regions of Kullu and Kangra District of HP.	Apparel wool	Small Animal With short Legs. Most animal polled, Ears are leaf-like.
7. Chokla/ Shekhawati	Churu, Bikaner Jaipur and Nagaur district of Rajasthan.	Carpet wool	Face is free of wool with broad forehead heavy prominent, Roman nose. Both sexes polled. Tail thin and of medium in length.
8. Magra/ Bikaneri	Bikaner, Nagaur, Jaisalmer and Churu district of Rajasthan.	Carpet wool	The head is without any wool but covered with white hair. White face with light brown patches around the eyes. Skin colour pink, Both sexes polled. Tail medium in length and thin.
9. Malpura	Jaipur, Tonk, Ajmer, Bhilwara &Bhundi district of Rajasthan.	Meat & carpet wool	Fairly well built animal. Long legs, face light brown, ear short and tubular, white fleece, both sexes are polled.
10. Marwari	Jodhpur, Udaipur, Pali district of Rajasthan & Jeoria Region of Gujrat.	Meat & carpet wool	Medium Sized animal, black faced, head is also covered with black hair with twisted small ears. Both Sexes are polled. Fleece is white and not very dense.
11. Sonadi	Udaipur, Dungarpur & Chittorgarh district of Rajasthan.	Meat & carpet wool	Per year 0.6 to 1.2 kg white fleece production, the breed also yields 1 to 1.5 kg of milk per day, goat like sheep. Ear generally have cartilaginous appendage, belly and legs devoid of wool.
12. Patanwali/ Kutchi/ Kathiawari	Saurashtra, Kutch, Kathiawar, Mehshana district of Gujrat	Carpet wool	Medium to large size animals. Typically roman nose, both sexes polled. Tail thin and short, good quality carpet wool known as "Joria", corkscrew shape horn.

Contd.

13. Deccani	Maharashtra, Andhra Pradesh and Karnataka.	Meat	Medium sized, colour black with white marking, ears medium in length, flat and dropping. Fleece coarse, hairy and open.
14. Mandaya or Bannur	Mandya district of Karnatka.	Meat	Compact body with typical U shape conformation from rear, ears long leafy and dropping, slightly roman nose, both sexes are polled. Shortest breed.
15. Nellore	Nellore and Ongole district of Andhra Pradesh	Meat	Spiral horns, two appendages of skin hanging from the throat. Tallest breed.
16. Madras Red	Chingalpet and Madras district of Tamil Nadu.	Meat & carpet wool	White markings on the forehead, inside the thighs and on the lower abdomen. Rams have corrugated and twisted horns while ewes polled. Body covered with fine small hairs which are not shorn.
17. Chotta Nagpuri	Nagpur district of Maharashtra.	Meat & carpet wool	Ears small and parallel to the head. Both sexes are polled.
18. Jaisalmeri	Jaisalmer district of Rajasthan.	Meat & carpet wool	Tail, well built animals, black or dark brown face, typical roman nose, long dropping ears, both sexes are polled.
19. Nali	Ganganagar & Jhunjhunu district of Rajashthan, Southern Hisar and Rohtak District of Haryana.	Carpet wool	Medium sized animals, face light brown, skin pink, both sexes are polled, ears large leafy, tail short medium and thin, forehead, belly and legs are covered with wool.
20. Pugal	Bikaner and Jaisalmer district of Rajasthan.	Meat & carpet wool	Fairly well built animals, black face with small light brown strips on either side above eyes. Lower jaw typically light brown. Both sexes are polled, ears short and tubular.
21. Hassan	Hassan district of Karnataka.	Meat	Small animals. White body with light brown or black spots. About 39% of males are horned. Females white, fleece

Contd.

			extremely coarse and open. Legs and belly generally devoid of wool.
22. Bellary	Bellary district of Karnataka.	Meat & carpet wool	Medium sized animals. Body colour ranging from white and various combinations of white and black. About 1/3rd males are horned. Tail short and thin.
23. Lohi	Punjab, Hariana, Chandigarh and Delhi.	Carpet wool & meat	Large body and white colour, face is brown, both sexes are polled, tail shot and thin.
24. Balaingiri	Orissa	Meat & carpet wool	Medium sized animals, ram are horned, face colour brown, black fleece colour.
25. Jalauni	Jalaun, Jhansi and Lalitpur district of UP.	Meat & carpet wool	Medium sized straight rose line. Both sexes are polled, ears large, flat and dropping, tail thin and medium in length, fleece coarse, short stapled and open, Generally white in colour.
26. Bhakarwal	Bhakarwal has no distinct home tract and animals are entirely migratory (J&K, UP, HP, Punjab, and UK).	Carpet wool	All animals are spotted, fawn or grey, rams are horned and ewes are polled, typical roman nose, ears are long and dropping, tail small and thin.
27. Mecheri	Salem & Coimbatore district of Tamil Nadu.	Meat	Medium sized animals. Light brown in colour, ears medium sized, both sexes are polled, tail short and thin, body covered with very short hairs which are not shown.
28. Changthangi	Changthong region of Laddakh of Jammu & Kashmir	Carpet wool	Animals are of strong built, large framed with good fleece cover. Fleece has extra ordinary long staple.
29. Coimbatore	Coimbatore, Madurai district of Tamil Nadu.	Meat & carpet wool	Medium sized animals, white with black or brown spots. Ears medium sized and directed outward and backward. About 30% males are polled. Fleece white, coarse, hairy and open.

Contd.

30. Kilakarsal	Ramnathpuram, Madurai, Thanjavir & Ramand District of Tamil Nadu.	Meat	Coat dark tan, with black spot on head, belly and legs. Ears medium sized. Tail small and thin, Most of the animals have wattle.
31.Vembur	Vembur, Nagalpuram, Melakhandhai, Achangullam villages of Tamil Nadu.	Meat	Coat dark tan with black spot on head, ear medium sized, males with thick twisted horns, Most animal have wattle.
32. Tiruchi	Tiruchy, Arcot, Salem district of Tamilnadu.	Meat	Small animal, body completely black, males are horned, ears short and directed downward and forward. Tail short and thin.
33. Ramnad	Ramand district of Tamil Nadu.	Meat	Medium sized animals, predominately white, some animals have fawn or black marking over the body, males have twisted horn.
34. Kenguri	Raichur district of Karnatka.	Meat	Medium sized animals, body colour mostly dark brown. Males are horned.
35. Shahabadi	Shahabad, Patna, and Gaya district of Bihar.	Meat & carpet wool	Animals are medium sized and leggy. Fleece is mostly grey and sometime with black spots. Tail is extremely long and thin. Both sexes are polled.
36. Ganjam	Koraput, Pholbani, and parts of Puri district of Orissa.	Meat & carpet wool	Animals are medium sized with coat colour ranging from brown to dark tan. Nose line is slightly convex, males are horned.
37. Bonpala	Sikkim.	Meat & carpet wool	Animals are tall, leggy and well built. Ears are small and tubular, sexes are horned, tail is thin and short, belly and legs are devoid of wool.
38. Tibetan	Sikkim and Arunanchal Pradesh (Kemang district).	Meat & Carpet wool	Animals are medium sized, mostly white with black or brown face, both sexes are horned, nose is convex, and ears are small broad and drooping. Belly and legs are devoid of wool.

Contd.

39. Karnah	Karnah district of Kashmir	Carpet wool	Animals are large, rams have large curved horns and prominent nose line. The fleece is generally white.
40. Nilgiri	Southern region of India, Developed by crossbreeding of Coimbatore, Southdown, Cheviot and Tasmanian merino sheep.	Carpet wool & Meat	The animals are medium sized, and white in colour. Face line is convex, males are horned, tail is medium and thin. Ears are broad flat and drooping.
41. Garole	West Bengal Plains.	Carpet wool & meat	The animals are medium sized, and white in colour. Tail is medium and thin. Ears are broad flat and drooping. Twining birth.
42. Munjal	Rajasthan	Carpet wool & Meat	The animals are medium sized, and white. Tail is medium and thin. Ears are broad flat and drooping.

Exotic Breeds of Sheep: Some of the important exotic breeds of sheep are shown in Table 3.4.2.

Table 3.4.2 : Important Exotic Breeds of Sheep with their unique features.

Breed	**Name of country**	**Utility**	**Identification**
1. Cheviot	Scotland	Fine wool, Annual yield of fleece 2.5 to 3.5 kg.	Erect ears, Clean white face, Ram Wheight 80 kg. and ewes weight 55kg, fleece light colour.
2. Hampshire	U. K.	Wool yield 2 to 2.5 kg., Mean fibre diameter 24 to 30μ.	Large ear black color, strong rugged head, black marks on legs and face, large size rapid growth.
3. Lincoln	England	Coarse wool, yiel 4.5 to 5 kg per year.	Larged sized, Mature weight of ram is 150kg, prolific breeder
4. Leicester	England	Average quality wool yield 3 to 4 kg per year, Dual purpose	Moderate sized, mature weight of ram is 100 kg, stylish in feature
5. Merino	Spain	Fine wool yield 4 to 7 kg per year	White in color, Ram with spirally twisted horns, ewes polled, folds in the skin, small sized, hardy in nature. Body weight 60-70 kg. Head and legs covered with wool.

Contd.

6. Rambouillet	France	Dual purpose, good quality wool yield 4 to 5 kg per year	Large sized head, white in colour, prolific, average weight of ram 125 kg, ewe weight 90 kg.
7. Southdown	England	Fine quality wool	Small sized, blocky in appearance, grey in colour, short round ears.

Indigenous crossbred strains of sheep: Some of the crossbred strains of sheep developed in India are depicted in Table 3.4.3.

Table 3.4.3 : Crossbred strains of sheep developed in India.

Crossbred Strains	Exotic inheritance	Crosses made with Place
1. Bharat Merino	75%	Chokla and Nali with Rambouillet and Merino at Avikanagar.
2. Avivastra	50%	Chokla and Nali with Rambouillet and Merino at Avikanagar.
3. Avikalin	50%	Malpura with Rambouillet at Avikanagar.
4. Avimans	50%	Malpura and Sonadi with Dorset and Suffolk at Avikanagar.
5. Indian Karakul	75%	Marwari, Malpura and Sonadi with Karakul at CSWRI, Bikaner.
6. Kashmir Merino	50-75%	Gaddi, Bhakarwal and Poonchi with Merino and Rembouillet.
7. Patanwadi Synthetic	50%	Patanwadi with Rembouillet and merino at GAU, Dantiwara.

Crossbred exotic sheep : Some of the exotic sheep breeds developed by crossbreeding are shown in Table 3.4.4.

Table 3.4.4 : Exotic sheep breeds developed by crossbreeding.

Breed	Crosses made
1. Columbia	Rambouillet x Lincoln rams.
2. Panama	Lincoln x Rambouillet rams.
3. Targhee	Rambouillet x Long wool breed.
4. Romnelet	Romney Marsh rams x Rambouillet ewes.
5. Suffolk	Dark face South down rams x Norfolk strain.
6. Corriedale	Merino ewes x Lincoln rams.
7. Karakul	It was also developed by crossbreeding of two local breeds.

Goat (*Capra hircus*) Breeds of India

In the plains, goats are mostly raised by the landless or marginal farmers. Goat produce lean and juicy meat which excellently meets the kitchen needs for all the Indian dishes. The non vegetarian in India, irrespective of cast, creed and religion, prefer goat meat (chevon) to beef and buffalo meat. The goat milk contains lower fat percent as compared to buffalo milk, with smaller fat globules, higher protein and lactose content and rich minerals. For babies and patients suffering from the cow milk intolerance and diarrhoea, the goat milk works well. Goat hairs have higher tensile strength preferred for rope making and mohair from the angora goats is suitable for fancy bags. Chegu and Changthangi goats of north Himalayan regions produce the finest under coat (Pashmina), useful for the manufacture of superfine fabric. The average pashmina production from the Indian breeds is low (about 100g in Chegu and 250g in Changthangi per year) but the fibre is superior in quality to the breeds of other countries. The leather produced from the goat skin is superior to those from other skins for fancy and high value products. The goat breeds have since long been described in terms of physical conformation, body type and coat colour and body measurements with subtle reference to production traits. There are as many as 23 well defined breeds of goats in India. Most of these breeds have been evolved over ages through geographical isolation and natural selection to suit specific agro ecological conditions. Some important goat breeds are listed below in the Table 3.5.1.

Table 3.5.1 : Goat Breeds of India with their unique features.

Breeds	Home Tract	Utility	Identification
1. Changthangi	Changthang (Ladakh),Hills of Leh.	Meat, Fiber, Pashmina.	Coat colour 50% white and rest brown, grey and Black, Ears small pointed outside, Large turning outward horn, Produce softest fibre (Pashmina) 250g/goat/year.
2. Chegu	Lahaul & Spiti, Yaksar (HP).	Meat, Fiber, Pashmina.	Coat colour is predominately white but greyish red and mixed colour may also be seen, small ears, twisted backward and outward horns, Pashmina production 120 grams/animal/year.
3. Gaddi	Kangra,Kulu, simla,Chamba (HP).	Meat, Fibre (long hair).	Each animal transports about 8 kg of merchandise. Both sexes have large horns, coarse long

Contd.

			hair measuring 7-10 inches (18-25 cm).
4. Barbari	Etawah,Agra, Aligarh(UP).	Milk & Meat.	Good milker, highly prolific, white with small light brown patches is most common, both sexes have twisted horns. Bucks have a large thick beard.
5. Jamnapari	Etawah, Agra, Mathura(UP).	Milk & Meat	Biggest & majestic breed. convex nose line with a tuft of hair known as "Roman nose" or parrot mouth appearance, Both sexes are horned with short and thin tail.
6. Beetal	Gurdaspur, Ferozpur, Amritsar(Punjab).	Milk, Meat	Both sexes have thick, medium-sized horns, roman nose. Male beard while females are beardless, udder is having big conical teats.
7. Jhakrana	Alwer,Nagaur (Rajasthan).	Milk, Meat	Predominantly black with white spots on ears and muzzle, and their skins are popular with the tanning industry.
8. Marwari	Jodhpur, Pali, Nagaur, Bikaner, Jalore, Jaisalmer (Raj).	Milk, Meat	Jet black colour, hair covering lustrous and grows at the rate of about 10-12 cm annually, male has a thick beard but absent in females.
9. Sirohi	Sirohi (Rajasthan).	Milk, Meat	Body color white and brown, body covered with dense coarse short hair, medium sized, compact body, long lope ears, well suited to stall feeding. Milk yield 0.9kg/day.
10. Kutchi	Kutch, Kathiawar (Gujarat).	Milk, Meat	Black color, medium sized, reddish mark on neck. Nose slightly Roman, cork screw horns pointed upwards. Milk yield around 1.5 kg. Hair annual yield about 500 gm when shorn twice a year.
11. Mehsana	Mehsana, Banaskantha, Baroda, Gandhinagar (Gujarat).	Milk, Meat	Large sized, coat colour is black with white spots at the base of the ear, udder is moderately developed having large and conical teats.
12. Surti	Surat, Nasik (Maharashtra)	Milk, Meat	Short stature, unable to walk long distance, reared in town-stall feeding. Adult female is heavier than male, white in colour with highly developed

Contd.

			udder, both sexes have small horns.
13. Osmanabadi	Around Osmanabad Maharashtra	Milk, Meat	Tall stature, Mostly black in color with brown or white spots. Ninety percent males are horned; females may be horned or polled.
14. Malabari	Malabar Hills of Kerala	Milk, Meat and hair	All males and a small number of females are bearded, occasional roman nose; skin is popular with the tanning industry.
15. Black Bengal	West Bengal	Meat, Skin	Excellent mutton and skin quality, small in size. Black colour, beard is found in both sexes. Prolificacy and fine leather. Meat of good quality.
16. Ganjam	Ganjam, Koraput and Phoolbani District of Orissa	Meat	Brown, black and grey colour, Body and legs compact, tall and leggy. Ears medium sized, straight and horizontal.
17. Kannaiadu	Ramnathpuram, Tirunelveli District of Tamil Nadu.	Meat	Body colour predominately black with white spots. Stout body, tall legs. Bucks horned and doe polled.
18. Zalawadi	Surendarnagar, Rajkot and Zalawad District of Gujrat.	Meat, Milk and Skin	Coat colour black with white spots. Body large sized. Tail short and curved. Twisted and Screw like horns. Ear long leaf like and drooping.
19. Bhakarwal	Hilly tract of Kashmir.	Meat, Hair and Skin	Coat colour white brown or black with white marking. Very hardy large sized animals, Long curved horns upward.
20. Gohilwadi	Bhawnagar, Amreli district of Gujrat.	Meat Hair and Skin	Coat colour black coarse and long hairs, medium sized animals, ear tubular and drooping, twisted horns backward.
21. Kangan	Hill of Himachal Pradesh.	Meat, hair and Skin	Coat colour black coarse and long hairs, Medium sized animals, Ear tubular and drooping.
22. Khasi (Assam Hill)	Khasi, Naga, Assam, Meghalya and Nagaland.	Meat hair and Skin	White with silver grey coat colour, short leged long body, Ears short and erect.
23. Sangamneri	Pune and Ahmadnagar District of Maharashtra	Meat, hair and Skin	Highly variable coat colour white, brown, black with spots of three colours. Medium sized body, Ear drooping, medium size horns turned backward and upward.

Important Exotic Goat Breeds: Some of the important exotic breeds of goats have been shown in Table 3.5.2.

Table 3.5.2 : Exotic breeds of Goat with their unique features.

Breed	Home Tract	Utility	Identification
1. Saanen	Saanen & Simmental valleys of Switzerland	Known as milk queen of the goat. Average daily milk yield in the tropics is 1-3 (kg) with fat % of 3.5	White or light cream colour body with black spots on the nose, pendulous udder, legs short in comparison to body, ears erect pointing forward, breed is polled. Produces about 100 kg milk per lactation.
2. Alpine	Nubia (Africa)	Average dairy milk yield in the tropics is 0.9-1.3 (kg) with fat % of 3.6	Colour varies from white to black, large animals, long lagged and hardy, erect ears and straight nose, scimitar (curved) type horns.
3. AngloNubian	Britain cross between Nubian and Jamnapari	Average daily milk yield in the tropics is 0.8-1.2 (kg) with fat % of 4.5, most outstanding dual purpose (milk and meat) breed.	Colour varies but brown and white are usual, long pendulous ears and roman nose and forehead, horns when present lie over the coat, largest and heaviest breeds among all European breeds.
4. Toggenberg	Toggenberg valley of Switzerland	Average dairy milk yield in the tropics is 1-3 (kg) with fat % of 3.5.	Light brown to dark chocolate body colour, small white ears pointing forward, white strips on both sides of face, these are large goats having long neck which are kept erect.
5. Angora	Asia Minor (Turkey)	Average yield of Mohair is 3 kg annually.	White creamy body colour, small in size with shorter legs. Horns are grey, spirally twisted and inclined backward and outward. Tail is short and erect.
6. Orenberg	Kazakhstan	Produce pashmina fibres of 16 microns diameter and 5.5 cm in length of about 300 gm per year.	

Pig Breeds of India

Indigenous domesticated pigs have no distinct breed features. Therefore, characteristics vary with topography and climatic conditions from region to region. These are raised traditionally by weaker society. However, some pig breeds has been evolved like Desi, Ghori and Ankamali.

The desi pig (Country Hog) which has been evolved from gradual domestication of wild pig (*Sus scrofa cristatus*) are small sized pigs with a body weight of 40-80 kg and are found in Uttar Pradesh, Bihar, Madhya Pradesh and Punjab state of India. Besides this, Ghori breed of pigs has evolved from *Sus salvanius hodsgon* is found in Manipur, Assam, Arunachal Pradesh, Nagaland, Mizoram, Sikkim and Tarai Himalayan region and are called as dome pig of pigmy pig. Ankamali pigs evolved from domestication of *Sus scrofa* and *amanuensis byth* and are found in Kerla, Karnataka, Tamilnadu and Maharashrta states of India. Besides *desi* breeds, recognised exotic breeds e. g. Large white Yorkshire, Duroc and Hampshire also found in India.

Table 3.6 : Characteristics of important breeds of pigs experienced in India.

Breed	Native Place Body features	Identification/ weight	Mature Body
1. Large White Yorkshire	England	Longer, face dished board snout, erect ears, long thin and forward head. Entire white, black pigment spots long levelled back, fine skin without wrinkles and smooth, prolific breeder.	Boar-300 to 400 kgSow-230to 320 kg. Principally bacon type.
2. Middle White Yorkshire	England	Short head and neck, jaw strait and light, unturned dishes face, cross between large and small Yorkshire, long levelled back white in colour, fine skin, medium sized, smooth skin, small bones.	Boar-250 to 340 kgSow-180 to 270 kgLess prolific than large white, Bacon type.

Contd.

3. Landrace	Switzerland	Longer snout, Drooping ears, medium sized, straight face and of medium length, short legs, lean carcass, white in colour with black skin spot, longer body, skin without wrinkles.	Boar-300-350 kgSow-200 to 250 kgBacon type.
4. Berkshire	England	Short unturned nose, dished face, erect ears, inclined forward, medium sized, legs of medium length, black in colour, good width and broad back, hams and shoulders are well fleshed.	Boar-270-220 kgSow-200 to 295 kgMeat is of good quality.
5. Hampshire	USA	Head and tail are black, erect ears. Black hog, with a belt encircling the body including front legs, short legs, smaller breed.	Boar-250 to 320 kg.Sow- 200 to 270 kg.Meat type midway between bacon and lard type.
6. Tamworth	U. K.	Longer head, long snout, golden brown in colour, thin shoulders, strong back.	Boar-200 to 320 kg.Sow- 180 to 280 kg.

Horse Breeds of India

Amongst various domestic animals horse is the only animal which occupies a unique position in Olympic games. Similarly, horse serve as pleasure animal to human through riding and racing. They are used in polo, racing, jumping and many other sports events. In India there are six well known distinct breeds of horses. These breeds have distinct breeding tract in various parts of the country including Northern, North-West and North-East India.

Table 3.7 : Horse breeds of India.

Breeds	Origin and Home Tract	Utility
1.Marwari	It has Arab blood. Marwar and Mewar region of Rajasthan	Riding, Sports and Safari
2. Spiti	Spiti Ponies originated in Tibet and found in spiti division of Lahaul, Kullu and Kinnaur distt. Of HP	Riding, Sports and Safari
3. Kathiawari	Evolved from wild horse of Kathiawar and Arabian horses. Saurashtra Province of Gujrat is the home tract	Riding and Pack animal
4. Zanskari	Leh and Ladakh area of J and K	Draught, Transport and Riding
5. Manipuri	These ponies are found in Manipur and Assam. Originated from Asiatic wild horse and Arabian horses	Beauty, fastness and hardiness
6. Bhutia	Sikkim and Darjiling. Resembles to Tibetian pony	Riding and Pack animal

Camel Breeds of India

The genus *Camelus* has two species, the heavily built; double humped camel (*Camelus bactrianus*) which inhabits the deserts of central Asia reaching up to Mangolia and Western part of China. The other species is single humped (*Camelus dromedarius*) widespread throughout the Middle East, India and North Africa. The Indian camels are mostly single humped and a very small numbers of double humped camels are also found in Ladakh area of Jammu and Kashmir. Indian camels termed as a plain camels and hill camels, according to the regions where they are bred. Plain camels include riverine (area where water supply through rivers and canals) and dessert varieties (inhabitants of arid sandy tracts). Hill camels and riverine camels are Baggage type (Capable of carrying loads), whereas desert camels are riding type (incapable of carrying heavy load but very good in travelling long distance). According to national commission on agriculture (1976) based on certain distinct characteristics camels can be classified in to distinct types and could be recognised as breeds: Bikaneri, Jaisalmeri, Kachchhi, Marwari, Mewari, Sindhi, Shekhawati, Revierine, Mewati, double humped camels. The Bikaneri breed is principal breed which has been defined by a committee constituted by Government of Rajasthan. Some workers have pointed out there are only 4 distinct breed of camel viz. Bikaneri, jaisalmeri, Mewri and Kachchhi and the rest are crosses of different major breeds or locally developed strains of these major breeds. Bikaneri camels are predominantly bred in Bikaner and nearby districts, such as Sriganganagar, Hanumangarh, Churu, Jhunjhunu, Sikar and Nagaur of Rajasthan and adjoining parts of Haryana and Punjab state. The breeding tract of Jaisalmeri breed encompasses the Jaisalmer, Barmer and part of Jodhpur district in Rajasthan. The Jaisalmeri camels are of active temperament and are quite tall with long and thin legs. The Kachchhi breed inhabits the Ran of Kachchh in Gujarat state. The major breeding tract encompasses the Kachchh and Bansakantha districts of Gujarat. The camels of this breed are generally brown to dark brown in colour with absence of hair on eyelids and ears. The Mewari breed of camel has derived its name from the Mewar area of Rajasthan and is well known for milk production potential. The major breeding tract of the breed encompasses the Udaipur, Chittorgarh, Rajsamand districts and adjoining Neemuch and Mandsour districts of Madhya Pradesh. According to the National Bureau of Animal Genetic Resources there are 8 breeds of camels in India, e.g. Bikaneri, Jaisalmeri, Jalori, Kutchi, Malvi, Marwari, Mewari and Mewati.

Indian Poultry (*Gallus domesticus*) Breeds

Poultry keeping in India is as old as its civilization. Red jungle fowl found in India and its neighbouring countries is considered to be the progenitor of all domestic breeds of fowl. India is the fifth largest egg producer in the world at present only next to China, USA, Japan, and Russian Federation. Indian Council of agricultural Research playing a vital role for poultry improvement in the country. It has established two poultry institutions, *viz.* Central avian Research Institute, Izatnagar and Project Directorate on Poultry, Hyderabad, to provide necessary research. ICAR also support a number of schemes covering all aspects of poultry production at the various agricultural universities in the form of All India Co-Ordinate Research Projects. Government of India has undertaken basic breeding programmes at its central poultry breeding farms located at Banglore, Bhubaneshwar, Mumbai and Chandigarh. Random sample test units has been established in different agro climatic zones, viz. Mumbai, Banglore, Bhubaneshwar and Gurgaon to evaluate the performance of the commercial stocks sold in the country both by public and private sectors. In the country, 20 breeds of chicken have been documented and some of them important chicken breeds are listed below:

Table 3.8.1 : Important poultry breeds of India.

Breed	Home tract	Distinctive features
1. Faverolla	Kashmir	Feathered comb.
2. Punjab Brown	Punjab & Hariana	Meat type with brown plumage.
3. Brown Desi	UP	Layer type single comb.
4. Ankaleshwar	Gujarat production.	Single comb, hardy with poor
5. Kadaknath/ kalamasi	MP	Black pigmentation on external and internal surface.
6. Assel	AP	Game bird with fighting quality and biggest in size among native breeds.
7. Denki	AP	Males (Cocks) good fighter with long neck and legs.
8. Ghagus	AP and Karnataka	Small size with small comb and wattles, thick neck and feathered shank.
9. Tellichery	Kerala	Small bird with black skin.
10. Haringhata black	W.B.	Good layer, small size black bird.
11. Chittong/ Malay	N.E.Region	Large Heavy bird with fighting nature.
12. Nicobari	Nicobar island	Short legs, small size, good layer.
13. Teni	All parts of country	Small all purpose bird, tolerates high temperature.
14. Frizzle fowl	Hot & humid costal region and N.E. region	Curved feather, better heat dissipation and adaptability in arid region.
15. Naked neck	Hot & humid costal region and N.E. region	Neck naked which makes them well adapted to high temperature due to heat dissipation.

Exotic poultry breeds: Some of the exotic poultry breeds have been discussed below

Table 3.8.2 : Exotic poultry breeds.

Breed	Home tract	Distinctive features
1. Leghorn(Mediterranean breed)	Spain	Number one egg producer, white in color with yellow skin and rose/single combed, heavy egg layers (250-300 eggs annually).
2. Minorca(Mediterranean breed)	Spain	Largest and heaviest Mediterranean breed, produces large white eggs, long wattles and large white ear lobes with black beak, shanks and toes.
3. Australorp(English breed)	Australia	Single combed with black plumage, beak, shank and white skin, good egg producer.
4. Sussex(English breed)	England	Single comb, white skin, good layers, small size breed.
5. Cornish(English breed)	England	White in color, heavy breed, excellent meat breed, pea type comb.
6. Rhode Island Red (American breed)	Rhode Island (USA)	Heavy breed, single and rose combed, dark red color, good strains for both egg and broilers.
7. Plymouth Rock (American breed)	America	Heavy breed with strains for both eggs and broilers, single combed with yellowish skin, dark tip which with the alternate dark and light bars, give a bluish cast. Barred plymouth rock: The male birds have black and white bars of equal width, where as in female the white bars should be as wide as the black bars.
8. New Hampshire (American breed)	America	Chestnut red plumage, single comb, good producer of large brown shelled eggs.
9. Wyandotte (American breed)	America	Dual purpose breed with rose comb, white or buff in colour.
10. Brahma(Asiatic breed)	Asiatic breed developed in India	Pea comb, have three varieties viz., light, dark and buff.
11. Cochin(Asiatic breed)	China	Massive bird with single comb.

Crossed poultry genotypes: Followings are the some of the poultry genotypes have been developed in India

Table 3.8.3 : Crossed Poultry genotypes evolved by the ICAR institutes and SAU's.

Name	Breeds involved		Purpose	Institute
Poultry genotypes	Male	Female		
1. Vanraja	Cornish	Synthetic population	Dual	ICAR, Hyderabad
2. Grama priya	Synthetic	White leg horn	Dual	ICAR, Hyderabad
3. Cari gold	Selective breeding of RIR		Egg	CARI, Izatnagar
4. Giriraj			Dual	USA, Bangalore
5. Krishna -J	Synthetic		Egg	JNKVV, Jabalpur
6. Yamuna	KadaknathNew Hampshire		-	
7. Kalinga Brown	Selective breeding of RIR		Egg	CPBF, Bhubneswar
8. Grama lakshmi	Australorp white leghorn		Dual	KAU, Kerala
9. Nandanam	RIR population			TNVASU, Chennai
10. Cari-Nirblik	Assel cross		190 eggs	CARI
11. Cari- Shyam	Kadaknath cross		210 eggs	CARI
12. Upcari	Frizzle cross		220 eggs	CARI
13. Hitcari	Naked neck cross		190 eggs	CARI

Other Poultry Birds and their Breeds have been mentioned below

1. **Duck:** Duck is the common name for a large number of species in the Anatidae family of birds, which also includes swans and geese. The ducks are divided among several subfamilies in the Anatidae family; they do not represent a monophyletic group (the group of all descendants of a single common ancestral species) but a form taxon, since swans and geese are not considered ducks. Ducks are mostly aquatic birds, mostly smaller than the swans and geese, and may be found in both fresh water and sea water. The ducks are generally monogamous, although these bonds generally last a single year only. Larger species and the more sedentary species (like fast river specialists) tend to have pair-bonds that last numerous years. Most duck species breed once a year, choosing to do so in favourable conditions (spring/summer or wet seasons). Ducks also tend to make a nest before breeding. Although mother ducks are very caring and protective of their young,

they are not above abandoning some of their ducklings if they are physically stuck in an area they cannot get out of or are not prospering due to genetic defects or sickness brought about by hypothermia, starvation, or disease. Ducklings can also be orphaned by inconsistent, late hatching where a few eggs hatch long after the mother has abandoned the nest. Ducks are mostly concerned in the eastern and southern states of the country (India), and reared for both egg and meat. Most breeds of domesticated ducks (*Anas platyrhynchos*) have descended from Mallard and wild ducks, and were first domesticated in Asia. The breeds of duck may be divided according to the purpose for which they are raised like meat, egg and ornamental type. The meat type ducks include Pekin, Aylesbury, Rouen and Sweedish breeds. **White Pekins** is most popular table duck, originated from China. Drake weight 4 kg and duck weight a little over 3.5kg at maturity. **Aylesbury** is the finest breed for table but is not as hardy as the Pekins. Plumage is spotless white in both the sexes. Drakes weight over 4.5kg and duck weight 4 kg. The egg type ducks include only one breed, The runner, commonly called **Indian Runner**. The Khaki Campbell may also be classified under the egg type. There are three standard varieties of Runner ducks the Fawn and white, The White, and the Penciled. All the varieties have the same shape but differ in plumage colour. Body shape is like a soda water bottle set at an angle of 50 to 60 degrees with no indication of a keel. Adult drakes weight 1.9 to 2.5 kg and ducks 1.5 to 2.2kg. Many of them lay 300 eggs annually. The **Khaki Campbell** breed was developed in England from the crossbred of Meat breed Rouen and Indian Runner. Egg production resembles with that of the Indian Runner but larger in size then Runner. The ornamental class includes the **Call, Crested white and Black** East India breeds.

2. **Turkey:** Turkeys (*Meleagris gallopavo*) are classed in the taxonomic order of Galliformes. Within this order they are relatives of the grouse family or subfamily. Males of both species have a distinctive fleshy wattle or protuberance that hangs from the top of the beak (called a snood in the Wild Turkey and its domestic descendants). They are among the largest birds in their ranges. As in many galliformes, the male (tom or gobbler) is larger and much more colorful than the female (hen). Genus Meleagris is the only genus in the subfamily Meleagridinae, formerly known as the family

Meleagrididae, but now subsumed within the family Phasianidae. Turkeys are mainly reared for meat production. There is no well defined breeds for Turkeys, however categorized into seven different varieties Viz. White Holland, Narrgansett, Bourban Red, Black, Slate, Bronze and Beltville Small white. All varieties have the same general shape, but differ mainly with respect to size, fleshing properties and plumage colour. Great variations in weight and type may be found among the strains of the same variety.

3. **Quail:** Quail (*Coturnix japonica*) is a collective name for several genera of mid-sized birds generally considered in the order Galliformes. Old World quail are found in the family Phasianidae, and New World quail are found in the family Odontophoridae. It is raised for eggs and meat. Quail are small birds with rapid growth and early maturity (5-6 week), have hight rate of lay (more than 250 eggs annually, egg weight 10-13gm per egg) and require less feed and space. It has short generation interval for which used as laboratory animal. In India, the quail is also known as bater. Quail attain about 90% of body weight by the market age of 6-7 weeks and start laying at this stage. Quails are maintained for one year only because the egg production, fertility and hatchability are reduced in second year. Quail eggs may be stored for nearly one month at 13 degree centigrade with 70% RH. The quail meat is tender and tasty and each quail may yield 100-150gm meat at market age of 5 weeks. The CARI, Izatnagar has evolved some crossbred genotype of quail such as Cari Shweta, Cari Uttam and Cari Ujjawal for meat purpose, and Cari pearl for egg purpose.

4. **Guinea fowl:** The guinea fowl (*Numeda meleagris galeata*) sometimes called guineahen are a family of birds in the Galliformes order, although some authorities (for example the American Ornithologists' Union) include the guinea fowl as a subfamily, Numidinae, of the family Phasianidae. The guinea fowl are native to Africa, but the Helmeted Guinea fowl has been domesticated, and both feral and wild-type birds have been introduced elsewhere. The breeds of guinea fowl are based on plumage colour variations. The main varieties recognized world over are Pearl, Lavander and White. In India, pearl guinea fowl are most common. They have dark grey feathers with white spots having pearl like appearance. The lavender birds have light grey spotted feathers. The feathers

of these two varieties are used for ornamental purpose. The white variety has completely white plumage. A new variety of guinea fowl has been developed at CARI, Izatnagar (UP) named as Guncari. It produces about100-120 eggs per year from March to October and hence seasonal layer. The egg weight is about 40gm and hatchability is about 70%, the body weight at 12 week is about 1 kg.

5. **Emu:** The emu (*Dromaius novaehollandiae*) is the largest bird native to Australia and the only extant member of the genus Dromaius. It is the Third biggest extant bird in the world after Ostrich & Cassowary. The birds have small wings and are incapable of flying. But they can run very fast, sometimes 40-50 kms per hour. The birds live in groups and can thrive under varying climatic conditions ranging 0 degree to 52^{0}C. These birds are omnivorous and eat leaves, vegetables, fruits, insects, worms. As emu can adjust to any type of weather, there is no need to prepare an artificial climatic condition to start its cultivation. The main thing required is a huge open land where emus can roam freely. The hatching of its egg requires hatchery and incubator machines. A three-room facility of 15-20 sq. ft area is needed to keep these machines. The eggs of emu are kept in hatchery machine for 49 days. After that they are kept in incubator machine for 4-5 more days. So a total 50-54 days are needed for the emu eggs to hatch under poultry process. While naturally it takes 57-60 days. So the time is consumed and the emus remain free for those days to produce more eggs. A year old emu is around five and half to six feet in height, its weight is 30-40 kg. Fully grown emu weighs 50-60 kg. When it grows it changes its colour thrice. In the beginning as it comes out of the egg up to 3 months there are stripes on its body. After 4 months it looks chocolate brown, after one year it again starts changing its colour. The feathers on the neck fall off. Then its neck looks bluish green and feathers exhibit a variety of colours. There is no distinction between a male and a female. Generally 18-month-old female is capable of reproduction; however in some emus this period can extend up to 2 years. As per available national data on an average during breeding period, female emu lays up to 32 eggs. The breeding period is 6 months, from October to March. Emu's eggs are dark green in colour. Compared to egg of a hen they are 10-12 times bigger. Female emu may lay eggs after every 3 days. But sometimes it lays eggs after 3 or 5 days.

Breed Descriptors

To develop breed descriptors of a breed requires the survey in the breeding tract. The purpose of breed descriptors is to facilitate valid comparison and classification of a breed with in a species. This means the information on habitat, physical and morphological characteristics, performances and genetic architecture of breed. The habitat of a breed should include the history of breed and geographical distribution (home tract), soil, topography, climate, feeds and fodder availability. These may be recorded from district headquarters of the breeding tract of the breed. The morphological and physical characteristics include the body size, the size and shape of different physical traits like horns, ears, tails, body colour etc. The data should be recorded on the performance include the information on growth in terms of body weights at different ages, production (milk yield, fleece, eggs, meat etc.) and reproduction traits. The genetic architecture of a breed may be known from blood groups, biochemical polymorphism and DNA markers.

Characterization: The act of describing distinctive characteristics or essential features in terms of breed descriptors. This includes the representation of all available knowledge which contributes to the reliable prediction of genetic performance in a defined environment.

Methods for breed characterization and its importance: There is need for undertaking systematic studies on the livestock resources by conducting survey and recording the information on these genetic resources. The recording of such observation on large and random population would help in describing norms of a breed, and descriptor for each breed can be prepared which can be subsequently used for recording the status and performance of the breed. This will not only help in gaining knowledge but will also be useful in planning breed improvement and conservation programme. Different approaches have been developed to analyze breed multiple features, including registration of their performances (phenotypic characterization) and molecular genetic as well as proteomic characterizations.

- **Phenotypic characterization:** A network project with National Bureau of Animal Genetic Resources, Karnal, India as the nodal agency for characterization and conservation of indigenous breeds of livestock and poultry came into operation in VIIth five year plan. This project aimed to develop breed descriptors, to characterize the breeds of phenotypic traits (qualitative and quantitative) and molecular genetic characterization. Three different levels of description can be used such as micro (phenotypic characterization as well as conservation), meta (phenotypic characterization) and micro (phenotypic and genetic characterization). The macro level consists of on farm survey based on stratified sampling procedure to collect information about herds and animals. The basic information about the location, village identification of the herd/animals details like type or breed, sex, age and some simple measurements (body weight, height etc.). The meta level based on monitoring of on farm husbandry practices. The information should be recorded on sampled animals on all performance traits viz. physical traits (linear measurements, body score), performance traits (milk yield, fleece yield, body weight, fertility, viability), health records, environmental conditions and management. Micro level is complementary to Macro and Meta studies for comparing and testing animal populations. The micro level is based on selected methods that can be used to gain knowledge of the genome of individual animals within a population.

- **Molecular Genetic level characterization:** Molecular characterization can play a role in uncovering the history, and estimating the diversity, distinctiveness and population structure of AnGR. It can also serve as an aid in the genetic management of small populations, to avoid excessive inbreeding. A number of investigations have described within and between-population diversity, some at quite a large scale. However, these studies are fragmented and difficult to compare and integrate. Moreover, a comprehensive worldwide survey of relevant species has not been carried out. As such, it is of strategic importance to develop methods for combining existing, partially overlapping datasets, and to ensure the provision of standard samples and markers for future use as worldwide references. Marker technologies are evolving: microsatellites analysis played and is still playing a fundamental role in molecular studies aiming to genetic characterization of breeds and populations, but it is likely that microsatellites will increasingly be complemented by SNPs. These markers hold great promise because of their large numbers in the genome, and their suitability for automation in production and scoring. However, the efficiency of SNPs for the investigation of diversity in animal species remains to be thoroughly explored.

Methods of data analysis are also evolving. New methods allow the study of diversity without a priori assumptions regarding the structure of the populations under investigation; the exploration of diversity to identify adaptive genes (e.g. using population genomics); and the integration of information from different sources, including socio-economic and environmental parameters, for setting conservation priorities. The adoption of a correct sampling strategy and the systematic collection of phenotypic and environmental data, remain key requirements for exploiting the full potential of new technologies and approaches. Characterization at the molecular genetic level based on molecular markers is undertaken mainly to explore genetic diversity within and between animal populations, and to determine genetic relationships among such populations. More specifically, the results from the 27 laboratory work are used to determine within and between-breed diversity parameters:

- Identify the geographical locations of particular populations, and/or of admixture among populations of different genetic origins.
- Provide information on evolutionary relationships and clarify centres of origin and migration routes.
- Implement gene mapping activities, including identification of carriers of known genes.
- Identify parentage and genetic relationships within populations.
- Support marker assisted genetic improvement of animal populations.
- Develop DNA repositories for research and development.
- In populations with limited or no information on pedigrees and population structure,

In the absence of comprehensive breed characterization data and documentation of the origin of breeding populations, molecular marker information may provide the most easily obtainable estimates of genetic diversity within and between a given set of populations. Furthermore, marker assisted selection offer new opportunities in AnGR management as DNA markers are useful in both basic (e.g. phylogenetic analysis and search for useful genes) and applied research (e.g. marker assisted selection, paternity testing and food traceability). Molecular markers are DNA locations presenting different detectable variants. The first markers used in livestock were the gene's coding protein polymorphisms (allozymes), but they present the limitation that both the number of loci that can be assayed and their polymorphism level are low. For this reason, as soon as DNA technologies were developed, markers at the DNA chain level (especially on nuclear material) quickly replaced allozymes as a tool to assess genetic diversity. Mitochondrial DNA (mtDNA) markers are very useful in phylogenetic analysis for several reasons. First, mtDNA is maternally inherited with no recombination and, therefore, the number of nucleotide differences between mitochondrial genomes reflects directly the genetic distance that separates them. Second, it mutates 5–10 times more frequently than nuclear DNA, thus allowing the study of the divergence between wild and domestic populations under the short time scale of domestication. For example, in sheep and goats there is one major geographical mtDNA lineage that probably represents initial domestication in the Fertile Crescent, with two more restricted lineages representing later independent domestication events.

There are several types of nuclear DNA markers. The AFLPs are dominant biallelic markers that provide an easy way to carry out a genome-wide screening of variation. They have the disadvantage of a reduced power to analyse within-breed diversity due to the dominant mode of inheritance, but could be very useful in analysing between-breed variation. One potential problem of AFLPs is associated to size fragment homoplasy, i.e. that different loci with the same fragment length will be contributing to the same band, thus biasing upwards the estimates of allele frequencies. This problem, however, is minimised if the number of loci analysed per primer combination is low. Microsatellites have been the markers of choice to study genetic variation in the recent years. Based upon sites in which the same short sequence is repeated multiple times, they present a high mutation rate and codominant nature, making them appropriate for the study of both within- and between-breed genetic diversity. Single Nucleotide Polymorphisms (SNPs) are point mutations in the genome sequence, predominantly biallelic and highly abundant throughout the genome. They have the potential to detect both neutral and functional genetic variation because, although most of them are located in non-coding regions, some correspond to mutations inducing changes in expressed genes. Large numbers of SNPs, however, are required for precision (it is said as a rule of thumb that about six SNPs are equivalent to one microsatellite). Sample collection is an important step in any molecular diversity study. The FAO Secondary Guidelines (FAO, 1998) required that at least 25 non-related animals per breed should be sampled and genotyped for at least 25 microsatellites with an effective allele number of two.

- **Proteomic level characterisation:** Aim of proteomics is the description of identity, quantity and state of all proteins in a cell under a specific set of conditions. Proteomics complements and extends the study of genomes and transcript data, reflecting the true biochemical outcome of genetic information. While genomic and transcriptomic data provide the "blueprint" for the possibility of cell function, they do not always inform on the actual protein content and thus the structural and biochemical effectors of a cell.

Avian proteome studies have been limited, and include muscle development, egg production, craniofacial disorders and the chicken lens using proteomic technologies. Proteomic approaches aiming to characterize breeds and to study differentiation have not been yet exploited. Advances made in avian genomics, particularly the publication of the chicken genome sequence, should improve confidence in the protein identifications provided by a typical proteomics experiment and provide the basis for further exploration of the protein component of avian species (Doherty *et al.*, 2007). The systematic study of protein structures, posttranslational modifications, protein profiles, protein–protein, protein–nucleic acid, and protein–small molecule interactions, and the spatial and temporal expression of proteins, are crucial to understanding complex biological phenomena. The number of different protein variants arising from protein synthesis (alternative splicing and/or post-translational modifications) is significantly greater than the number of genes in a genome. Mass spectrometry in combination with chromatographic or electrophoretic separation techniques, is currently the method of choice for identifying endogenous proteins in cells, characterizing post-translational modifications and determining protein abundance. Two-dimensional gel electrophoresis is unique with respect to the large number of 29 proteins (>10 000) that can be separated and visualized in a single experiment. Protein spots are cut from the gel, followed by proteolytic digestion, and proteins are then identified using mass spectrometry. However, developing high-throughput technologies would be useless without the capacity to analyse the exponentially growing amount of biological data. These need to be stored in electronic databases associated with specific software designed to permit data update, interrogation and retrieval.

Chapter - 4

Fodder Production and Grassland Management

K B Kore

Livestock farming is an integral part of Indian agriculture system and vital source of livelihood particularly in the rural areas of the country. This sector has been the primary source of energy for agricultural operations (for land preparation, sowing, pulling bullock carts etc.) and major source of animal protein (mutton, chevon, chicken, pork, beef, eggs etc.) for majority of population. India has been habitate of major draught, milch and dual-purpose breeds of cattle as well as buffaloes. Besides their potential our indigenous as well as crossbred animals are unable to produce maximum output due to unavailability of adequate and nutritious feed and fodder throughout the year. The quantity and quality of forage production is directly proportional to the livestock health, production and productivity. There is colossal gap between demand and supply of good quality nutritious fodder. Besides, the availability of lesser area under cultivated fodder crops, the grazing lands also becoming smaller due to the degradation of the land, urbanization, industrial development and increased area under food production to fulfil the demand of increasing human population. To meet the nutritional requirement of such a huge population of indigenous, crossbreds as well as genetically upgraded animals, proper management and conservation of fodder and grasslands is necessary.

History

The systemic research on forage crops was initiated in Maharashtra at Mumbai and later extended to Chennai, Madhya Pradesh and West Bengal. Indian Council of Agricultural Research (ICAR) and Government of India is the main funding agency for carrying out extensive research on forages in India. First time in 1952-53, ICAR started scientific research on grassland and fodder under the recommendations of FAO. Later on necessities for extensive research was gained substantial importance and in third five year plan Indian Grassland and Fodder Research Institute (IGFRI) was established in 1962 at Jhansi (Uttar Pradesh) to conduct, collate and co-ordinate research on all aspects of grasses, grasslands and fodder crops. Further to strengthen the specific research focus in 1947 'Agrostology' section was established at ICAR, New Delhi for studies on forage and fodder crops.

Common Livestock Feed Resources

The most common livestock feed resources are

1. **Crop reduces:** like straw of paddy and wheat, stovers of maize and jowar, haulms etc.
2. **Green grasses:** grasslands of alpine, sub-alpine pasture lands, cut and carry grasses etc.
3. **Community lands:** common property resources and wasteland.
4. **Cultivated fodder:** legumes (like lucern, berseem, stylo, cowpea), cereals (like maize, jowar, bajra) and grasses (like, anjan, napier etc).
5. **Forest lands:** wild natural grasslands, forest tree leaves, bushes and shrubs.
6. **Agro-industrial by-products and unconventional feeds:** like cotton seed cake, neem seed cake, jatropa cake or meal, mango seed kernel etc.
7. **Top feeds:** tree leaves like banayan, pipal; various fodder tress like subabul, gliricidia etc.
8. **Cereal grains:** jowar, bajra, maize, ragi etc.
9. **Oil meals and cakes:** soybean meal, mustard cake, groundnut cake etc.
10. **Cereal and legume by-products:** rice bran, wheat bran, rice polish, gram chuni, pulse chuni etc.

11. **Animal protein sources:** fish meals, bone meals, meat and bone meal etc.

Forage & fodder crops : Crops which are used in the form of whole plant (except roots) as pasture, hay, silage or green chops for feeding of animals is known as forages. The term forage is broadly used for all the plant materials that are eaten by herbivores. The term includes wild as well as cultivated plants that are used as livestock feed. The cultivated plant species that are utilized for feeding to the animals in the form of soilage (forage cut green and fed in fresh condition), silage (green fodder preserved under anaerobic condition) and hay (dehydrated green forage) are called fodder crops. Forage and fodder crops are synonyms terms but later are referred to cultivated forage crops which may either be cereals or legumes. Forages or roughages are coarse bulky feeds with more than 18 % crude fibre (CF) and low in digestible nutrients (less than 60 % total digestible nutrients, TDN).

Supply and Demand of Livestock Feed and Fodder

There is tremendous pressure of livestock on available feed and fodders, as land available for fodder production has been decreasing. Currently there is huge gap between supply and demand of livestock feeds (Figure 4.1.1 & Figure 4.1.2). The availability and demand of green roughage, dry roughage and concentrates is 395.2, 451.0, 50.0 and 1061.0, 589.0, 135.0 million tonnes, respectively with net deficit of 62.76, 23.46 and 63.0 % respectively. By 2025, the demand will reach to 1,170 million tonnes of green fodder, 650 million tonnes of dry forage and 152 million tonnes of concentrate feed. At the current level of growth in forage resources, there will be 65% deficit in green fodder and 25% deficit in dry fodder. It is obvious that deficit in green and dry fodder is increasing every year, while for concentrate gap is almost static.

Fig. 4.1.1 : Demand and supply of green fodder.

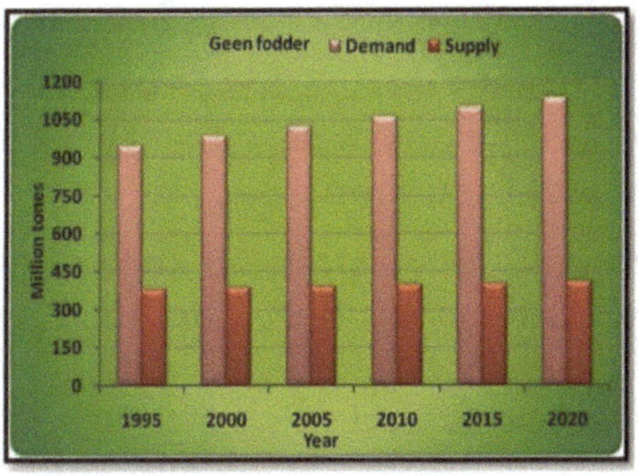

Fig. 4.1.2 : Demand and supply of green fodder.

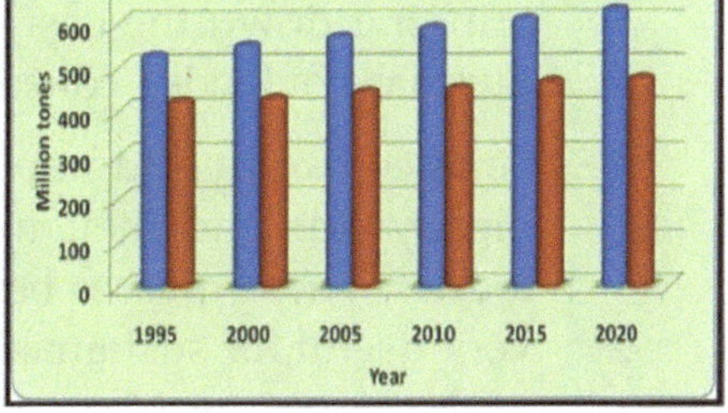

Importance of Cultivated Fodders and Grassland in Livestock Production

Supply of feedstuffs is a dominant variable on any livestock farm and most often due to poor planning coupled with inefficient production practices and adverse weather conditions, basic feed supplies are inconsistent and inadequate. It is not economic to meet these deficits of feedstuffs with concentrates, as these may lead to increased cost of production. The aim of fodder production management is to match the production capabilities of the farm with the animals' requirements in order to obtain the greatest margin over feed costs, within safe limits of natural resource utilization. Fodders and grasslands are important component of efficient livestock production due to following reasons.

1. Feed and fodder make up about 60-70% of the total costs of livestock production, so it is necessary to give much attention to supply of good quality, nutritious and least cost feeds and fodders.
2. The fodder crops meet the requirements of ruminants in terms of nutrients, satisfy the hunger and fill of rumen, fulfil the needs of rumen microbes very effectively and hence are important for ruminant production system.
3. Fodder crops provide most of the critical elements like highly digestible protein, carbohydrates, fats and minerals. Green fodders are a very good source of β-carotene (precursor of vitamin A).
4. Common cereal fodder crops like maize, sorghum and oats are rich in energy and the leguminous crops like lucerne, berseem and cowpea are rich in proteins as well as a good source of major and micro minerals, so critical for rumen microbes and animal system.
5. The green fodder crops are known to be cheaper source of nutrients as compared to concentrates and hence useful in bringing down the cost of feeding and reduce the need for purchase of feeds/ concentrates from the market.
6. Grasses like hybrid napier and guinea grass are known for high yields and their new varieties produce good quality fodder. These grasses being perennial and high yielding are very useful for sustainable livestock production.

7. Shrubs and small trees (like *Gliricidia, Desmanthus, Leucaena, Sesbania* spp.) are very good and cheap sources of protein and minerals and can be introduced between farm plots and have multipurpose utility.

Farm Machinery and Equipment

To carry out timely operations and optimum output, agricultural practices require farm mechanization or machine operated various farm activities including grasslands and fodder management. The improved agricultural and production practices, high yielding seed varieties along with increased cropping intensity putting pressure on traditional animal power operated farm equipments. These traditional practices are deficient in meeting the high demand of modernized farming system. Hence, to fulfil the ever increasing needs of farm management mechanization or provision of various modern machineries for carrying routine farm practices has become imperative. In the beginning of 20th century the development of agricultural engineering in the country was started based on modern technologies being used in developed countries. Various farm machinery and equipments are listed in Table 4.1.1.

Advantages of farm mechanization: Some key advantages of farm mechanization are:

1. Increase in crop intensity and yield thus ensuring better returns to the farmer.
2. Reduction of weather risk and risk of non-availability of labour thus minimizing post harvest wastages.
3. Improved working conditions and enhanced safety for the farmer.
4. Conversion of uncultivable land to agricultural land through advanced tilling technologies.
5. Saves time of farmer.
5. Shifting wasteland for feed and fodder cultivation.
6. Farm mechanisation also provides different streams of employment related to handling of farm machines thus resulting in increased rural employment.

Table 4.1.1 : Various farm machinery and equipments.

Sl. No.	Operations	Traditional equipment	Improved equipment
1.	Seed bed preparation	Spade, country plough, *bakhar*	Disc-plough, mouldboard plough, cultivator, disc-harrow, leveller, rotavator, roto-tiller, puddler, chisel plough, *Patela*harrow, pulverizer roller etc.
2.	Sowing and fertilizer application device	Broadcasting, dibbling, transplanting, line sowing behind country-plough	Seed drill, seed-cum-ferti drill, till-planner, fertilizer broadcaster, ammonia applicator, potato planter, groundnut planter, sugarcane planter, rice transplanter, pneumatic planter, inclined plate planter, multicrop planter, vegetable transplanter etc.
3.	Interculture	*Khurpi*, spade plough, blade harrow (*dora*)	Hand weeders, cultivator, wheel hoe, rake etc.
4.	Irrigation	Swing basket, *dhenkli*, *pur/jhori*, percian wheel	Rotary tiller, ridger, furrower, sweep centrifugal pump, submersible pump, sprinkler, drip irrigation etc.
5.	Plant protection	Summer ploughing, solar radiation, sieve, rat trap, bird scarer	Sprayer, duster, light trap, heat, solar radiation, net, biotechnology etc.
6.	Harvesting and digging	Sickle,*khurpi*, Spade	Serrated sickle, power, reaper, combine harvester, digger.
7.	Threshing, shelling and decortications	Beating, rubbing, animal treading, *olpad*thresher	Pedal thresher, power thresher, decorticator, combine harvester etc.

Adopted from Handbook of Agriculture, ICAR, New Delhi

Farm Power and Agro-Energy

Indian agriculture is characterised by overwhelmingly small holdings coupled with unabated land fragmentation. Major sources of farm power include both animate (humans and draught animals) as well as inanimate sources such as diesel engines, tractors and electric motors. There has been a rapid change in the way agriculture operations are carried out like- cropping system, water resources, diversification of crops, intensification of agriculture, increasing use of mechanical power versus bullock power, transformation from sustenance farming to market oriented farming and changing food habits. All this has its impact on animal husbandry practices. Human, animal, diesel engine, electric motor, tractor and power tiller are the major power sources for field operations in agriculture. Due to awareness and farmers education the animal power which contributed about 60 % of the total farm power in 1971-72 come down to only 17-18 % in 2000-2001 with increased contribution from machine power (~82-83%) for various agricultural operations.

Soil and Water Conservation

Soil-and-water conservation measures are extremely important for fodder and grassland management especially in dry areas and on sloppy degraded lands. There are many practices for soil and water conservation and some of which includes bunding, biological measures like cultivation of legumes, water storage, agro-forestry etc.

1. **Bunding:** This is quite effective in sloppy and undulating lands. Along the contour lines and periphery, various types of bunds are made depending upon the topography of the area. In dry degraded areas, bunds are required to be stabilized with suitable grass species. In areas having better soil and rainfall, technique of graded bunds along with grassed waterway and drop structure for safe disposal of water has been found to be better than contour bunds.

2. **Biological measures:** Various grass and legume species are known to play an important role in soil and moisture conservation by reducing run-off and soil loss. In excessive windy areas, a combination of trees may be planted as wind/ shelter belt in specific designs to check wind impact. The spreading of dry grasses or shrubby materials on bare ground along the contours or terraces has been found to check soil and water loss.

3. **Water storage/discharge structures:** These are extremely useful in ravines and watersheds, water storage/ discharge structures, *viz.* check dams, anne-cuts and embankments of desired engineering specifications, are constructed for efficient conservation and utilization of already scarce water resources in dry areas. In some ravinous and hilly areas, gully plugging may also require erecting structure of various engineering specifications.

4. **Agroforestry:** It has positive impact on soil and water conservation. "Agroforestry is the science of designing and

developing integrated self-sustainable land management system, which involves introduction or retention of woody components such as trees, shrubs along with agricultural crops including pasture/animals simultaneously or sequentially, on the same unit of land and at the same time, to meet the ecological as well as socio-economic needs of local people". Effects of fodder tree-crop interactions in agro-forestry systems are more positive. The woody component acts as semi-permeable barrier to water movement and mulch from trees, reduces the raindrop impact and thus, helps in soil conservation. The radiation balance and surface wind pattern influence plant water uptake and thus the conversion efficiency increased in the agro-forestry system. The agro-forestry systems are relevant for soil and water conservation are: multistorey tree garden, perennial crop combination, contour hedges and shelter belts/wind breaks.

Irrigation and Drainage for Fodder Production

Irrigation of fodder crops and water management aims to provide suitable moisture to the fodder crops to obtain optimum yields, commensurate with maximum economy in irrigation water and maintenance of soil productivity. The limited availability of water enables only 30% of country's cropped area under irrigation and that too major share goes to food/cash crops. In view of rising human and livestock population pressure, the forage farming is faced with twin problems of meager allocation of land and water resources. However, water is required in huge quantities for producing high tonnage of succulent biomass of forage species. It is in this context irrigation management requires entirely different approaches in forage production systems.

Development of suitable irrigation schedules offers an important approach to water saving because timely supply of water to fodder crops in adequate quantity is the root of the efficient water management for ensuring optimum yield and quality of produce. The optimum soil moisture regime for berseem, oat, maize, teosinte, sorghum, hybrid Napier and cowpea has been found to be 75% available soil moisture (ASM). Lucerne, cluster bean and barley may be irrigated at 50% ASM. The detailed irrigation requirement for individual forage production is given in following sections.

Agronomical Practices for Production of Leguminous and Non-leguminous Fodders in Different Seasons

Important Characteristics of Fodder Crops

1. Grown in closer spacing and high seed rate with short growth period.
2. Prevention of soil erosion and dense stand to smother weeds.
3. Improves soil health through addition of higher amounts of organic residues in the soil.
4. High persistency and regeneration capacity will reduce the need for frequent sowing and tillage.
5. Specific crop management which differs with the purpose of growing forages and mode of their utilization.
6. Adaptability to wide range of condition with capacity to grow under stressful environment.
7. Multicut nature with capacity to provide regular income and employment.
8. Easy for storage, transport, processing and conservation.
9. They should not be shy seed producer, poor harvest index and narrow seed multiplication ratio.

A. Non Legume/Cereal Fodder Crops

Cereals are the crop plants belonging to the grass family 'Graminae' which are grown for their edible starchy seeds, botanically known as 'caryopsis'. The important characteristics of cereal forages includes

1. They are rich in starchy (energy rich) material and carbohydrates.
2. They are determinate in growth habit.
3. Forage quality starts decreasing after flowering.
4. Dependent on soil for their nitrogen requirement.
5. They are less sensitive to climatic variations and soil variations compared to legumes.
6. They can be more conveniently conserved in the form of silage (preserved under anaerobic condition).

Some of the common non legume/cereal fodder crops have been discussed in detail in the following paragraphs

1. Maize (Fig. 4.2.1)

Botanical name: Zea mays, Synonym: Corn, Makka etc.

Varieties: African tall, Ganga-5, Ganga safed- 2, Vijay composite, Jawahar, APFM-8 etc.

Distribution: It is widely grown in Jammu and Kashmir, Himachal Pradesh, Punjab, Haryana, Utter Pradesh, Madhya Pradesh, Bihar, Maharashtra, Rajasthan, Karnataka, Andhra Pradesh, Tamil Nadu and parts of North Gujarat.

Characteristics: It is almost an ideal cereal forage and grain crop because of its quick growing, high yielding, palatable and nutritious qualities. It is the most suitable crop for silage making. It is a principle *Kharif* crop in India and other Asian countries. Maize can successfully be grown as *kharif, rabi* and *zaid* crop.

Soil: Fertile, well drained loamy soils and silty clay loam are well suited. Slightly acidic to neutral soils are suitable.

Sowing and manuring: In summer sowing between June-July. Seed rate should be 50-60 kg with spacing of 30×10 cm. It is preferred to sow 10-15 days before start of rain which will give 15% higher yield. Nitrogen fertilizers are good for maize with a dose of 120:60:40 of N P K (kg/ha). It is also advisable to apply 20 kg zinc sulphate/ha along with basal dose of fertilizer.

Irrigation: For winter and summer crops irrigation at an interval of 8 to 10 and 10 to 12 days is necessary, however *kharif* maize do not require irrigation.

Harvesting and yield: It is harvested between 50-80 days of growth from teaselling stage up to milk stage of grain maturity. For silage, late dough stage is preferred. An average green fodder yield is 40-50 tonnes/ha in *kharif* and 50-60 tonnes/ha in winter and summer seasons.

Nutritive value (% on dry matter basis): CP-6.74, CF-5.6, EE-2.09, NFE-47.5, Total ash-8.2, Ca-0.52, P-0.28, DCP-4.14, TDN-68.

2. Sorghum (Fig. 4.2.2)

Botanical name: Sorghum bicolour, Synonym: Great millet, Jowar, Durra etc.

Varieties: MP Chari, CSH- 13, HC 171, Safed moti, Jumbo, CSV 15 and CSH 20 MF etc.

Distribution: It is grown in Madhya Pradesh, Maharashtra, Gujarat, Rajasthan, Karnataka, Andhra Pradesh, Tamil Nadu, some parts of Uttar Pradesh and Haryana.

Characteristics: Sorghum is one of the most important fodder crops of low rainfall areas. Forage sorghum is characterized by quick growth, high biomass accumulation, dry matter content. It is also suitable for silage and hay making.

Soil: It is grown in different types of soil except very sandy soils, preferably well drained sandy loam to black soils are suitable.

Sowing and manuring: It is mostly grown in *kharif* as rain fed crop and during summer under irrigation, however less suitable for *rabi* season (September). Seed rate of 12-15 kg/ha for single cut and 20-25 kg/ha for multicut sorghum is required. The recommended dose of NPK is 80:40:40 kg/ha. FYM can be applied at the rate of 10 tonnes/ha for dry region and 25 tonnes/ha with 100 kg ammonium sulphate at planting in irrigated area.

Irrigation: Kharif sorghum does not require irrigation but during summer irrigation is required once in 8-12 days. For single cut 4 to 5 irrigations whereas for multicut 7-8 irrigations are required. Frequent irrigation can reduce the cyanide content.

Harvesting and yield: It is harvested at 50% of flowering stage with average yield of green fodder is 35-40 tonnes/ha for single cut and 100 tonnes/ha in 3-4 cuts for multicut varieties. In dry region the yield may be 10-30 tonnes/ha. There are improved yielding on an average 50 tonnes/ha in single cut varieties and up to 70 tonnes/ha in multi cut varieties.

Nutritive value (% on dry matter basis): CP-7.75, CF-32.30, EE-1.73, NFE-50.0, Ash-8.5, DCP-3.5, TDN- 54.0.

Toxicity: In early or young stage up to 40-50 days, it contains Hydrocynic acid (HCN) in the form of Dhurrin, which leads to cyanide poising if consumed by the animals. Hence, proper care has to be exercised during harvesting for avoiding HCN poisoning.

3. Bajra (Fig. 4.2.3)

Botanical name: Pennisetum typhoides, Synonym: Pearl millet, African millet etc.

Varieties: Rajco, Giant Bajra, MB72, BJ 104, HB 3, HB 4 etc.

Distribution: It is grown in central, southern and north westerns states except in the areas of high rainfall like Assam. It is grown mainly in Rajasthan, Maharashtra, Gujarat, Uttar Pradesh, Haryana, Karnataka, Tamil Nadu and Andhra Pradesh.

Characteristics: It is an outstanding summer growing fodder crop suitable for tropical and subtropical regions due to its drought and heat tolerance with high photosynthetic efficiency and also adoptable for varied climatic conditions.

Soil: Crop grows on a wide range of soils from very light soils from sand dunes in Rajasthan to red loams of Karnataka and Maharashtra. However, red chalky, black and clay soil is suitable.

Sowing and manuring: It can be sown at any time of the year in all seasons with sufficient irrigation facility. However, it is preferably sown in June to September (Kharif), October to January (Rabi) and during summer (January to May) season in north India. The seed rate of 8-10 kg/ha for single cut and 12-15 kg/ha for multicut. The recommended dose of NPK is 60-80:35:26 kg/ha. FYM can be applied at the rate of 20 tonnes/ha with supplementation of 30 kg nitrogen, 20 kg superphosphate and 20 kg muriate of potash.

Irrigation: Kharif bajra requires irrigation at the time of long spell of rain only. While in summer bajra, irrigation should be apply at an interval of 10 to 15 days according to soil type.

Harvesting and yield: It is harvested at 50% of flowering stage or 30-45 days from sowing with average yield of 35-40 tonnes/ha as fresh herbage for well managed farm and 15-20 tonnes/ha in rain fed condition.

Nutritive value (% on dry matter basis): CP-7.0, CF-31.50, EE-1.50, NFE-50.0, Ash-11.0, Ca-0.15, P-0.07, DCP-4.25, TDN- 60.

Toxicities: Traces of oxalic acid (1.19-2.16 % on DM basis) not harmful at this level.

4. Oat (Fig. 4.2.4)

Scientific name: Avena sativa, Synonym: Jai, Ganer, Togi Koddi etc.

Varieties: Kent, Flaming gold, Radar, JHO-851, Algerian, IGFRI-3021, OS-6, 7 etc.

Distribution: It is grown in Punjab, Haryana and Uttar Pradesh as a Rabi crop. It is also being grown as fodder crop in milk-shed areas.

Characteristics: It is commonly grown as a winter crop, tolerates hot and dry conditions. Next to wheat and barley it is an important cereal fodder crop of the temperate regions. It is an ideal fodder as a green chop, silage and hay mostly in the lean period of the year.

Soil: It is suited to all types of soils. However, fertile and well drained loam to clay loam soils is best.

Sowing and manuring: Late September to late November is a suitable period for sowing . Seed rate is 80 kg/ha for small seed variety and 100 kg/ha for large seed variety. For single cut, 80 kg N and 30 kg P_2O_5 per ha is required. Apply 40 kg N and 30 kg P_2O_5/ha at the time of sowing, while remaining 40 kg N should be given at 30 days after sowing.

Irrigation: One pre-sowing irrigation followed by subsequent irrigations for every four weeks interval depending upon the soil. Four to eight irrigations are required for two cutting system.

Harvesting and yield: It can be harvested at 50% flowering or 70-80 days for single cut while for two cuts, first cut is harvested 50 days after sowing and second cut at 50% of flowering stage. The average green forage yield varies from 4 to 5 tonnes/ha for single cut and in case of two cuts it may vary from 5 to 6 tonnes/ha.

Nutritive value (% on dry matter basis): CP-9.5, CF-26.5, EE-2.25, NFE-51.0, Ash-11.0, DCP-7.0, TDN-70.

Toxicity: Over application of nitrogen fertilizer above recommended dose leads to increase in nitrogen content of forage to a toxic level in the form of nitrates. The nitrates are reduced to nitrite in rumen ecosystem. This nitrite when absorbed in blood streams converts haemoglobin to methaemoglobin due to oxidation of Fe^{+2} to Fe^{+3}.

5. Ragi

Scientific name: Eleusine coracana, Synonym: Finger millet, Rajika, Mandua etc.

Varieties: Annapoorna, Kaveri, B1, CO-1, CO-2, AKP1, Hansa etc.

Distribution: It is grown in almost all areas of the country.

Characteristics: It is a drought resistant crop grown in rain fed as well as in irrigated areas. It is grown as *Kharif* (North India), *Rabi* and summer season (South India) crop.

Soil: It is suited to all types of soils. Loamy, sandy and black cotton soils are better than other type of soils.

Sowing and manuring: It can be sown throughout the year, however early June is good for short duration variety. Heavy seed rate of about 10-15 kg/ha is suggested in broadcasting method. In rain fed and irrigated condition nitrogen can be applied 40 and 160 kg/ha respectively.

Irrigation: It requires frequent irrigations, at the interval of one week 11 to 13 irrigations are required.

Harvesting and yield: The crop can be harvested 3 months after sowing. The average yield is about 10-15 qt/ha and 40-50 qt/ha for rain fed and irrigated areas, respectively.

Nutritive value (% on dry matter basis): CP-3.7, EE-1.1, CF-33.0, NFE-51.2, Ash-10.7.

Fig. 4.2.1 : Maize

Fig. 4.2.2 : Sorghum

Fig. 4.2.3 : Bajra

Fig. 4.2.4 : Oat

B. Leguminous Fodder Crops

The word legume is derived from the Latin word 'legre' (together) because the pods have to be gathered or picked by hand. The plant belonging to family Leguminouseae and has nitrogen fixing nodules on their roots. Important characteristics of leguminous forage are

1. They are partially independent of the soil for their nitrogen requirement.
2. They are high in protein content compared to cereals.

3. Indeterminate in growth and maintain quality traits over longer period.
4. They are more sensitive to variations in climatic and soil conditions than the cereals.
5. Improve soil fertility because of nitrogen fixing ability.
6. They can be more conveniently conserved in the form of hay (air dried or dry forage).

Some of the common leguminous fodder crops have been discussed in detail in the following paragraphs

1. Berseem (Fig. 4.2.5)

Botanical name: Trifolium alexandrinum, Synonyms: Egyptian clover.

Varieties: Mescavi, Pusa giant, Varda, Khadarvi, UPB-103 etc.

Distribution: Berseem is the prominent legume fodder crop of *rabi* in entire North West, Zone. It is widely grown in Punjab, Delhi, Uttar Pradesh and also in Rajasthan however also grown in Hill Zone and part of Central and Eastern Zone of the country.

Characteristics: It is known as "King of fodder" and established as one of the best *rabi* fodder crop with high nutritive value and palatability. Berseem makes most digestible and palatable green fodder to the cattle and especially milch animals. It is also a very good soil builder.

Soil: Although all soils are suitable, it grows well on heavy, dry and medium soils. Light sandy soils and water logging conditions are not suitable for berseem. The crop can be grown successfully on alkaline soils having good water retention capacity.

Sowing and manuring: The ideal time of sowing berseem is first to third week of October in north India and from last week of September to first week of December in North West to Eastern and Central India. The optimum seed rate is 25 kg/ha, which may be increased up to 35 kg in early or late sown conditions. For obtaining good yield, 20 kg N and 80 kg P_2O_5/ha should be applied as basal dose. N:P:K ratio is 25:30:40.

Irrigation: First two irrigations should be given at 5-6 days interval. Subsequent irrigations may be given at an interval of 10-12 days in winter and 8-10 days in summer. Thus, about 12-15 irrigations will be needed during the entire crop season. Normally the crop should be irrigated after each cutting.

Harvesting and Yield: It is harvested at 50-55 days after sowing. The subsequent cuttings should be taken at 30-35 days interval. A good berseem crop can give 100-120 tonnes/ha green fodder and 15-20 tonnes/ha dry fodder. An average fodder yield is 45-55 tonnes/ha.

Nutritive value (% on dry matter basis): CP-17.9, CF-26.0, EE-2.3, NFE-36.0, Ash-20.0, Ca-1.65, P-0.33, DCP-12.5, TDN-60.

2. Lucerne (Fig. 4.2.6)

Botanical name: Medicago sativa, Synonym: Alfalfa.

Varieties: CO-1, Anand-2, 3, GAUL-2, RL-88, T-9 etc.

Distribution: It is an important winter fodder crop in Rajasthan, Gujarat, and parts of Tamil Nadu, Kerala and in Leh area of Laddakh.

Characteristics: It is also known as "Green gold" and Queen of fodder. It is very important leguminous grown as an annual in hot humid region and as a perennial in dry regions. It is persistent as well as heat and drought resistant. It supplies green fodder for a longer period (November - June) in comparison to berseem (December - April).

Soil: Well drained deep loamy to clayey soil rich in calcium, phosphorus and potash produces high yield of lucerne. It cannot thrive on alkaline soils but can be grown on acid soils with liberal application of lime.

Sowing and manuring: Mid-October to early November is the best time for sowing of lucerne. In north, early October while that in east and south late December is the preferred period for sowing. The seed rate is 20-25kg/ha for broadcast method, whereas in line sowing 12-15 kg/ha can be used. If lucerne is grown for the first time in a field, the seed are to be inoculated with *Rhizobium mellilotus* culture for beneficial effects. N:P:K ratio is 20:100:40. Farmers growing lucerne for green forage are advised to apply 20 kg N, 50 kg P_2O_5/ha and 50 kg K_2O/ha as a basal application before sowing.

Irrigation: The crop needs very frequent irrigation during its early growth period at an interval of about one week but once the plants are established, subsequent irrigations are provided at an interval of 10-12 days during Rabi and 8-10 days during spring and summer seasons.

Harvesting and yield: For green forage production, first cut should be harvested two months after sowing and the subsequent cuts may be taken at 28-30 days during winter and 20-25 days during summer seasons. On an average 7-8 cuts will be obtained in case of perennial while 4-5 cuts in case of annual lucerne. Annual lucerne yields green fodder of 60-75 tonnes/ha while perennial crop may provide 80-1,100 tonnes/ha.

Nutritive value (% on dry matter basis): CP-20.0, CF-29.5, EE-1.81, NFE-34.68, Ash-14.0, DCP-16.0, TDN-58.0.

3. Cowpea (Fig. 4.2.7)

Botanical name: Vigna unguiculata/sinensis, Synonym: Lobia, Black eye pea, Cherry bean, Barbati.

Distribution: It is grown in Central, North as well as in South India.

Varieties: Russian giant, UPC-287, EC- 4216, IGFRI-457, 450, CV-1, GFC- 3.

Characteristics: It is an important leguminous fodder crop after lucerne and berseem. It is quick growing, high yielding with high protein content. It is bushy tailing, tap rooted legume fodder mainly cultivated in *Kharif* and summer season having considerable importance in dry land farming.

Soil: It can be grown on wide range of soils from heavy to red sandy loam. Water logged and heavy drought is not tolerable but tolerant to heavy rainfall than any other pulse crop.

Sowing and manuring: For monsoon crop, it should be sown in month of June-July with onset of rain. The crop is sown in the first week of July in the hills and in the second fortnight of March in lower hills and in October in plains. For summer cultivation, the sowing should be done in the month of February. The row to row spacing is 30-45 cm with a seed rate of 20-25 kg/ha. Farmyard manure may be applied @ 10 tonnes/ ha at the time of land preparation. About 120 kg N and 80 kg P/ha are recommended. Half the nitrogen is applied as basal dose and half for top dressing.

Irrigation: During monsoon it doesn't requires irrigation. In plains, 3-4 irrigations are required. In summer season, the crop is irrigated at an interval of 12 to 15 days.

Harvesting and yield: Generally it is ready to harvest in 70 to 75 days. For obtaining two cuts first cut is to be harvested at 50-55 days and second cut at 50% flowering or harvest at 40-45 days after first cut.

Nutritive value (% on dry matter basis): CP-15.3, EE-1.1, CF-34.8, NFE-30.0, Total ash-9.0, Ca-1.43, P-0.30 DCP- 10.3 & TDN- 50.5.

4. Stylo

Botanical name: Stylosanthes hemata, Synonyms: Caribbean stylo and Verano stylo.

Distribution: It is a native of the Islands of West Indies, Brazil and found generally adjacent the coastal regions of North and South America, and introduced in many tropical regions of Australia, Burma and India.

Characteristics: It is used extensively in agro-forestry systems mainly in horti-pastoral system that combines with perennial grasses like guinea grass and napier. Being nutritive and palatable, it is used as feed for all types of animals in the form of hay, silage.

Soil: It is adapted to a wide range of soil types and is drought resistant.

Sowing and manuring: In a well prepared field the seeds are broadcast or sown in line at 50 cm apart @ 5- 6 kg/ha in pure and 3-4 kg/ha in mixed pasture during the early rainy season. Before sowing, the seeds should be scarified or treated with hot water for 1-1.5 minutes. At the time of field preparation and before sowing 5-8 tonnes/ha FYM + 20 kg and 40 kg P_2O_5/ha and 40 kg K_2O/ha are applied. From second year onwards, 30 kg P_2O_5 and 15 kg N/ha are sufficient.

Irrigation: It grows well in areas with rainfall of minimum 800 mm. *Stylosanthes guianensis* has special adaptation for low rainfall. This crop requires irrigation at regular intervals.

Harvesting and yield: First cut is available after 60 days when crop is sown in monsoon and after 70 days when crop is sown in winter, subsequent cuts are available at 40 days intervals.

Nutritive value (% on dry matter basis): CP-12-18, EE-1.3, CF-38, NFE-42, Ca-0.61-1.72, P- 0.10-0.4 and Total ash-7.0 -14.2.

5. Sunhemp

Botanical name: Crotalaria juncea, Synonyms: Sunnhemp, Bombayhemp, Sunn, Saha etc.

Distribution: It is distributed all over the tropical and subtropical regions of the world including India.

Characteristics: It is an erect shrubby annual with much branched coarse stems and simple or digitate leaves. It is considered as multipurpose fodder species because it serves as a green manure plant, as a forage crop and for fibre production. It is hardy, quick growing and fairly drought resistant.

Soil: It is adapted to all type of soils but thrives well in light well drained loams.

Sowing and manuring: It is usually sown in June as *Kharif* crop. In some parts of Maharashtra and West Bengal, it is sown in October as *rabi* crop. Seed rate of 30-40 kg/ha in row sowing and 90-100 kg/ha in broadcasting method. Usually fertilizers are not required, however 40-60 kg of P_2O_5 per hectare can be applied as in the form of single superphosphate.

Irrigation: As a rule it is neither irrigated nor intercultured but one or two irrigations during growth phase will enhance the yields.

Harvesting and yield: The crop can be harvested in about 50-60 days after sowing. The average yield of green fodder is about 12-15 tonnes/ha of land.

Nutritive value (% on dry matter basis): CP-11.30, EE-1.1, CF-38.9, Total ash-6.4.

Toxicity: Seed and pods if fed along with forage may lead to toxic effects like bloat or tympanitis due to high protein content (35% in seed) and high cytoplasmic proteinaceous substances.

Fig. 4.2.5 : Berseem

Fig. 4.2.6 : Lucerne

Fig. 4.2.7 : Cowpea

Fig. 4.2.8 : Anjan grass

C. Grasses

Some of the common grasses have been discussed in detail in the following paragraphs

1. Anjan Grass (Fig. 4.2.8)

Botanical name: Cenchrus ciliaris or *Pennesetum ciliare,* Synonym: Buffel grass, Anjan, Dhaman Baiba and African foxtail grass.

Varieties: S 262, S 358, BundelAnjan 1 & 3, CO 1

Distribution: In India, it is a natural species largely found in Rajasthan, Haryana, Punjab, Gujarat and parts of western Uttar Pradesh and Tamil Nadu.

Characteristics: It is indigenous perennial tufted pasture common on the plains particularly in dry areas. It is mostly used as cut and carry fodder but also well suited to grazing for all classes of livestock.

Soil: It comes up best on well drained, light (red coloured) to medium textured soils and calcareous in nature. It can withstand drought and can also grow very well under irrigation. It is good soil binder and hence used as a cover crop on bunds for soil and water conservation purposes.

Sowing and manuring: Sowing can be done in all seasons by broadcasting method at the rate of 5 to 6 kg/ha of seed. In the first year, a basal application of 5 tonnes of farmyard manure (FYM) along with 40 kg N and 20 kg P_2O_5/ha is to be mixed thoroughly in the soil at land preparation. Afterwards, 20 kg N/ha is top-dressed at one-month crop stage.

Irrigation: Regular irrigation with an interval of 10-15 days during summer sustains the grasslands.

Harvest and yield: The first cutting should be taken at 60 days and subsequent cuttings at 30-45 days intervals depending on the rainfall. Under irrigation 12 cuts are possible in a year. Yield varies 90-110 tonnes/ha in arid tract with less than 300 mm rainfall to 225-275 quintals/ha/annum in the regions receiving 400 to 700 mm rainfall.

Nutritive value (% on dry matter basis): CP-9.1, CF-29.2, EE-1.6, NFE-44.2, Total Ash-6-8 , Ca-0.43, P-0.28.

2. Dinanath Grass

Botanical name: Pennisetum pedicellatum, Synonym: Deenabandhu grass.

Varieties: Jawahar Pennisetum 12, Bundel 1 & 2, IGFRI 43-1, CO 1, TNDN 1.

Distribution: It is widely grown under dry conditions in many states of India. Although it was reported to be growing wild in Chhota Nagpur and Gaya districts of Bihar and Ganjam district of Orissa.

Characteristics: It has good ability to spread naturally by self seed set therefore has a tendency to become a notorious weed in certain areas. The grass is good for monsoon season or kharif season in tropical and semi-arid areas. The grass is mostly used as cut and carry fodder. It is also very good for hay making.

Soil: It grows well on fertile loam soil but with manuring can be grown in sandy soils. The grass can tolerate both acidic and alkaline soils and marginal stony, gravely, undulated and hilly lands are suitable.

Sowing and manuring: Sowing is usually done just before rainy season when one or two good showers are received. Sowing may also be done during January to February with irrigation. The seed rate is 1-2 kg/ha, a basal application of 5 tonnes of farmyard manure (FYM) along with 40 kg N and 20 kg P_2O_5/ha. Afterwards, 20 kg N/ha is top-dressed at one-month crop stage.

Irrigation: During prolonged drought periods 1-2 irrigations are required. Summer sowing requires one presowing irrigation, later at an interval of 10-12 days depending upon soil nature.

Harvesting and yield: It has good quality forage with production up to 45 tonnes/ha under medium to high rainfall conditions.

Nutritive value (% on dry matter basis): CP-7.4, CF-22.2, EE-2.8, NFE-49.0, Total Ash-18.6.

3. Napier - Bajra Hybrids (Fig. 4.2.9)

Botanical name: Pennisetum glaucum × P. purpureum

Synonym: Bajra-Napier hybrids, N-B Hybrids, King grass, Elephant millet, Cumbu-napier hybrids etc.

Varieties: Yeshwant (RBN 9), Pusa Giant, CO 1, CO 2, CO 3, Swetika (Hybrid napier 3), NB 21, NB 37.

Distribution: Distributed in many part of India

Characteristics: These are tall growing (200-300 cm), erect, stout, deep rooted, perennial hybrid grass derived from inter specific cross between *Pennisetum glaucum* and *P. purpureum.*

Soil: It grows well on deep, retentive soils of moderate to fairly heavy texture and also grows on light textured with sandy loam to loamy soils. For better response to management, loamy soils with good drainage are preferred.

Sowing and manuring: One rooted slip or stem cutting is planted at a depth of 3-5 cm on one side of the ridge at 75 × 30 cm spacing at the rate of 40,000 rooted slips or stem cuttings/ha in the month of mid-February to July. In irrigated situation, it can be planted throughout the year. A basal dose of 5 tonnes of FYM/compost: nitrogen: 50 kg/ha (urea 110 kg); phosphorus 50 kg/ha (super phosphate 310 kg); potassium 40 kg/ha (muriate of potash 65 kg) should be applied followed by a top dressing of 50 kg N/ha after each cut.

Irrigation: The fields should be irrigated 3rd day after planting and as and when required thereafter. The irrigations need to be frequently applied especially during spring and summer.

Harvesting and yield: First cut should be taken 60-75 days after planting and subsequent cuts once in 45 days. Among grasses, N-B hybrids are the highest green forage yielder in a unit time and space. The normal yield levels range from 150-160 tonnes/ha/year. CO_2 variety yielded 300-350 tonnes green fodder/ha in southern India when cut at 10-15 cm from the ground at regular intervals of 45 days.

Nutritive value (% on dry matter basis): CP-10.2, CF-30.5, EE-2.1, NFE-41.0, Total Ash-16.2.

4. Para Grass (Fig. 4.2.10)

Botanical name: Brachiaria mutica, Synonym: buffalo grass, California grass, giant couch grass, water grass and paniwali ghas.

Distribution: It is native to Brazil and is widely distributed in tropical countries.

Characteristics: It is a hairy and perennial grass, spreading rapidly by surface runners, profusely rooting at nodes. The grass is used as green fodder, soiling and even for hay. It can withstand moderate grazing.

Soil: It performs well in moist clayey/silty/peaty/sewage farm soils. It grows well in water logging conditions on river and canal banks and can withstand prolonged flooding. It is highly profitable to grow on submerged or low lying as well as saline soils where nothing else survives.

Sowing and manuring: In a well prepared and levelled field, shoot bits of 30 cm length each with 2-3 nodes are planted in lines at about 50 cm distance; under irrigated conditions best time of planting is in March while under rain fed conditions at the onset of monsoon. A basal dose of 5 tonnes of FYM/compost is given with 40:50:50 kg NPK/ha followed by a top-dressing of 40 kg N/ha after each harvest.

Irrigation: The crop needs liberal irrigations at fairly close interval. The land should always be kept moist.

Harvesting and yield: Generally, the first cut is taken in 3 months from planting and subsequent cuts at monthly intervals. The green forage production ranges from 100-140 tonnes/ha.

Nutritive value (% on dry matter basis): CP-6, CF-35, EE-2.1, NFE-45.0, Total Ash-11.2, Ca-0.76 and P-0.49, DCP- 3.0, TDN-50.

5. Guinea Grass (Fig. 4.2.11)

Scientific name: Panicum maximum

Varieties: Makueni, Riversdale, Hamil, PGG-4, FR-600, Haritha, and Marathakom.

Characteristics: Guinea grass is a popular fodder grass of the tropics. It can be profitably grown as a component of agro-forestry systems. It is a perennial bunch grass. The grass tolerates shade and grows under trees and bushes and is best suitable as an intercrop in coconut gardens.

Soil: The grass is adapted to a wide range of soils. It usually grows on well-drained light textured soil, preferably sandy loams or loams, but is better suited to medium to highly fertile loams. It cannot tolerate prolonged water logging.

Sowing & manuring: The best season of planting is with the onset of southwest monsoon during May-June. As an irrigated crop planting can be done at any time of the year. Seeds are used (3 kg/ha), it should be sown in nursery and the seedlings transplanted in the main field. FYM should be applied along with phosphorus and potassium fertilizers. A basal dose of 10 tonnes of FYM, 50 kg P_2O_5 and 50 kg K_2O per ha (applied in trenches) is recommended.

Irrigation: At planting two irrigations are required within seven to ten days for quick establishment. Usually irrigation once in 7-10 days is required.

Harvesting and yield: The first cut is usually ready in 9-10 weeks after planting and subsequent cuts are taken at 45 to 60 days intervals. About six to seven harvests can be made in a year. Approximately 80-100 tonnes/ha of green fodder is obtained per year.

Nutritive value (% on dry matter basis): CP-8.0 to 14, CF-38, EE-1.7, NFE-40-48, Total Ash-12-15, DCP- 4.4, TDN-55.

Fig. 4.2.9 : Napier - Bajra hybrids

Fig. 4.2.10 : Para grass

Fig. 4.2.11 : Guinea grass

Fig. 4.2.12 : Subabul

D. Fodder Trees

A major constraint to livestock farming in the tropical country like India is the seasonal fluctuations in forage yield and quality of the grasses. In many tropical countries trees have been used as sources of fodder, fuel wood, and timber. Trees and shrubs can provide high quality fodder in the dry season. Now-a-days more attention has been given to the species which fix atmospheric nitrogen (*Leucaena* spp., *G. sepium*and *Acacia* spp). Further, they can be used in pastures leading to the development of tree based forage production systems such as alley farming system, intensive feed gardens and scattered trees on pastures or fodder bank. For small holder farmers using traditional livestock rearing systems tree grass combinations have great potential for improved management systems. Tree fodders contain high levels of crude protein and minerals and many show high levels of digestibility. Most of the tree leaves generally contain 30-45 % dry matter and 9-25 % crude protein. The ether extract appears to be fairly high in tree leaves but not as digestibility in most cases in cultivated fodders. The crude fiber in tree leaves is complex and highly lignified in the mature stage. Though they contain a lower percentage of crude fiber than grasses and hays and they are poorly digested.

Importance of fodder trees in livestock production: Trees and shrubs have the advantage over herbaceous plants in as much they are persistent, produce more edible dry matter (mainly during the dry season) and usually retain their leaves during prolonged drought periods. Fodder tree species offer the advantage of integrating crop and livestock production and a potential for improving the overall productivity. Further, desirable agronomic characteristics of fodder trees are- ease of establishment, good competitive ability, highly productive and persistence under repeated cutting or grazing, ability to adapt to asian and pacific climatic edaphic conditions, require no fertilizer (low input system), resistant to local pests and diseases, ability to produce seed or be reliable for vegetative propagation, have good nutritive value and resonable palatability to animals. Some of the common fodder trees are as follows

1. Subabul *(Leucaena leucocephala)* (Fig. 4.2.12)

It is a perennial hardy evergreen shrub also known as miracle tree or leucaena or ipilipil. It had its origin from Mexico and is now widely spread throughout the tropical and subtropical countries of the world.

Varieties: Hawaiian type- K-340, 341, Salvador type- K-8, 28, 29, 67, Peru Cunningham type-a cross between Salvador and Peru types.

Characteristics: Subabul is best suited for warm regions. It has strong and deep root system and is highly drought resistant.

Soil: It cannot withstand water logging. It requires a deep well drained neutral soil and can tolerate saline and acid soil.

Sowing and manuring: Planting of seedlings can be done with the onset of rains in May-June or Sept-October. A seed rate of 3-4 kg/ha is recommended. Sowing is preferably done during February March in a nursery or polythene bags or *in situ* at 2-3 cm depth. A basal application of N: P_2O_5: K_2O at the rate of 20:50:30 kg/ha is recommended.

Harvesting and yield: First cutting is done after 5-6 months at a height of 70-80 cm from the ground level at a time when the plants reach a height of 1.5- 1.75 m. Subsequent harvests can be made at 50-60 days interval depending on the re-growth.

Nutritive value (% on DM basis): CP- 16, EE-7, CF-13, NFE-52, ASH-13, DCP-12, TDN-58. Subabul is a highly nutritious leguminous tree fodder with 27-34 per cent protein. The fodder is rich in carotene and vitamin A. Pro-vitamin A content is the highest among all plant species.

Toxicity: The foliage contains mimosine (amino acid), which is toxic to non-ruminants at levels of about 10% of the diet. Main toxic effects are listlessness, body coats become rough and stary and within 2 -3 weeks, the animals lose hair form different parts of the body especially on tail, pizzle and eustacheon.

2. Hedge Lucerne *(Desmanthus virgatus)*

It is a small shrub, 2 to 3 m tall and roughly resembles leucaena. It is a native of tropics and subtropics of the new world, palatable, aggressive, persistent and tolerant to heavy grazing. It has 22 per cent protein in leaves and 10-15 per cent in stems. It is highly productive, yielding about 40 to 70 tonnes of green fodder per ha per year. No poisonous principle is observed in the foliage. Because of its pithy stem, the harvesting is easier. It is an ideal plant for wasteland development.

3. Agathi *(Sesbania grandiflora)*

The outstanding feature is its extremely fast growth rate, especially during the first three to four years. Average wood yield of 20 to 25 m^3/ha. Prolific nodulation and extremely large nodules are its characteristic features. Cattle relish both its leaves and pods. It has also been viewed as a source of pulp for paper industry. Leaves contain 36 per cent crude protein.

4. Shevri *(Sesbania aegyptica)*

It is a promising perennial fodder shrub for both dry and wet areas. The leaves and young twigs form nutritious forage to the livestock. When cut at 50 cm height at 60 days interval, green fodder yield of 12 tonnes/ ha per year can be obtained.

Limitations of tree fodder: Interference of trees is a major constraint in the integration of trees with field crops, as trees mature and canopy is formed the intensity of light at the ground level decreases and forage productivity may decline. A further disadvantage of the trees is the high proportion of inedible woody tissue that these plants produce as well as anti-nutritive and toxic factors which can produce toxic effects in ruminant animals, as well as their slow establishment.

Table 4.2 : Fodder trees which are being used for fodder banks.

Species	Major eco-zone	Countries
Albizia lebbeck	Humid/ sub-humid	India, Nepal
Dalbergia sisso	Tropical highlands	India, Nepal
Derris indica	Arid/ semiarid	India
Gliridia sepium	Humid/ sub-humid	Panama
Leucaena leucocephala	Humid/ sub-humid	India, Philippines, Mexico
Prosopis cineraria	Arid/ semiarid	India
Sesbania bispinosa	Humid/ sub-humid	India, Pakistan
Sesbania grandiflora	Humid/ sub-humid	India, Indonesia
Sesbania aegyptica	Humid/ sub-humid	India

Source: Banerjee (2000).

Harvesting and Post-harvest Techniques for Fodder Preservation

In forage production, the time taken in harvesting of the crop, its transportation, chaffing and other handling operations are the important factors that influence the total yield or production of fodder. Delayed harvesting leads to over maturation of the fodder crops which may reduce their palatability, nutritive value as well as digestibility. For proper management of fodder harvesting, the order in which farmer will harvest the fodder areas should be outlined and process should be limited to the extent necessary to provide fodder for that particular period and to adjust for round the year continuous supply of fodder. For harvesting forage harvester tools or equipment's are required which includes improved sickle or cutter grass slasher or grass cutter, manual chaff cutter, power operated chaff cutter with safety gadgets, tractor mounted fodder harvester and fodder block making units etc. Followings are the methods of fodder harvesting

a. **Lopping:** this method is suited for small unit of fodder production.

b. **Cutting:** best suited to lower harvesting rates and/or where fodder species are less abundant. Fodders are cut selectively or in strips or blocks.

c. **Pushing or chain pulling:** better suited to higher harvesting rates and/or where fodder species are abundant. Pushing can be preferred in soft bed area whereas chain pulling may be preferred in hard bed area.

d. **Breaking, pushing or chain pulling:** better suited where larger quantities of fodder are required or where fodder species are abundant. Push or pull in strips or blocks.

Followings are the method of harvesting fodder in large area/ natural grasslands and pastures

i. **Strip harvesting:** refers to harvesting in a strip aligned across the slope *e.g.* by pushing or pulling. Strips must retain vegetation along both sides of the strip. Strip harvesting can be carried out in three ways (1) *Parallel* strips: strips can be cleared in parallel direction having uniform width without curves. (2) *Curved strips*: strips may be curved so that each segment of the strip is aligned across the slope to avoid land degradation. (3) *Variable width strips*: the width of strips can vary to make best use of the fodder available.

ii. **Block harvesting:** refers to clearing (e.g. pushing or pulling) of patches of fodder vegetation up to 4 hectares in area. Each block must be surrounded by retained vegetation. Within the block, you must retain non-fodder trees higher than 4 meters.

iii. **Random harvesting** (combination of strips and blocks)**:** Under the circumstances where the distribution of fodder is patchy, various combinations of strips and blocks can be used to best utilize the available fodder.

Post Harvest Technology

Post-harvest technology is inter-disciplinary "Science and Technique" applied to agricultural produce after harvest for its protection, conservation, processing, packaging, distribution, marketing, and utilization to meet the food/feed and nutritional requirements of the people/animals in relation to their needs. It is important because it has capability to meet food/feed requirement of growing population by eliminating avoidable losses making more nutritive food/feed items from low grade raw commodity by proper processing and fortification. Various institutions (IGFRI, IARI) or private organizations have developed many types of post-harvest processing equipment. Some of them are forage densifying machine, feed processing mill, feed grinders, pelletizer, forced air batch type hay drier, bailing machine, mechanical hay drier are some of the equipments used for improving quality of feed and fodder resources after harvesting.

Conservation of Fodder

Availability of nutritious fodder throughout the year is very essential for profitable livestock farming however there are fluctuations in availability of fodder from season to season. Therefore, it is necessary to conserve the surplus fodder in the form of silage or hay. The surplus forages of the two seasons *i.e.* rainy season (August-September) and winter season (February-mid March) could be conserved and easily be carried over to the succeeding lean periods (May and mid July and November-December) for the fodder supply.

Silage

Silage is defined as a material produced by controlled fermentation of crops under anaerobic conditions. The process of preserving feed in the form of silage is called as ensiling. The green fodder harvested at proper stage is stored, packed and compressed in silo and it is then tightly covered to prevent the contact with fresh air. The receptacle or structure in which silage is being prepared is called as silo; it is generally air tight/semi-airtight structure. In India at most of the farms *pit silos* are more popular than other types.

Important Points to be considered during silage preparation

1. The crops at the time of ensiling should have about 30-35 per cent dry matter and should have sufficient soluble sugars for acid production during fermentation.
2. Crops of high moisture should be ensiled with the addition of preservatives and additives.
3. The crop harvested in bloom stage is the best for silage preparation.
4. If a crop is deficient in soluble carbohydrates (e.g. legumes), then a carbohydrate source like molasses can be added.
5. The crop should be properly trampled/pressed to remove the air out of the silo.
6. Crop should be chopped before ensiling. For good silage, the shorter the chop length, the better is the quality.

Silo : It is an airtight structure designed for the storage and preservation of high moisture feed as silage. It is a hole/pit in the ground, trench or a tower where green fodder is stored for preparation of silage. While preparing a silo some precaution should be taken. It should be airtight to prevent growth of moulds, the walls of silo should be smooth, perpendicular and strong so as to allow better compression, height and depth of silo depends on demand of silage, water level in the area, cost of construction and machinery available for filling the silo, the top of the silo thoroughly covered with a water proof material like plastic sheet, concrete slabs etc. Normally the height of a cylindrical silo is taken double of its diameter. There are different types of silos; the classification is depicted as follows

Figure 4.3 : Classification of silo.

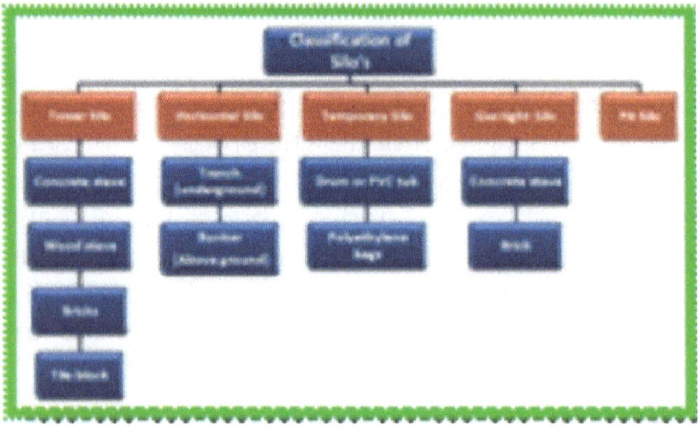

Crops suitable for silage making: Crops having thick, solid stems and rich in soluble carbohydrate are best for silage preparation *e.g.* maize, sorghum, bajra etc. Silage can also be prepared from oats, berseem etc. after wilting to 35-40% DM.

Stage of crops suitable for ensiling: Maize-dent stage; oats, sorghum, bajra- milk or dough stage; berseem, lucerne- 20-25% bloom stage; natural grasses- early flowering stage.

Method of Silage Preparation

1. First step for silage making is preparation of silo pit at desired location and levelling the pit.
2. The crop suitable for silage making is first cut into small pieces of about 2-4 cm length and wilted up to 35% DM.
3. Silage additives like molasses (@ 3 to 3.5%) can be added to provide soluble carbohydrates for efficient bacterial fermentation.
4. To improve palatability and nitrogen content, salt (0.5%) and urea (1%) can be added, respectively.

5. The whole material is thoroughly mixed and then filled tightly in the silo. Silo should be filled rapidly and should not be left open. It should be sealed as soon as possible and pressed so that no air pocket is left in the silo otherwise chances of mould formation leads to spoilage.

6. After filling, silo should be covered with polythene sheet followed by that of a layer of soil or mud to make it airtight.

7. Removal of silage can be done after 45 days of ensiling and used for feeding of animals. Care should be taken in removing the silage from silo. It should not be allowed to deteriorate after the silo is opened for feeding. Covers should be kept firmly in place as long as possible and the minimum face should be exposed at one time.

Characteristics of good quality silage: Silage quality is determined mainly by the odour, physical state, pH, ammonical nitrogen, volatile acids and lactic acid. For desirable fermentation, the forage should be rich in water soluble sugar (more than 5% on dry-matter basis). It should have Colour: green, yellow or golden brown, Smell: pleasant or vinegar type, Texture: firm, pH: 3.5-4.2, Lactic acid: 5 -15 %, Butyric acid: No/ traces (<0.2 %), Ammonical nitrogen : < 10% of total N.

Advantages of Silage Making

1. It is palatable, slightly laxative in nature, provide succulent feed during scarcity.

2. Nutrient loss is minimum as compared to other methods of preservation.

3. Less storage area is required as compared to hay.

4. It helps in biological control of pests and insects by preventing them to complete their life cycles due to early harvest of crops.

5. It can be prepared in all seasons.

6. Silage can be prepared from plants which have thick stems and thus are not suitable for hay making and also when the weather does not permit for hay making.

7. The organic acids mainly VFA produced in the silage are similar to those produced in rumen and therefore are utilized in the same manner.

Disadvantages of Silage Making

1. Silo construction is costly.
2. Loss of nutrients may be very high, if silos are not properly prepared.
3. Due to fermentation, there is 5-20% loss of dry matter.
4. If air enter silo, carotene loss is much.

Hay

Conservation of high-quality forages with minimum loss of nutrient by drying is termed as hay making. The principle of hay making is to preserve nutritional value of forages through drying it to a level (<12-15 % moisture) at which the activity of microbial decomposers is inhibited. In India, sunlight is available in abundance, which enables farmers to dry the green forage in open sunlight and thus making hay more economical. The hay making leads to reduction of moisture content to 10-15%, which inhibits the enzyme activity in the plant to be conserved. Various methods of drying includes drying on wire, roof tops, tree tops, galvanized tin sheets, tripod stand, tetra pod stands, field curing etc.

Crops suitable for hay making: Thin stemmed cereal crops like sorghum, oat, range grasses, range legumes particularly *Sylosanthes, Siratro,* cowpea, lablab, and all the cultivated legume fodders like berseem, lucerne, and cowpea are suitable for hay making. Leguminous forages have high buffering action and high nitrogen content, and hence are more suitable to be conveniently conserved as hay. Cultivated fodders like hybrid napier, para grass, guinea grass, teosinte are not suitable for hay making.

Method of Hay Making

Hay making is relatively more convenient and easy for Indian farmers. It can be done as follows.

1. Cut leguminous crops (berseem or lucerne) in the pre-blossom stage in order to ensure conservation of protein and available energy to a great extent.
2. Chop the forage while still moist (fresh or wilted) with a chaff-cutter. Thin stemmed crops including legumes can be dried without chopping while thick stemmed fodders like sorghum,

maize and *bajra* (pearl millet) require chopping or crushing before they are allowed to dry.

3. Chopping need not be too fine. The best length of the cut is about 5 to 8 cm.

4. Spread the wet chopped forage in the sun on a smooth hard surface in a thin layer not exceeding 12 to 15cm in height. The usual threshing floors, roof tops, polythene sheet etc. can be used for drying of forages.

5. Stir the drying forage every 2-3 hours during the day to speed up the drying process under exposure to the sun and the air.

6. When thoroughly dry (usually) after 2-3 days, depending on the frequency of stirring, the intensity of the sun and the movement of the air, gather the mixture of dried stems and leaves to store or market. When hay balers become available, the chopped and dried forage can be baled. Baling reduce the storage space and facilitates the transport of the forage to the market.

7. After it is well dried (dry-matter content at the time of storing should be around 85-90%) the forage can be stored at the farm as hay in the same way as wheat *bhusa* in thatched or mud-covered stacks or in buildings normally used for storing wheat *bhusa* or rice straw.

Characteristics of good quality of hay: Good quality hay should contain moisture between 12-15%, should be highly digestible, leafy, pliable, palatable, free from undesirable weeds and moulds. It should be free from dust, green in colour and should have minimum loss of nutrients while curing.

Advantages of Hay Making

1. It is most suitable method of conservation of green fodder for small holders.

2. Less expensive to prepare a high quality conserved form of feed.

3. It is possible to maintain more stock on a certain area of land (66 kg/m^3).

4. Many undesirable things present in a fresh crop are eliminated after it is converted into hay.

Disadvantages of Hay Making

1. In making hay from high-quality forage, the biggest drawback is the loss of valuable leaves in handling.
2. Some nutrients are always lost in field curing of hay.
3. Drying of green forages at ordinary temperature reduces its digestibility.
4. Losses of vitamins due to bleaching and fermentation

Storage of Feeds and Fodders

In order to feed to retain its freshness, storage facilities should be kept cool, dry, and pest-free. A good storage structure provides for effective fumigation, proper ventilation, protection from rodents, birds, rain water leakage, fire and floor moisture seepage and ease of loading/ unloading. Concrete walls for the storage of feeds and feed ingredients must be watertight. The roof, walls, doors, windows and floor must be leak-proof. The floor must not transmit water vapour from the soil. Doors, windows should be sealable in order to permit control of ventilation. Building must have devices to protect against the entry of rats and mice and birds. Pipes, shafts, ducts etc. should be fitted with wide metal guards outside and netting inside. Storage containers such as garbage cans with tight-fitting lids can help keep grain safe from elements and rodents. Plastic containers are preferable to metal ones because they are less likely to produce condensation. All storage bins, regardless of the source or type, should be cleaned before being used and maintained on a regular basis.

Table 4.3 : Storage structures for feed ingredients and fodder.

S.No.	Storage structure	Description
Traditional storage structures		
1.	Muc-bin	Made by bricks and mud or by straw and cow dung. These are usually cylindrical in shape with varying capacity.
2.	Bamboo reed bin	Made up of bamboo splits plastered with mixture of mud and cow dung
3.	Thekka	These are made up of gunny or cotton cloth wound around wooden support and generally in rectangular shape
4.	Metal drums	Made up with iron sheets in cylinder and square shape with various sizes

Contd.

Improved storage structure		
1. d	Improved bins e s i	Different organisms developed and g n e d improved storage structures for scientific storage of feed-grains, which are moisture resistant and rodent-proof. Eg. Pusakothi, PKV bins, Chittore stone bins etc.
2.	Brick-built godowns	These are made by brick-walls with cemented flooring for storing jowar in bulk and bags.
3.	CAP (Cover and Plinth) Storage	It is an economical way of storage on a large scale. The plinth is made by cement concrete and bags are stacked in open and covered by polythene cover.
4.	Silos	These are made from concrete, bricks and metallic materials with loading and unloading equipment's

Scarcity Fodders

There is usually adequate forage of fair to good nutritive value in the wet season, but in the dry season available forage from natural pastures are usually inadequate both in terms of quality and quantity to meet even the maintenance requirements of livestock. Differential natural calamities like droughts, floods, cyclones and fire quite often affect the crops in different parts of country. Suffering of animal is generally more than the humans because livestock owner, incapable to avoid the feeding of their large animals, leave them stray. In serious situation of scarcity of prolonged nature livestock relief campus are organized by government agencies for protection of livestock.

Sources of feed stuffs during scarcity: Feeds and fodders for the feeding of farm animals under scarcity may be obtained from the following sources. These include both conventional and non-conventional organic sources, which may be fed to animals during scarcity.

1. **Crop residues and grain byproducts:** These include dry crop residues like wheat straw, paddy, finger millet (ragi) and many minor types of millet; stovers of sorghum, pearl millet (bajra) and maize; and natural grasses collected from reserve forests. Straws and stalks of pulses and oil seeds are generally less palatable and digestible but can be fed during scarcity as a survival feed. Sugarcane tops and trash are available in plenty in some areas.

2. **Partly damaged crop left after flood:** Partly damaged crop is mostly unfit for grain production. These damaged crops of flood affected areas are usually coated with a layer of mud silted during stagnation of flood water. This may be washed off the mud layer may be removed after sun drying the plants followed by mild beating of dried plants with a small piece of bamboo manually.

3. **Oilseed processing residues:** Cakes available after extraction of oil by different extracted processes are valuable animal feeds and include mustard cake, ground nut cake, linseed cake, sesame cake, soybean meal, sunflower cake, cotton seed cake, etc.

4. **By products of sugar and jaggery industry:** These include bulky roughages like bagasse and different kinds of sugar rich syrups like screens, molasses and sludge. After harvesting the sugar cane the green tops available as a waste can be used for the feeding of cattle and buffaloes. Some quantity of cane tops is converted into hay at some places, while good quantity of it goes waste, which can be preserved by ensiling. Sugarcane trash mostly used as fuel for the preparation of jaggery, may also be use to supply part of the roughage requirement after chaffing and enriching with more palatable and nutritious feeds. Bagasse is available in sugar factories and crushers after extraction of juice. The palatability and nutritional value of bagasse for the livestock (cattle and buffaloes) are much better than the rice hull.

5. **By products of forests and forest based industries:** These include mature jungle grasses, fodder tree leaves, fallen tree leaves, wild fruits and flowers, berries and kernels, wood shavings, saw dust, etc. These are mostly much less palatable and poorly digestible. Besides common fodder, shrubs and herbs like pipal, neem, saura, tara, mango, kathal, etc. other non-toxic tree leaves may also be fed to farm animals to supply part of their nutritional requirements. Saw dust and papers are the important sources of feed during acute scarcity.

6. **Aquatic herbage:** Several aquatic herbages may be fed to maintain the animals particularly during scarcity caused by floods. These are water hyacinth, water spinach, water lilies, duckweeds, *azolla, hydrilla* and several algae and their by products available after the extraction of agar-agar. Besides

supplying protein and energy they are rich sources of carotenes.

7. **Herbages of arid zone:** The non-conventional herbages fed during scarcity in semi-arid and arid zones are several varieties of cacti like opemtia (nagphani) and euphorbia (sehud), acacia pods, juliflora pods, some wild melons etc.

8. **Animal organic wastes:** A large amount of animal organic wastes contributed by excreta of farm animals and poultry, waste materials from slaughter houses, dead animal carcasses etc can be processed and used as scarcity feed. The animal excreta are richer in crude protein content but their use is limited due to the presence of pathogenic micro-organisms and ova of different parasites.

9. **Fish and other aquatic animal's wastes:** These are trash fish, fish soluble, ensiled fish processing wastes, frog meat residue and crabs and lobsters processing wastes etc.

10. **Miscellaneous organic matters:** These include silkworm pupae meal, snail meal, hotel wastes, brewing wastes and penicillin mycelium wastes, etc.

11. **Fruit factory waste:** The waste materials of fruit factory processing plants like pine apple wastes, orange peel, tomato pomace are found to be abundantly available can be utilized as animal feed. These can form a part of the diet of livestock after processing through ensiling.

12. **Unconventional cakes and seeds:** The utilization of deoiled sal seed meal, treated neem seed cake, nahar seed meal, tapioca waste, extracted tea leaves has already been tested. These feeds may be incorporated to supply about 10-30% dry matter requirement of farm animals. These unconventional feeds can also be used for the feeding of simple stomached pig and poultry during scarcity of costlier conventional feeds replacing limited proportion of conventional ingredients.

13. **Utilization of top feeds:** Top feeds collected from the tree tops like subabul, avisa, acacia, etc. during cyclones and floods and used as forage supplements. During the cyclones fallen tree leaves and twigs are collected which are good in their nutritive value. Because of their higher feeding value they can be used in placed of conventional fodders. Jamun, biul, kachanr and other leaves have high digestibility and are also good protein supplements.

Newly Formed Feeds during Scarcity Period

1. **Urea molasses liquid feed:** Liquid molasses containing 2-3% uniformly mixed urea fortified with minerals and vitamins has been named as "Liquid feed". Liquid urea molasses feed can form an economic system of animal production not only during scarcity but also during the normal periods also.
2. **Urea-molasses nitrogen supplement:** Mixture contains 10-15% urea and is preferably the hot process. Calculated quality of this mixture is fed after mixing with concentrates or chaffed straw to meet about 50-60% of the dietary nitrogen requirement during scarcity. This supplement contains urea nitrogen equivalent to about 30-45% crude protein and may be much low rate (500g to 1kg per head adult bovine daily) than the 3 to 6 kg urea insoluble liquid feed.
3. **Urea-molasses-minerals-blocks:** These are solidified like blocks prepared in 1-5 kg blocks. The mixture contains about 10% urea, 10% complex mineral mixture, 5% common salt, 10% mustard or any other cake, 25% de-oiled rice bran and 40% wheat bran. De-oiled rice bran helps in keeping the blocks soft and for increasing hardness powdered lime may be added or de-oiled rice bran may be replaced with flour of damaged wheat, broken rice, sorghum, finger millet or pearl millet. This should be fed gradually and in limited quantity to avoid harmful incidence due to sudden excessive intake of blocks.

Feed and Fodder Management for Individual Animals

Fundamental to the whole fodder production plan is an estimate of the amount of feed that is needed. Herd requirements are a function of the herd structure, which means taking into account the average number of animals the herd, size of animals, average butterfat production, inter-calving period, heifer retention policy (sell surplus heifers as calves, or rear them all and sell the surplus animals when heavy in calf), and age of heifers at first calving. Other major factors affecting the amount of forage to be produced are roughage quality and wastage of forages. The dry matter requirements of each group of animals can be determined. The energy and protein requirements could also be estimated, to ensure that they will be met.

Calculation of Feed Requirement of Individual Animals

1. To calculate the feed requirement for individual animals we need to know
   a. The number and class of stock in the farm/paddock.
   b. The length of time that you plan to graze the paddock *i.e.* days
   c. The size of paddock in hectares

$$\text{Animal days per hectare (ADH)} = \frac{\text{Number of stock} \times \text{number of days}}{\text{Area of paddock (inhectares)}}$$

2. Then calculate the area of land that must feed to one animal for one day

$$\text{Area of land} = \frac{10000\ \text{m}^2}{\text{ADH}}$$

If there is insufficient feed in the square to feed one animal for one day, you can either reduce stock numbers or reduce the time of grazing by spelling the paddock.

Feeding Management of Individual Cattle or Buffalo

The cows' requirements must be met from two sources: roughages and concentrates. Since the two types in combination must supply the total nutritional requirement within a fixed dry matter package, it follows that the amount of concentrate needed depends on the quality of the roughage.

1. Cattle and buffalo require daily dry matter @ 2.5 kg and 3.0 kg/ 100 kg of body weight, respectively.
2. About 2/3rd part from dry and green fodder as roughages and 1/3rd part from concentrate mixture out of total dry matter requirement of the animal.
3. If sufficient leguminous fodder is present 1/4th of total roughage should be provided through legumes and 3/4th from non-leguminous fodder.
4. Breeding bull should be provided 20-30% additional ration above to maintenance need.

5. For growth, milk production and pregnancy, extra ration in the form of concentrate mixture should be offered.
6. About 1 kg concentrate mixture for production of each 3 kg milk in case of cattle and each 2 ½ kg milk in buffalo should be given per day.
7. Above 6 months of pregnancy, 1 kg and 1 ½ kg concentrate mixture per day should be given to cow and buffalo respectively.
8. At least 5 kg green fodder per day must be included in the feeding of milch animal to maintain the milk production.

Fodder Production for Small Units through Inter-Cropping or Back Yard Cultivation

Mixed Intercropping System of Forage Production

It consists of growing together two or more plant species with different growth habit, canopy structure, rooting pattern and offering little or no mutual competition. Mixed cropping of cereals/grasses and legumes is commonly practiced to improve herbage quality, increase biomass production and economize on fertilizer usage. Besides, it also ensures efficient land utilization, suppression of weeds and insurance against aberrant weather conditions. In India with warm climate, most of the grasses and cereals do not have enough protein to support animal growth and production. It was found that by associative cropping of graminaceous and leguminous forages the nutritive value of the mixed herbage could greatly be improved. Thus, besides enriching the soil productivity and ensuring better utilization of plant nutrients from different soil layers, mixed cropping produces nutritious herbage since the legume component contains higher amounts of protein, calcium and phosphorus. Investigations on intercropping systems showed that sorghum/maize/*Pennisetum pedicellatum*/ pearl millet with cowpea/ rice bean (*Vigna umbellata*) and oat with *Melilotus alba*/pea in paired rows recorded good forage yields along with higher forage protein.

Integration of perennial forages on bunds and boundaries: Bunds around fields are a common feature on cultivated lands and occupy 2-10% of the cultivated area. Bunds are formed to demarcate field boundaries, conserve water and soil and to use as pathways. If not properly protected, the bunds are eroded by rains and damaged by animals. Growing of appropriate forages/bushes stabilizes the bunds through protective cover and soil binding action of roots, besides providing considerable amount of forage for feeding animals. Experiments conducted at the IGFRI, Jhansi, showed that in rice fields, *Sesbania* grown on field boundaries yielded maximum forage among perennials. Similarly, under rain fed conditions planting of hybrid Napier+Stylo along with fruit trees like *aonla* (Indian gooseberry), *ber* (*Ziziphus*spp.) and *karonda* recorded maximum forage yields.

Incorporation of fodder trees and shrubs into farming system: Tree and shrub fodders usually sustain a long life-cycle. Competitive land use between crops and livestock exerts considerable pressure against utilization of arable land for planting fodders and pastures. In this situation, a number of approaches have been suggested in order to incorporate fodder shrubs and trees without competing with crops. Fodder shrubs such as *Leucaena leucocephala, Gliricidia sepium, Sesbania grandiflora* and *Artocarpus heterophyllus* etc can be grown as living fences which provides not only human food and fuel wood but also animal feed.

Livestock Waste Utilisation and Recycling

Farm animals consume forages and concentrates (energy, proteins, minerals, and vitamins) to produce meat, milk and eggs. Livestock waste and manure are inevitable by-products of this process. Wastes produced due to livestock's activities are dung, urine, feed refusals, feed processing wastes and animal droppings. It can also include slaughter house by-products. Animal production is developing rapidly in the country and hence the wastes and manure produced by it.

Livestock manure represent a valuable resource that, if used appropriately, can replace significant amounts of chemical fertilizers. However, unless animal waste and manure is managed carefully to minimize odour, nutrient losses and emissions, it becomes a source of pollution and a threat to aquifers and surface waters, it can also be a direct threat to human and livestock health. Thus, suitable management of livestock waste including manure is essential to avoid various ill effects due to its mismanagement. A livestock waste management is a plan to specify how, when and where animal waste will be handled and disposed of safely or a system that store, stabilize, transport or apply animal waste to land or recycle it. Recycling is the bioconversion of biomass into organic fertilizer, animal feed, clean fuel, or chemical intermediate compounds. In many developing countries including India there exists a whole sector of recyclers, scavengers and collectors to salvage 'waste' material out of livestock industry and reclaim it for further use.

Advantage of Livestock Waste Utilization and Recycling

1. There are social, economic and environmental benefits of utilization and recycling of livestock wastes.
2. Successful recycling of livestock waste reduces production cost of livestock and allied industry including agricultural with minimum environmental pollution.
3. Recycling of livestock manure as a bio fertilizer helps to increase in both quantity and quality of crop yields while reducing the use and impacts of pesticides, and chemical fertilizers.
4. Manure contains the undigested fraction of the organic matter in the diet, and generally more than 70% of dietary nitrogen, and 65% of dietary phosphorus. Hence, manure is an important source of organic matter for agricultural soils and may provide a major contribution to biological, physical and chemical soil quality.
5. Soils with low natural fertility can be improved by application of animal organic waste.
6. Helps to control and reduce spreading of livestock diseases from animals to humans.
7. Ensures a sustainable use of nutrients and mitigation of environmental impacts, including odour and NH_3 emissions, greenhouse gas emissions, hence reduction in global warming.
8. The integrated use of animal manures, in combination with inorganic fertilizers, is likely to increases crop and forage production, resulting in economic benefit for farmers, as well as environmental benefits.
9. It can also be used as a source of fuel in the form of biogas.
10. Recycling will contribute with plant nutrients to crop fertilization, thereby reducing the need for nitrogen, phosphorus, potassium and micro nutrients applied to the fields in mineral fertilizers.

Application of Livestock waste as a fuel in the form of biogas: Biogas is obtained in the process of biodegradation of organic materials under anaerobic conditions. Therefore, the bio-gas technology *i.e.* production of methane gas from livestock waste under anaerobic condition is the best alternate source of energy from livestock organic waste which is abundantly available in rural and semi-urban areas. Biogas can be used as an excellent fuel for many applications including power generation. It is used as fuel for cooking and lighting purposes. It can also be used in diesel engines to substitute diesel-oil up to 80 per cent. The left over decomposed slurry of biogas plant is a good source of manure for agricultural lands. Thus biogas plants help in total recycling of organic wastes in an environment-friendly manner. Cooking through biogas not only reduces consumption of fuel wood, dung cake, agricultural wastes, coal or kerosene but also reduces work load for women who in rural areas of the country spend considerable time to collect fuel for cooking. Use of biogas as cooking fuel provides an efficient way of converting fuels to heat and reduces indoor air pollution. The other environmental benefits include improved waste management, reduced emission of methane to the atmosphere and reduction in deforestation.

Calculation of Economic Aspects of Fodder Cropping and Procurement of Feed

An amount of expenditure on cultivation of particular fodder on specified area is its cost of cultivation while total money or expenditure spend to produce defined quantity of fodder is called as its cost of production. Different expenditures on cultivation of fodder include cost for preparation of land, sowing, manuring, irrigation, harvesting and post-harvest processing etc. Production cost includes all the inputs added for getting a unit amount of fodder from particular defined fodder land. The cost of fodder production can be calculated by dividing total green fodder yield (tonnes/ha) by cost incurred per tonne of green fodder.

Grassland Management

Grazing plays a significant role in livestock feeding. After crops residues, natural grasses are an important source of nutrition for livestock and cultivated fodder crops plays very meagre role in livestock production due to land scarcity. Grasses belong to a group of plants that have more than 10,000 species. They are found everywhere in deserts, in snow covered areas, in mountainous regions, on the coasts and even in water. Natural pastures, waste and fallow lands, forests, tree groves etc. are major natural feed resources for grazing of livestock's which can also be called as grasslands. Grassland can also be an area where the annual rainfall is insufficient to support a luxuriant growth of trees, but is still high enough so that deserts are not formed. Grasslands are however not restricted to low rainfall areas. Rangeland is a broader term than grasslands, including regions where even woody vegetation is dominant. It is a term looking at the land from the viewpoint of livestock production. It also serves as a habitat for wildlife. The main floral component of rangelands is grass or grass like vegetation. However due to our ever increasing population these grassland are being converted into agricultural lands through the development of dams and canals. Hence, it is imperative to take stringent steps to conserve and manage the grassland to meet the demand of livestock feeds.

Grassland resources of India : Presently livestock production is primarily based on rangeland grazing. The grazing activity is mainly dependent on the availability of the grazing resources from pastures and other grazing lands viz. forests, miscellaneous tree crops and groves, cultivable wastelands and fallow land. Such lands are about 40% of the total geographical area of the country.Grassland ecosystems form Himalayan pastures, the terai grassland of the foothills, semi-arid grasslands of Western and Central India, scrublands of the Deccan Plateau and in the Shola forests of the Western Ghats, Nilgiri and Annamalai ranges. Thus depending on the quantity of rain, there are tall, medium and short grasses. Natural grasslands have a variety of grasses, forbs (small plants) and herbs. These differ according to the different grassland types. The grazing intensity in the country is as high as 12.6 adult cattle units (ACU)/ha as against 0.8 ACU/ha in developed countries. Therefore, the task to deal with such situation is two-fold (i) improvement of pastures, and (ii) judicious implementation of grazing management. The problems of grazing lands are related to ecological and socio-economic causes. In arid areas, water is the limiting factor, in semi-arid areas, proper land utilization and excessive grazing are the main problems and in high rainfall zones, it is the lower nutritive value of the herbage. Another serious problem is low proportion of legume component in the herbage.

In India, presently out of 328.8 million hectares of land, 11.04 million hectares are under permanent pastures grasslands. The grass cover of India is majority situated in semiarid and arid tract on undulating and hilly areas. Grazing based livestock husbandry continues to play an important role in rural economy of the country as around 50% animals depend on grazing in forests and other grazing areas in many parts of the country. Pasture lands constitute the main grazing resources of the country, available over an area of 12 million ha (3.94% of the geographical area). Nearly 30 pastoral communities in northern and western parts of the country depend on grazing based livestock production. Based on the practice followed by these pastoral communities in various regions, the grazing systems may be categorized either on the basis of methods of grazing or patterns of migration. Considering methods of grazing the pastoral communities change the site of grazing after its utilization. The example of such type of grazing system is *Kharak*in Uttarakhand and *Goals* in the desert area of Rajasthan.

Classification of Grasslands

According to the Indian Council of Agricultural Research (ICAR) the grass cover of India has been classified into five major types:

a. **Sehima-Dichanthium type**: This cover type is spread over the whole of Peninsular India and are spread over the Central Indian plateau, Chota-Nagpur plateau and Aravalli ranges, covering an area of about 17,40,000 km^2. There are 24 species of perennial grasses, 89 species of annual grasses and 129 species of dicots, including 56 legumes. Dominant perennial grass species are *Dichanthium annulatum, Sehima nervosum, Bothriochloa pertusa, Chrysopogon fulvus, Heteropogon contortus, Iseilema laxum, Themeda triandra, Cynodon dactylon, Aristida setacea, Cymbopogon* spp.etc.This is also a rich wildlife area, with a large number of protected areas, especially forest protected areas (sanctuaries and national parks).

b. **Dichanthium-Cenchrus-Lasiurus type**: This type is associated with sub-tropical arid and semi-arid regions. These are spread over an area of about 436,000 km^2, including northern parts of Delhi, Aravalli ranges, parts of Punjab, almost whole Rajasthan, and Gujarat, and southern Uttar Pradesh. There are 11 perennial grass species, 43 annual grass species, and 45 dicots including 19 legumes. The principal perennial grass species are *Cenchrus ciliaris, C. setigerus, D. annulatum, Cymbopogon jawarancusa, Cynodon dactylon, Eleusine compressa, Lasiurus sindicus, Sporobolus marginatus, Dactyloctenium sindicum, Desmostachya bipinnata* etc. These grasslands are extremely important for the survival of certain bird species.

c. **Phragmites-Saccharum-Imperatatype**: These types of grasslands cover about 2,800,000 km^2 in the Gangetic Plains, the Brahmaputra Valley and the plains of Punjab and Haryana, North-Wastern states, West Bengal, Bihar and Uttar Pradesh. Important species of grass cover includes *Imperata cylindrica, Saccharum arundinaceum, S. spontaneum, Phragmites karka,*

Desmostachya bipinnata. Bothriochloa intermedia, Vitevaria zizanioides, Imperata cylindrica, Chrysopogon aciculatus, Panicum notatum, etc. There are 10 perennial grasses, 26 annual grasses, and 56 herbaceous species, including 16 legumes. Some wet grasslands survive in protected areas of the *terai* region and the Brahmaputra flood plains. These wet grasslands harbour many globally threatened wildlife species.

d. **Themeda-Arundinella type**: These grasslands cover about 230,000 km^2 of entire Northern and North Western mountain tract, includes the states of Assam, Himachal Pradesh, Jammu and Kashmir, Manipur, Uttar Pradesh, Punjab, Haryana and West Bengal. There are 37 major perennial grass species, 32 annual grass species, and 34 dicots including 9 legumes. The principal species of this grass cover are represented by *Arundinella benghalensis, A. nepaolensis, Bothriochloa intermedia, Chrysopogon fulvus, Cymbopogon jwarancusa, Cynodon dactylon, Heteropogon contortus, Themeda anathera, Euloliopsis binata, Ischaemum barbatum.*

e. **Temperate and Alpine cover**: These are spread across altitudes higher than 2100 m and include the temperate and cold desert areas and high hills of Uttarakhand, Himachal Pradesh, Jammu and Kashmir, Uttar Pradesh, West Bengal and the north-eastern states. There are 47 perennial grasses, 5 annual grasses and 68 dicots, including 6 legumes. The principal perennial species are: *Agropyron conaliculatum, Chrysopogon gryllus, Dactylis glomerata, Danthonia cachemyriana, Phleum alpinum, Carex nubigena, Poa pratensis,* and *Stipa concinna.* These high altitude grasslands harbour wildlife not generally found in other parts of the country. Depending upon the biotic influences and local variations in topography and soil structures, these five broad categories can still be subdivided into several grass associations.

Some of the other grass covers of India are

i. Himalayan pastures

ii. Terai grasslands

iii. Shola grasslands of the Western Ghats

Management of Grasslands

Protection, development and sustainable use of grasslands are very important for the rural economy and livestock. India has more than 500 million livestock, more than 50% of the fodder for this livestock comes from grasslands. Grasslands evolved under a system of grazing, drought and periodic fire and almost all the existing grasslands are maintained by either of these or a combination of all these factors. Tropical grasslands, which are in the mid succession stage, are largely maintained by annual or biannual burning in most of the protected areas (sanctuaries and national parks) whereas in unprotected areas they are maintained by livestock grazing and other biotic factors. Hence, a careful planning and management of these grasslands is necessary.

Steps to improve grasslands: It is recommended that specific crucial grasslands should be selected for effective conservation, as part of the Protected Area network, or as a part of watershed management.

1. Grazing would have to be regulated and fires prevented.
2. The emphasis being on harvesting grass rather than grazing it, which would result in augmentation of both the generation of grass as well as its nutrition value.
3. Efforts should be enhanced to improve cattle quality, as it is proven that improved varieties tend to be stall-fed and sent less to free-graze on rangelands.
4. The provision of a sustainable supply of fuel is undertaken by a newly created Fuel wood Mission. Not only will this mitigate the drudgery of millions of women who have no option but to forage for every possible scrap of fuel, but also will reduce pressure on trees and shrubs whereby our remaining forest and trees will be well-protected
5. Under poor soil, water and nutrient situations where cropping is not possible *Silvipasture Systems*, integrating woody perennials and pasture species, can serve the twin purpose of forage and firewood production and ecosystem conservation.
6. In order to reduce grazing pressure, the concept of horti-pasture, utilizing land in the orchards by developing pasture stands should be encouraged.
7. Higher priority on generation of information on temperate grasslands with emphasis on (i) low input, clover based sheep grazing system (ii) ideal pasture for mixed grazing systems; (iii) inventory of grazing routes and grazing systems; (iv) designing of suitable production system for migratory grazers of Himalaya and the *Thar* desert.

8. A matching approach on fodder production, agro-waste-use and fodder trees should be brought under one umbrella in form of a national fodder mission.
9. Development of Common Property Resources (CPRs) available with village *Panchyats* through improved pasture/ silvi-pasture systems is undertaken by dairy/livestock cooperatives/associations.
10. The practice of stall-feeding should be encouraged among livestock owners in order to prevent over grazing consequent depletion of available forest fodder resources.
11. There is a need of capacity building at various levels for the rangeland development and seed production of range species. Demonstration of range improvement and management technologies at different locations.

Grazing Management

There are many systems of grazing by the livestock to manage the grassland efficiently. However, regardless of the grazing system used, there are several key principles of grazing management that should be applied for efficient use of grasslands includes adoption of a planned approach to grazing, know the plants that grow on your property and understand their response to grazing, manage to increase perennials, avoid overgrazing, manage total grazing pressure, provide evenly distributed and reliable water supply and keep good records and monitor rangeland condition.

The choice of grazing system will depend on the natural resources and infrastructure (water and fencing) on the farm. Native grasslands and shrub lands respond well to a period of rest, and paddock spelling should always be adopted as normal grazing management. Set stocking, crash, continuous, cell, strip, deferred, rotational, time controlled, block, mob and rational grazing are all terms used to describe different types of grazing systems. Each of these grazing systems is based on the manipulation of stock numbers and stocking density in the context of feed availability and plant recovery from grazing. The different types of grazing systems are described with their advantages and disadvantages in the following Table 4.6.

Table 4.4 : Classification of grazing systems.

System of Grazing	Advantages	Disadvantages
1. Continuous grazing		
• This is an extensive system of grazing in which the stock remains on the same pasture area for prolonged periods i.e. animals are grazed within one enclosed area over a long period of time. • Offers great opportunity for feed selection by animals. • Continuous grazing is a normal practice on range lands and tropical grasslands where fencing is economically impracticable.	• Least management input of all the grazing systems. • Least cost of all the systems. • Least disturbance to animals. • Easy to keep grazing records. • Area selection leads to inefficient pasture • At light stocking rates good production per animal can be expected. • The stocking rate is relatively low.	• The "precise" stocking rate needs to be known time to time, as the pasture growth rate varies. • Seasonal fluctuations in yield are difficult to cater. • Season growth rates and inefficient pasture utilization. • A serious disadvantage is built-up of tick and nematode infestation
2. Rotational grazing		
• In this system, pasture is subdivided into a number of paddocks and animals are allowed to graze in a regular sequence before returning to graze again in the first paddock after some regrowth period. • A farm paddock system and a rotational cycle of about 30 - 40 days are used. • Livestocks are moved every 1-3 day through a large number of paddocks (up to 20 or 30) in intensive rotational grazing systems.	• Animals are quieter and easily managed. • Flexible rotational grazing based on plant growth can boost pasture utilization. • Allow for better weed control. • Unhealthy animals can be spotted easily. • Herbage of the desired quality can be offered to animals by adjusting the number of paddocks or the period of regrowth following utilization	• More expensive as it requires more capital for fencing materials, water troughs and gates etc. • More manpower is required for moving cattle. • Pasture that has not been grazed from some time grows tall and carries leaves in the top portion of the pasture plant. • Once grazed these leaves are removed allowing sunlight to strike the soil surface and initiate weed germination.

Contd.

<table>
<tr><td>• Common names for intensive rotational grazing systems include high density grazing, short duration grazing, block and strip grazing, planned grazing and cell grazing.</td><td>• Excess growth can be used for hay, silage or set aside for forage.</td><td></td></tr>
<tr><td colspan="3">3. Strip grazing (close-folding, ration grazing)</td></tr>
<tr><td>• Intensive method of rotational grazing based on use of electric fence.
• A moveable electric fence is placed across a grazing paddock and is moved forward once or twice each day.
• The amount of herbage offered to the animal is restricted, selective grazing is minimized.</td><td>• The pasture is uniformly consumed.
• Less wastage by trampling and defecation.
• Valuable on highly nutritious pasture.</td><td>• More expensive & requirement of manpower is also high.</td></tr>
<tr><td colspan="3">4. Deferred grazing</td></tr>
<tr><td>• In this system the grassland is divided into three major divisions. One out of three is used for seeding and the other two divisions are used for grazing alternatively.
• Under deferred grazing, paddocks are regularly spelled (rested). At any one time, more of the property will be subject to grazing than spelling.
• The leafier grasses are the most suitable for standing hay e.g. Giant star grass (Cynodon plectostachyus) and Guinea grass (Panicum maximum).</td><td>• Used as means of improving range pasture by allowing plants to mature before grazing, their vigour is built up.
• Root systems are allowed to develop, and self-sown seedlings become established.
• This system of grazing also allows seed production.
• As livestock are deferred to different compartments into the grazing area, it prevents overgrazing.</td><td>• More space/land requirement
• Proper fencing is required.</td></tr>
</table>

Contd.

5. Deferred- Rotational grazing		
• In this system the whole pasture is divided into three equal compartments and fenced. Each compartment is grazed continuously for 1/3 of the total grazing season and protected through remaining 2/3 of the season for seed setting and maturity. • By the time animals are allowed to graze in the 3rd compartment, the seed would have been shed. When they go back to first and second compartments seed then would have also ripened and got shed. • In the second year, rotation will start from the second compartment and in the third year from the third compartment.	• It provides opportunity for a systematic rotation of pastures in which grazing is either delayed or discontinued to provide for plant reproduction, establishment of new plants, or restoration of the vigor of existing plants. • One or more pastures are grazed during the spring, while the remaining one or more pastures are rested until after seed ripening of key species and then grazed.	• More space/land requirement • Proper fencing is required.
6. Soiling or zero-grazing		
• Soiling refers to feeding of cut crops (including grass) to housed stock. • Animals reared on the farm without the necessity of grazing. • It is made possible by the evolution of modern machinery e.g. forage harvester.	• Higher animal production owing to more efficient herbage utilization • No losses from fouling and trampling • Pasture may be harvested at a more advanced stage of growth • Herbage intake is uniform • A greater measure of bloat control is possible especially • Animals are protected from heat and flies	• Cost of equipment is high and it restricts the practise to herds of at least 30 cows • Need to provide bedding for the housed animals • Manure disposal may present a problem. • Permits an increased stocking rate and an intensification of management • Need for fences and pipe water supplies

❑❑❑

Chapter - 5

Livestock Housing and Sanitation

H. N. Pandey, Rana Ranjeet Singh & Mukesh Singh

Functions of Livestock Housing

The most important purpose of livestock production is to increase the conversion efficiency of feed energy into the livestock products like milk, eggs, meat, wool, hair, hides and skins, draught power and manure (fertilizer). Scientific housing of livestock is necessary in order to get higher productivity from them. A proper housing design/structure should alter/modify the micro-climate inside the house so that animals are least exposed to the extremes of climatic conditions. Moreover, it should also provide better control of diseases & parasites, desirable working conditions for farm staff & better living conditions for the animals, better integration of various routine farm practices, opportunity for higher labour efficiency at farm, protection from predators & thieves, opportunity to express species specific natural behavior patterns and it must not compromise the welfare of the animals. Some of these functions are discussed in subsequent sub-headings.

1. **Protection from inclement weather conditions:** The climatic parameters i.e. the environmental temperature, humidity, rainfall, wind velocity and wind direction vary a lot in different seasons and months of the year particularly in tropics. It is experienced that for the livestock the temperature ranging from

25^0 to 35^0 C and the atmosphere humidity ranging from 55 to 65 are most conducive. The comfort zone for temperate animals and their crosses are lower than the native livestock of tropical region. The temperature range of the comfort zone for different temperate breeds of cattle is 9^0 to 21^0 C. But the range of these variables for different species and breeds are still narrow based on their comfort zone. A minor deviation from the respective comfort zone does not adversely affect the growth and production but as the deviation increases, the reduction in performance also increases. Animal housing is required to minimize the deviation of microclimate from the comfort zone of animals. It also helps in maintaining the other variables in a narrow range in different seasons and in different climatic zones. If the livestock is exposed much to a high or a low temperature for a longer duration, its performance may decrease drastically, and sometimes it becomes fatal also. Proper housing protects the livestock from these hazards.

2. **Minimizing the energy losses:** When the livestock are exposed to significantly high or law micro temperature, they have to waste some energy to maintain their body temperature and which affects the production, growth and the reproduction. The channels for loss of energy in both the seasons are different.

    a. **Winter:** During winter when the micro temperature is lower than the comfort zone the tropical breeds utilize the body energy for maintaining their body temperature. This energy loss is proportional to the differences of existing micro temperature and the comfort zone of the species and breed concerned. Though there is a slight increase in dry matter consumption per unit body weight but their available energy is reduced when the micro temperature is significantly reduced. This condition is not applicable in temperate cattle breeds reared in tropics and to some extent their crosses with tropical breeds as their comfort zone is nearly similar to the tropical micro temperature during winter. Hence their growth and production increases during winter.

    b. **Summer:** When the micro temperature goes significantly higher than the comfort zone, and the condition prevails for longer duration, the animals feel difficulty in dissipating the extra heat generated by the body due to minimizing or stopping the heat loss from body surface by conduction and evaporation particularly when atmospheric temperature and humidity both are increased. In this situation the feed consumption per unit body weight is reduced and hence the production and reproduction both are affected particularly in temperate breeds and their crosses with tropical breeds.

3. **Protection against loss of appetite and body weight:** During hot and humid summer, if housing is not proper and animals are not protected against the exposure to high temperature, the appetite is drastically reduced and accordingly the body weight of adult cows, growth rate of growing stock and production and reproduction of producing stock is affected significantly. The tropical cattle are not much affected but the temperate and crossbred cattle and buffaloes are much affected. For maintaining body weight and performance, proper housing facilities are required.

4. **Protection against theft and predators etc:** Proper fence or boundary wall is constructed and enclosures are provided for safeguarding the livestock. If such facilities are not provided the small and young stock can be attacked and harmed by the invading animals causing the heavy loss.

5. **Opportunity to express species specific natural behavior patterns:** A proper housing design provides an opportunity to the animals to express their natural species specific behavior as far as possible indicating that their welfare standards are taken care properly. Hence they are less stressed and more productive. Significance of behavioural need of the animals in a housing design usually increases with level of production and degree of confinement.

6. **Grouping**: The proper housing is also required for housing the animals in required group size and category for proper management.

In addition, for proper feeding, milking and other functions, the housing is needed. The insemination, treatment, segregation and parturition also proper housing is needed. Similarly for loading and unloading of animals appropriate facility is also needed.

General Principles of Design and Construction of Housing

The basic factors affecting the design and construction of livestock housing are mentioned below:

- Species.
- Breed.
- Productivity level.
- Climatic variables.
- Herd size.
- Availability of local materials etc.

The species for which the housing is proposed to be built, decides its size, strength and material to be used. Buffaloes, cattle, horses and ponies and camels require larger space, while small ruminant (sheep and goats) and swine needs comparatively smaller space. Cattle and buffaloes require strong wall and structures than the small ruminants, while the swine need strong impervious and non-slippery floor and strong solid wall up to the height of 60 to 90 cm depending on breed. Cattle and buffaloes need impervious and non-slippery and washable floor while sheep and goats prefer dry, absorbent and sandy floor. The poultry and birds can easily be kept in houses with AKRA walls.

Breed indicates the size of animals within the species and housing structure particularly the width, height and the depth of feeding trough, area of stand floor and paddock need to be provided as per the size of animals.

Productivity level of animal decides the special facility to be provided for maintaining the productivity as these are very sensitive to the housing and microclimate and comfort. Providing the sands on the floor and cool place during the hot days to high yielding buffaloes and regular water sprinkling on the body of high yielding crossbred cows during hot days are the best example of such housing design.

Agro-climate of the place is the most important factor for deciding the type of housing and structures. The closed and low ventilation houses with properly bedded floor is preferred for temperate climate, while the open, high and well ventilated houses are preferred for hot and dry areas. The well ventilated, open and low height houses with good roof gradient are used in hot and high rain fall areas. Other minor alterations are suggested for livestock housing as per the local climatic conditions.

The herd size decides the type of housing to be followed. If the herd size (cattle & buffalo) is large (usually more than 50 heads), the loose housing system can be followed otherwise for smaller herds (less than 50 heads) the conventional housing should be adopted. Similarly in case of sheep and goats the animals in smaller flocks can be tied individually while in large flocks they are kept loose.

If the herd size is less than 20, the one way of conventional byre is economical and for less than 5 the backyard housing is sufficient.

The locally available housing materials should be used for constructing the animal houses for economic reasons. Many such materials like stones, thatch, wood and bamboos are locally available and can be efficiently used for animal housing. They sometimes also decide the type of housing.

Points to be Considered for Selection of Site

The animal keeping can be operated safely, economically and with less disease occurrence and less parasitic load if the following mentioned points are considered in selection of site for animal houses.

1. Topography
2. Drainage
3. Soil type
4. Marketing facility
5. Transport facilities
6. Water availability
7. Labour availability
8. Climate

The topography of the place is an important factor in selection of site for animal farm. The animal houses should be constructed at the land of high level. It helps in quick and efficient disposal of animal wastes, washings and also prevents the parasitic infestation due to water logging and high moisture.

Animal houses should be constructed at the place where the efficient drainage is available. High volume of livestock waste and washings are daily to be disposed off (particularly in dairy farms) which requires an efficient drainage system.

There is no much scope for site selection based on soil type but where ever possible a sandy loam soil should be preferred. The basic soil consideration is that the land having good and fertile soil suitable for fodder production, should not be used for animal house construction, if other options are available.

The basic economic consideration in site selection of an animal based unit is the efficient and appropriate marketing facility. The animal products are usually of perishable nature and need quick and remunerative disposal. Though the technological interventions like chilling, pasteurizations and use of permissible preservatives in animal products have solved the problem to certain extent but cost and risk is involved in the processes. Hence it is always beneficial to establish any animal unit in the area where the efficient and remunerative price market or marketing system is available.

The marketing facility is also related with the availability of good quality and cheap inputs like feed, fodder and labour.

Availability of efficient transport is another factor to be considered. Rail or road transport should be available near but not adjacent to the farm. The near and efficient transport facility reduces the marketing cost and spoilage of products but the farm should not be situated very near to a railway track or station. It creates noise problem and unrest to the animals particularly poultry.

Clean and potable water must be available regularly to the farm. A large volume of water is required for drinking and washing of animals, sheds and utensils and other purposes. Water contaminated with chemicals, drugs etc. should not be used for animal houses.

The labour cost comes next to the feed cost in livestock keeping. Hence a place where laborious, efficient and willing workers are available at appropriate wages, should be preferred from economic and management point of view.

A place having a climate suitable for animal keeping should be preferred for establishing a livestock farm. It improves the production and reproduction of animals, reduces the managemental cost and the decease occurrence. In the areas having unsuitable and uncertain weather conditions in most part of the year, the maintenance cost of animals is very high.

There are some other factors which should also not be ignored while selecting the site for animal farm. These factors are usually associated with the local sociology and local rules and regulations. Law and order, peace, working nature of workers, and cooperation of local administration etc. are also important considerations while establishing a livestock farm.

Type of Animal Housing

The housing arrangement for different type of animals is also different. Various housing types for cattle buffaloes, sheep goats and swine are discussed below

A. Housing for Cattle & Buffaloes

For cattle and buffaloes usually two types of housing systems are followed.

(1) Conventional/tie/stanchion barns, or closed housing.

(2) Loose housing.

These housing systems are followed depending on environmental conditions and size of holding. These systems can also be modified depending on climatic conditions. In cold places and for small herd size, conventional barns should be preferred, while for hot dry or hot humid conditions and large herd size, the loose houses (with appropriate modifications) should be used.

(1) **Conventional barn**: In conventional barns, animals are always tied inside the house on a platform and secured at neck by neck chains/ neck ropes, stanchion except for some time during cleaning of house when the animals are let loose in an open area. Cows are fed in troughs at the front of the stall and water is provided in bowl or bucket. They are milked inside the barn. Barns may be partially or fully enclosed with proper arrangements of ventilation.

Advantages of Tie Barn

- Less land is needed.
- Useful for small holdings.
- Individual animal care is easy.
- Disease spread is slow.
- Feeding level of individual animals can be observed easily.

Disadvantages of Tie Barn

- Costly housing structures.
- Not suitable for large herds.
- Difficult to identify the sick and animals in heat.

- Not suitable for hot humid regions of tropics.
- Number of animals is fixed for a given space and any additional animals cannot be accommodated.
- Due to restriction in movement certain leg disorders may develop.

The tie barn system having two rows can be structured in two types

a. **Head to head or face to face:** In this system face of animals is towards the middle of the building and the hind quarter of animals is towards the outer wall (Figure 5.1.1).

b. **Tail to tail:** In tail to tail system, the head of animals is towards the walls where feeding manger and feeding space is provided but the hind quarter of animals of both the rows is towards the middle of the house where the gutter and the central (inspection) passage is provided (Figure 5.1.2).

Advantages of Tail to Tail System

- Milking can be supervised and organized easily.
- Disease spread is slow.
- Cleaning is easy.
- Animals in heat can be detected easily.

Disadvantage of Tail to Tail System

- Feeding of individual animal is difficult to observe.

Advantages of Head to Head System

- Feeding and its observation are easy.
- Better if comparatively lesser space is available.

Disadvantages of Head to Head System

- Milking operation cannot be observed and managed conveniently.
- Since the head of both rows is nearer the risk of disease spread is more.
- Cleaning is not convenient.

(2) **Loose housing:** Animals are always (day & night) left loose in open paddock or pasture in loose housing system except when they are brought for milking, insemination and any treatment or vaccination (Figure 5.1.4). The open paddock is provided with shaded area under which animal takes rest in the event of adverse climatic conditions (too hot/cold/rain). This shaded or covered area has common feeding manger. A common watering is also provided. Concentrates are fed at the time of milking in the milking barn. This type of housing is suitable for most part of the country except for temperate and heavy rainfall zones. However, this system of housing can suitably be modified as per the climate of the place.

Advantages of Loose Housing System

- Low cost of construction.
- Flexible in nature. If required for a small period more animals (up to 10 %) can be accommodated in a given space.
- Feeding watering cleaning and milking is easy.
- Animals in heat and sick animals can easily be identified.
- There is no emergency situation in watering during summer because the water is always kept available in common water trough and the drinking water is not provided individually.
- Social structure is built up among the animals.
- Best suited for large herds.

Disadvantages of Loose Housing

- More land is required.
- Different kinds of houses for different purposes are needed.
- Risk of disease spread is more.
- Not suitable for small holdings.
- Individual feeding and its observations are difficult in this system.

In this housing system various kinds of houses are built to accommodate various type of animals like milking byre, milch cows shed, dry cows shed, pregnant animal shed, down calver shed, calf shed, heifer shed, young male shed and bull shed etc.

Figure 5.1.1 : Head to head type barn.

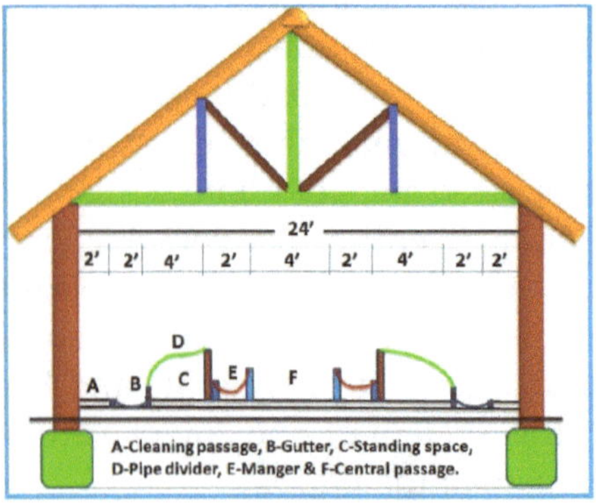

Figure 5.1.2 : Tail to tail type barn.

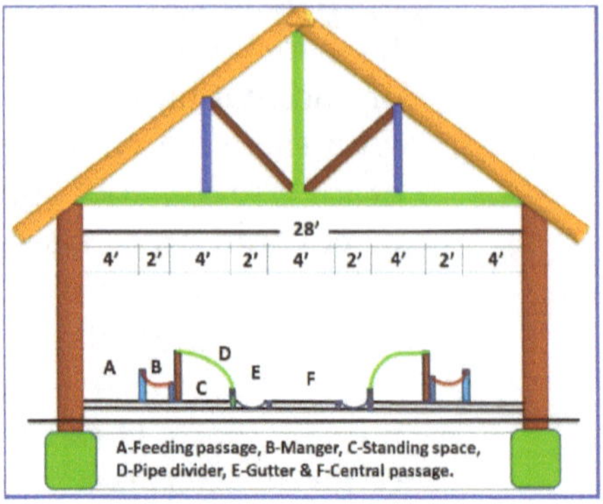

Figure 5.1.3 : Backyard cattle housing.

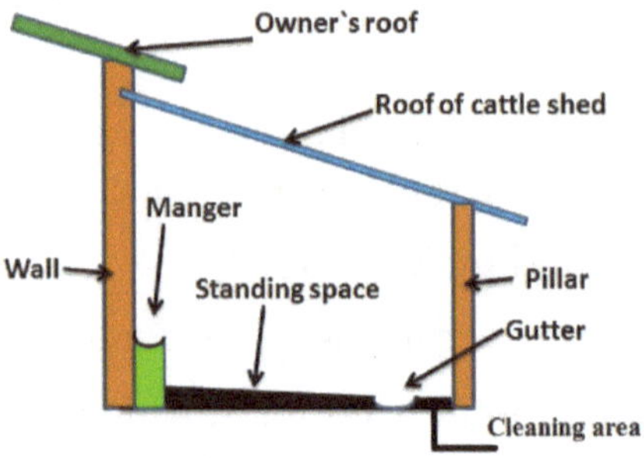

Figure 5.1.4 : Loose housing system.

Milking barn: Milking barn is a place where cows are milked. Milking barn should be located at a central place with other sheds arranged around it. It may be fully enclosed or side walls may have open spaces at suitable intervals. In former type, provision of windows and ventilators are made at suitable places while in later type wire netting of small mesh size is fixed. Individual stalls/standings are constructed in one or two rows. Each of these stalls is separated by bent iron pipes. Double rows of stalls can be constructed either in head to head or tail to tail arrangements (Figure 5.1.5). While practicing hand milking, each stall can be used for 2-3 milch animals. Hence, number of stalls required on a farm is about half to one third number of the milch animals. The length and width of stalls should be around 1.5-1.7m and 1.05-1.2m while width of central passage should be about 1.5-1.8m and there should be 35 cm wide U shaped drain on either side of central passage. There should be feeding manger and feeding passage (75 cm) on either side along the heads of the stalls.

Figure 5.1.5 : Milking barn for cattle & buffaloes.

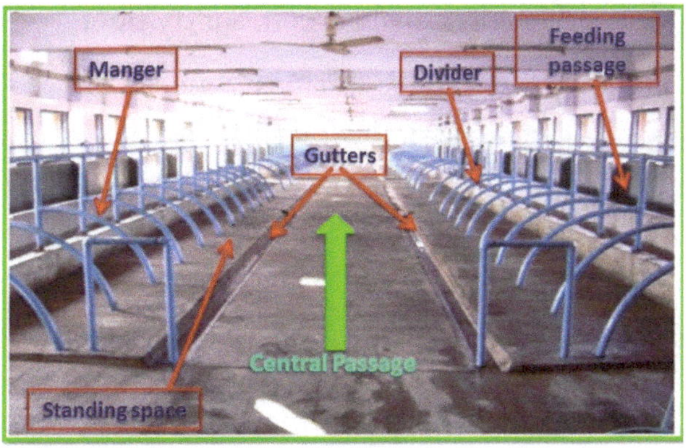

Milking parlour is a special type of milking shed in which stalls are organized for milking two or more batches of cows at a time. Depending upon herd size, production level, time of milking, availability of labour and degree of automation, stalls can be designed in many ways and shapes. It may be abreast, herringbone, polygon rotary parlour etc.

In addition to these houses, ancillary structures like milk house (for collection, weighing and storage of milk for a brief period), stores (for keeping hay, straw, concentrate etc), silo pit and handling yard may also be constructed at an organized dairy farm.

B. Housing for Sheep & Goats

These species require the low cost housing and are left loose in enclosed area except the does and ewes near the parturition (kidding and lambing) and some period (a week) after the parturition for taking care and suckling of young ones. During this period the females are housed individually in cubicles framed by using the expended metal or weld mesh. One female occupies one cubicle for about a week at for one kidding/lambing.

The animals of different age and stage are group housed separately and these partitions should be framed by using the expended metal or the weld mash as per the requirement and later it can be dismantled or modified as per the need. Feeding and watering facilities are provided with different dimensions for different categories of stock. The housing structures that can be constructed on an organized sheep/goat farm are shed for general flock, ram/buck, lamb/kid, sick animal, lambing/kidding, shearing hall, wool store, handling and weighing yard, stores for feed and fodders, foot bath and dipping tank etc.

The houses for these species can be of two types:-

- Ground level with absorbent sandy floor.
- Raised and slatted floor.

Advantages of Ground Level Shed

- Low construction cost and easy to construct.
- Low labour cost in routine clearing.

Disadvantages of Ground Level Shed

- Risk of disease spread is more
- After long use the sand needs to be replaced.

Advantages of Raised Slatted Floor

- The risk of disease spread is less.
- The high level of sanitation can be maintained.
- The excreta can be utilized in integrated farming directly.

Disadvantages of Raised Slatted Floor

- Costly to construct.
- More labour cost for routine maintenance and cleaning.

C. Housing for Pigs

The basic principle of pig housing is that the floor and the wall up to 75 to 95 cm height (depending on breed) should be made impervious and hard. The floor should not be slippery, but the inner surface of wall should be cement polished so that it can be washed and cleaned easily. The pig house should be airy and well ventilated. It should also have the provision of protection against cold winds during winter where the minimum temperature goes below 12^0C.

The piggery should have separate houses for growers, fatteners, gilts and the dry sows and the boars. Separate farrowing facility should be provided to the pregnant sows/gilts 15 days before the expected date of farrowing and till they are dried and piglets are separated. This house must be well equipped with the facility so that the newly born piglets can suckle the dam's teat but can protect themselves from crushing due to overlying of the dam (Figure 5.1.8 & 5.1.9). Heavy mortality is experienced in piglets if the farrowing facility is not proper. This house should also have the facility for feeding the creep ration to piglets separate from dam's ration, but the dam should not be permitted to consume the creep ration. The farrowing pen should also be equipped with some heating arrangement if the farrowing takes place during winter where the temperature goes down, otherwise heavy piglet mortality takes place.

Facility of clean drinking water and feeders must be provided to all the categories of animals in all the seasons. The house should be cheap in construction by using the materials locally available.

The floor or the corners of the house should not have the pits or damage parts because due to rooting habit the pigs will damage the floor and there will be difficulty in cleaning and sanitization and also there will be risk of injury on the legs and body of pigs which takes long time to cure. The piggery should have good slope (1:40) and a good drainage and efficient waste disposal system.

Figure 5.1.6 : Goat house with raised and slatted floor.

Figure 5.1.7 : Goat house with raised and slatted floor.

Figure 5.1.8 : Farrowing pen.

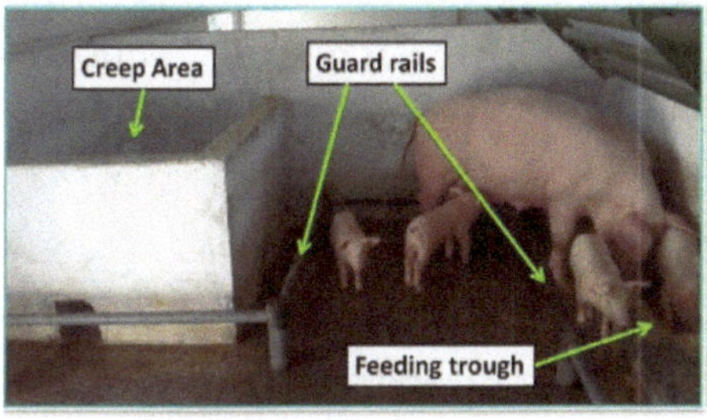

Figure 5.1.9 : Farrowing crate.

There are two types of farrowing facilities

a. Farrowing crate

b. Pen with guard rail.

These facilities are having different advantages and disadvantages.

The piggery should also have some auxiliary facilities like dipping tank, loading ramp and crush for vaccination and treatment.

Advantages of Farrowing Crate

- The piglets move freely.
- A multipurpose hall can be converted into farrowing hall and after the season is over and the piglets are weaned, the hall can be used for other purpose.

Disadvantages of Farrowing Crate

- The sow is not permitted to move freely.
- The crates have to be prepared based on breed (size) of sow.

Advantages of Guard Rail

- The sow and the piglets are permitted to move freely.
- Both sow and the piglet have free access to feed and water.

Disadvantages of Guard Rail

- The room is permanently converted into farrowing pan.

Creep Feeding Facility for Piglets

The piglets are provided with high nutritive value and palatable concentrate mixture from the second week of life, which is termed as creep ration. The piglets should be allowed to eat the creep ration freely but the sow is not permitted to eat it. There are two types of arrangements for this purpose-

- Creep feeding chamber.
- Rings and hooks on wall above the creep feeding trough area.

Any of these methods can be followed, depending on farm practices and farrowing facility.

Arrangement of Farm Buildings

This issue deals with the location of various buildings within the farm. Proper location of farm buildings reduces the labour cost, unnecessary movement of animals, maintains animals social satisfaction. Proper location improves the utilization of farm buildings. This also helps to keep the farm clean.

The farm office or operators room should be situated in front of the farm, so that the outside persons coming to farm may not have to roam in the farm for reaching and contacting the right person. The calf shed should be placed near the milking barn (if suckling), dry cow shed/ breedable heifer shed may be situated at one side, calving shed/ down calving shed should be kept at a side but near to the operators place so that it can be supervised regularly. The breeding bulls and young male shed should be placed at the back of the farm but it should be at a place where the sound of females can be heard and females can be seen also. This is helpful in maintaining the breeding ability of males.

The chaffing place, silage pit and *bhoosa* godown should be at the side or back of the farm but not at a long distance. These buildings should not be placed within the animal vicinity. Utensil cleaning and sterilization place should be nearer to milking place. The reproduction lab/place/ room should be placed near the breeding male shed. The veterinary unit/health unit should be placed at the back side of the farm. The foot bath and cattle crunch should be placed near the veterinary unit.

Construction Details of Various Structures

There are many structures in livestock houses like floor, walls, roof, manger, water trough, farm gate and passages, fence, foot bath and race. These structures have different construction details for different species of livestock.

A. Floor

The floor types are different for different species. Cattle buffalo and pigs need a solid, non-permeable and non slippery floor while sheep and goat need absorbent and sandy floor.

a. **Floor structure for cattle, buffalo and pigs:** For constructing the floor for cattle, buffalo and pigs two types of constructions are used (i) for covered area and (ii) for run or open area.

i) **Floor of covered area:** The floor of covered area is usually constructed by the concrete, cement mortar and is finished by the 5 cm thick layer of cement, sand and fine gravel. It should never be cement polished otherwise it will become slippery. The first step is to level the ground as per the need. The slope in the floor of covered area should be one in forty or one in fifty. Then the floor should be pressed well to make it compact. An 8 to 10 cm thick well mixed layer of the following composition should be spread uniformly and then should be pressed to make it compact.

Ingredients	**Parts**
1. 40 mm broken stone:	10
2. Coarse sand:	5
3. Cement:	1

Finally, the floor should be finished by 5 cm. thick layer of the fallowing composition.

1. 20 mm gravel :	4
2. Coarse sand :	2
3. Cement :	1

ii. Floor for open area: The open area is used by the animals for various activities like sitting, sleeping, lying, walking, etc. It should be non-slippery, impervious and cheaper in cost. The usual types of floor for open area are mentioned below

1. *Kacha* or earthen,
2. Brick on edge,
3. Cement concrete and
4. Stone.

The earthen floor is cheapest in cost but it has some basic disadvantages

1. Needs frequent maintenance.
2. It is not impervious and it may act as a source of various diseases particularly mastitis.
3. Cannot be washed.
4. Not suitable in high rainfall areas.

Therefore, the earthen floor is not recommended for cattle and buffaloes farms. The brick on edge is generally accepted type of floor for open area of cattle and buffaloes. Though it is costlier than the earthen floor but is comfortable, impervious and can be washed and cleaned daily and its life is 10 to 12 years in regular use. To construct this floor first the ground has to be leveled giving required slope (1:40 to 1:60). The location of surface channels and main channel may be planed. The 10 cm thick layer of cement, concrete and broken stone of 1: 5: 10 ratios should be spread and made compact. Finally, the brick of first grade is installed on edge giving about one cm space in between which should be filled properly by 1:5 to 1:6 cement and coarse sand mixture in order to make the floor impervious. Cement concrete floor is recommended for covered area only where the animals stand for longer time but it becomes very costly for the open area and hence it is not recommended for this purpose. The stone floor is used only in the areas where it is available locally and at a cheaper cost. The stone slabs are fitted in the same manner as brick on edge. Though it has got a longer life but becomes slippery and undulated in the long use. Floor space requirements for different species of livestock have been given in Table 5.1.1.

b. **Floor for sheep and goat:** Small ruminants prefer the floor having following characteristics

   (1) It should be absorbent to absorb the urine, so that the floor remains dry.

   (2) It should not be slippery or muddy.

For preparing this kind of floor, first the place should be well leveled and made compact. It should not have channels and drains. A 10 to 12 cm thick layer of coarse sand should be spread over this leveled floor. This sand layer needs replacement after about three year of regular use. If the sheep or goats are reared on raised and slatted floor, the floor should have the slat of about 1 cm. If it is more it will cause injury in hooves and if narrow, the cleaning will be difficult. Slatted floors can be made by using either weld mash of 4 mm iron rod or expended metal supported on metallic structure or slatted RCC slabs or wooden materials.

Table 5.1.1 : Floor space requirements for different species of livestock.

Species/Category	Floor space (m^2)		Maximum number of animals per pen	Height of roof at eaves
	Open	Covered		
1. Cattle & buffaloes				
a. Young calf up to 3 months of age	1.0	1.5	30	175 cm in medium and heavy rainfall areas while 220 cm in semi-arid &arid areas
b. Young calf between 3-6 months of age	1.0	2.5	–	
c. Older calf	2.0	4.0	30	
d. Cows	3.5	7.0	50	
e. Buffaloes	4.0	8.0	50	
f. Down calvers	12.0	12.0	01	
g. Bulls	12.0	120.0	01	
2. Sheep & goats				
a. Lamb/kid	0.4	–	75	220 cm in heavy rainfall areas and 300 cm in dry areas.
b. Ewe/nanny	1.0	–	60	
c. Milch doe	1.4×1.2	–	One stall for each	
d. Ram/buck	3.4	–	–	
3. Pigs				
a. Weaner Fattening pigs	0.9-1.8	0.9-1.8	30	200-250 cm
b. Dry sow/gilt	1.8-2.7	1.4-1.8	03-10	
c. Farrowing sow	7.0-9.0	8.8-12.0	–	
d. Boar	6.0-7.0	8.8-12.0	–	

The above dimensions are based on suggestion of Indian Standard Institute (ISI).

Drains and channels: The livestock farms produce a huge amount of dry, semidry or liquid waste daily which has to be disposed off daily at a proper place in a proper way. These wastes are very useful for plants and crops but very deleterious to livestock if it is retained in the animal farm for longer duration. The liquid wastes like urine, animal and shed washings and utensil washings etc are disposed off in two ways

(1) Surface channels system.

(2) Underground drainage system.

The surface channels may be *kacha* or made of RCC. or brick and cement sand mortar. Both the systems have got some advantage and disadvantage.

Advantage of Surface Channels

1. Low construction cost.
2. Easy in cleaning.

Disadvantage of Surface Channels

1. Occupy much space.
2. Not hygienic.
3. Require frequent cleaning.

Advantage of Underground Drainage

1. Does not occupy much space of farm land.
2. It does not need frequent cleaning.
3. It keeps the farm hygienic.
4. Needs no frequent repairs.

The underground drainage system has got only one disadvantage that it is very costly to construct as compared to surface channel. Hence, most of the livestock farms prefer the surface channel system. Both types of drainage system should have the capacity to drain out the liquid waste depending on the animal type, number and area of the farm. They should also have the recommended slope so that the material may flow smoothly. The *kacha* surface channel should have 1:40, *pacca* channel 1:50 to 60, while the under-ground channel should have 1:50 slope. There are some terms used in underground drainage system

1. **Catch pit or gully trap:** It is the start of underground channel where the washings are dropped into the drainage system. Its size depends on width of the saucer shape surface channel which carries the washings to the pit. Its top is slotted having

1 cm slots and the bottom should be about 20 cm lower than the opening of the underground drain.

2. **Manhole**: It is the junction point or opening in under-ground channel provided for entering a parson or equipment for opening/cleaning the channel. This hole or pit has got a solid cover.

The gutters in the floor of milking byre and the conventional barns should be U shaped and the depth should not be more than 15 cm at any place and the channels in the open area of loose house should be saucer shaped and should not have the depth of 15 cm or less.

B. Pillars & Walls

In tropical areas the roof of animal house is mostly supported on the pillars. In loose houses, there is least scope of any wall. Since the roof is based on pillars only, these pillars have to be sufficiently strong to bear the load of roof and thrust of animals. The materials used for construction of these pillars are (1) W.I. pipe of 10 cm diameter (2) Woolen poles of 15 X 15 cm (3) Brick posts laid down into sand cement mortar supported by iron rods 24 X 24 cm. (4) Stone 10 X 10 cm (5). Reinforced cement concrete poles of 20 cm diameter. Pillars must have sufficient formation at the ground and strong jointing arrangement with trusses.

C. Roof of Animal Housing

Roof of animal house is very important structure which protects the animals and other housing structures from rains, sun rays and the wind. There are two types of roof- flat & sloppy. There are many kinds of materials which are used for roof making.

1. Thatch,
2. Tiles *(Khaprail)*,
3. Slates and stones,
4. Wooden planks,
5. R.C.C. (*pacca* roof),
6. C.G.I. sheets and
7. Asbestos sheets.

1. **Thatch roofing:** It is made of dry grasses like *moonj, kans,* paddy straw etc supported by bamboo frame. It is made mostly by

the locally available materials. Since, it is bad conductor of heat hence most comfortable to the animals in both the seasons-winter and summer. It is also very cheaper in construction. But it is having limited use in small scale animal holding only because it has got many limitations. It needs very frequent replacement (life is just two years) and is also prone to fire.

2. **Tiles:** This kind of roof is common in villages and suburban areas where the holdings are limited up to 50 heads of large animals. These are prepared by the soil and are cured in fire after drying. This roof is supported on bamboo or wooden structure. The life of this roof is about 5 years. It is also comfortable to animals in both the seasons.

3. **Slates and stone:** This roof is used only in places where the slates and the stone slabs are cheaply available. This roof is also supported with the bamboo, wooden or metallic structure. This is not comfortable to animals in any season particularly in summer at a hot place and in winter at cold place. Its life is about 5 to 6 years.

4. **Wooden planks:** This roof is comfortable to animals but is in limited use due to the cost of planks. Hence, it is used in places where the planks are available at a cheaper rate. This kind of roof is common only where the animal number is up to about 40 heads.

5. **R.C.C. roof:** R.C.C. roof is generally flat in shape and is limited to the research farms and the government organizations due to its higher construction cost but its life is also long (20 years or more).

6. **Corrugated galvanized iron (C.G.I.) or Asbestos sheet:** These types of roofs are constructed by the corrugated sheets-one is galvanized iron and the other is asbestos sheet. The corrugated galvanized iron sheet (C.G.I.) is stronger than asbestos sheet. Both are supported on wooden or the metallic structure with formation of trusses of different shapes. The life of these sheets is about 15 years. The C.G.I. sheets get very hot during summer and very cold during winter season. It has to be painted white at the top during summer particularly in hot arid zone and should be provided with the false roofing of wooden planks, straw or other dry materials at inner side. Plantation of creepers on such roofings may also solve the problem to some extent.

In almost all the roofing system (except the R.C.C. roofing) trusses (pre-fabricated, triangulated wooden structures) have to be provided for support. These trusses decide the slope of roof. There are various deigns of trusses as

(1) Crown / king post truss

(2) Collar beam truss

(3) Closed-coupled truss

Figure 5.2.1 : Single post truss.

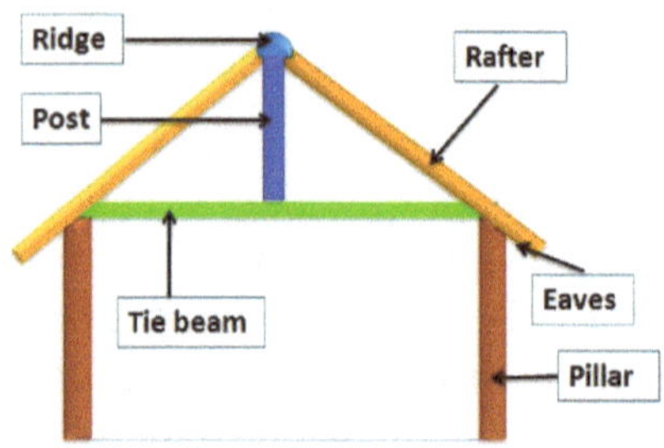

Figure 5.2.2 : Close couple truss.

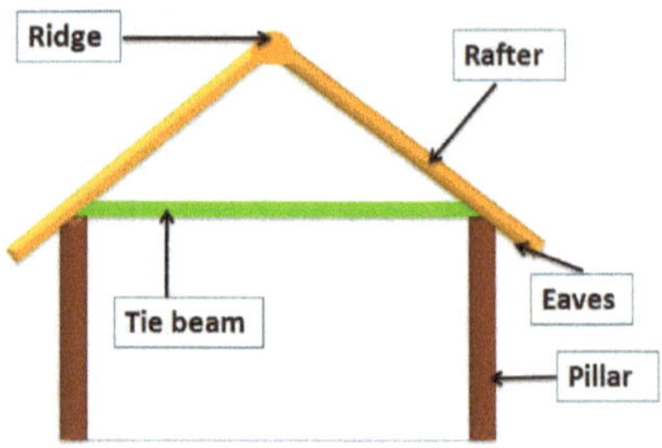

Figure 5.2.3 : Collar beam truss.

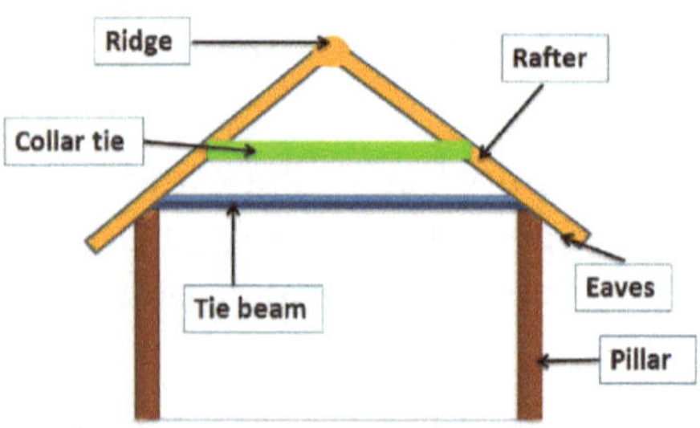

Figure 5.2.4 : King post truss.

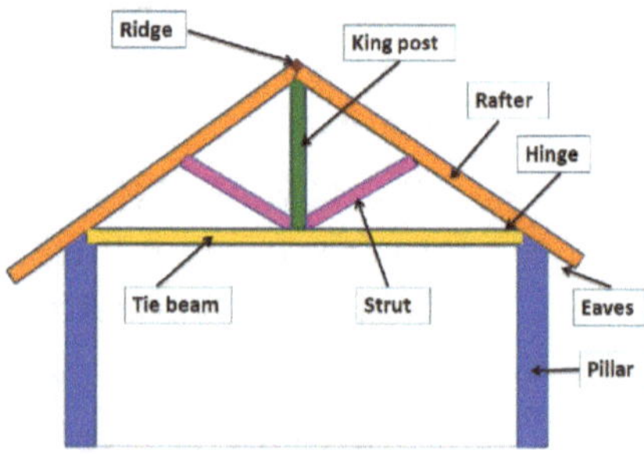

D. Feeding & Water Trough

The feeding troughs are provided to the animal houses for feeding concentrate and roughage. Their height, depth, width depend upon specs and size of animals. When feed is offered almost all animals come to the feeding trough. In tie barns the feeding trough is provided to every animal separately but in loose house, common feeding through is constructed for all the animals in the paddock (Figure 5.3.1). Usually 60 to 75 cm length of feeding through is provided for each adult cow/ buffalo which further depends upon the body barrel diameter and the horn condition of animals. The length of feeding through per animal can be minimized up to 50 cm without affecting growth and production if feeding frequency is increased. In this case all animals do not come to feeding trough at a time. The dimensions of feeding trough for different age groups are different. The dimensions of feeding trough for different species of livestock have been mentioned in Table 5.1.2.

Figure 5.3.1 : Side view of feeding trough for cattle.

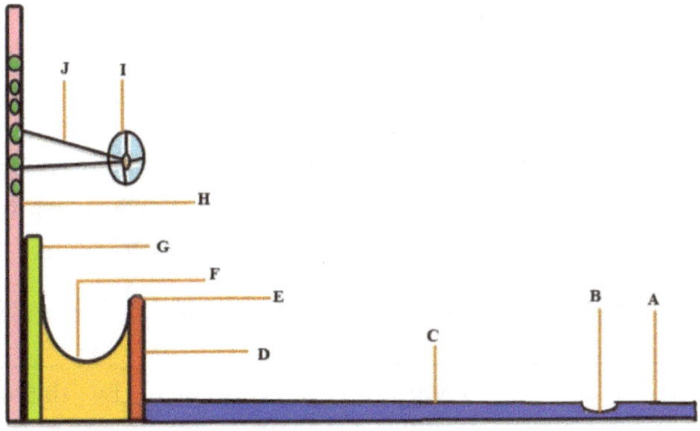

A-cleaning area, B-Gutter, C-Standing floor, D-Inner wall, E-Mettaloid top, F-Bottom, G-Outer wall, H-Pillar, I-Reel, J-Support

The height of inner wall of feeding trough depends upon the height of the animals. It should usually be 5 cm higher than the elbow point of animals, which comes to 45 to 55 cm. This height is most convenient for animals to take various feeds. The upper most corner of the inner wall should be metallic and half round being comfortable to the animals.

The outer wall in loose houses is not important from the view point of animal comfort; however, it is important from farm management view point. It should be 10 to 15 cm higher than the inner wall to minimize the feed wastage. The width of feeding trough for adult cattle and buffalo for single row should be 55 to 60 cm. Depth of feeding trough should be as per the face length of the animals and for adult cattle and buffalo it should be about 40 cm. The bottom should be U shaped to avoid the wastage of feed. There should be lengthwise slope (1:40) in the bottom and drainage facility for washing and cleaning the feeding trough time to time. There should be a reel or moving roller fitted above the feeding trough so that the animals may not put their fore legs in the feeding trough and spoil the feed. The distance between the reel and inner edge should be such that the animals can take feed easily but may not spoil it. It can be adjustable also. The feeding trough for the calves and heifers of various age groups can also be constructed considering their body size.

Water trough is provided in the open paddock of cattle and buffalo. The total length of water trough is kept $^1/_{10}$ of the feeding trough assuming that maximum $^1/_{10}$ of the animals kept in paddock come for drinking water at a time. There are many designs of water trough depending on the source of water pressure and frequency of water supply. In all the designs, there are some common constructions. The stand floor of 1.20 meter width having $^1/_{30}$ slope and a saucer shape shallow drain must be provided at the site where animals have to stand for taking water. Proper arrangement must be provided for routine cleaning and drainage of water from water trough. The water should fall in to the trough from a height of few centimeters, so that the water supply source may not get contaminated when the pressure stops or becomes negative. The water supply may be regulated or automated by providing float valve at a safe place. It can also be regulated by the siphon system.

Table 5.1.2 : Dimensions of feeding trough for different species of livestock.

Category of animals	Height of inner wall (cm)	Width of manger (cm)	Depth (cm)	Length per animal (cm)
1. Adult cattle & buffalo	50	60	40	60-75
2. Calves	20	40	15	40-50
3. Adult sheep & goat	35	50	30	40-50
4. Lamb & kid	25	50	20	30-35
5. Adult pigs	25	50	20	60-75
6. Growing pigs	20	30	15	25-35

The above dimensions of the feeding trough are based on suggestion of Indian Standard Institute (ISI).

Pigs are in the habit of eating from the ground or even by digging the ground (rooting). Hence the feeding troughs for pigs are always low in height and should preferably be made of cement concrete mortar. The feeding trough for a boar should be 50 cm long, 40 cm wide and 20 cm deep. Its inner wall should be 25 cm high above the floor level. Its bottom should be 5 cm above the floor and should have good slope and drain hole for frequent cleaning and washing. All the feeding troughs should have half round bottom and slope and drainage hole.

The sows and gilts are housed in group of about 10 individual hence they are provided continuous feeding trough. The sows and adult gilts should get about 30 cm length per head. The feeding trough should be 40 cm wide and 20 cm deep and its height should be 25 cm as in case of boar. The young piglets should get 10 cm length of creep feeder having 15 cm width and 6-7 cm depth. The feeding arrangements for the weaners and growers may range between 20 to 30 cm in width, 15 to 20 cm in depth and per head length should be 20 to 30 cm.

The water trough can be constructed in a corner of the living area having the size of 40 X 40 X 20 cm.

E. Farm Gate, Passage and Fences

The main gate of the farm should be in two parts and well hinged with wall or pillar built strong. The total minimum width should be 2.5 meters. It can be made of hard wood or iron pipe or angle and rods. Its height from the ground should be 1.5 meters. It is better to have one small passage of 60 X 75 cm in one of the parts to permit the entry of workers only when the gate is locked.

The passage for the movement of tractor and trolley along the feeding trough should be 2.2 meters for putting the fodder into the feeding trough. It should be metalloid road and have proper turning space where ever needed. The passage for animal movement should 1.2 meters wide. The passage gate should be single, made of hard wood or iron pipes of proper size. The gate should be hinged with wall or railing pillar in such a way that it can be fully opened on either side.

The outer fence of the farm protects the animals and other properties from the intruders; it isolates the property from the other neighbours. It should be 1.5 meters high and made of barbed wire and poles, wooden poles and rails, brick wall, live thorny bushes, depending on the locality and availability of the material at a cheaper cost. The fences coming in contact of animals should never be made of barbed wire. The height depends upon the kind of the stock housed.

Modifications in Loose Housing System for Cattle and Buffaloes

As discussed in earlier paragraphs, the standard loose house is a basic structure which can and should be modified depending on the prevailing climatic condition of the place, herd size, species and breed of animals to be housed. Common modifications are mentioned here for major agro-climatic zones.

a. Hot-Dry Zone

- Roof should be high, about 3.5 meter in the centre (ridge) and 3.0 meter at the side (eaves) so that the heat coming through roof may not reach to the animals housed under it.

- Roof should be painted white at the top so that the heat coming from sun rays may be reflected and the heat absorption by the roof may be minimum.
- The roof should be extended about one meter on both sides of covered area and this extension should be slanting having about 45^0 slope to protect the animals from sun rays and the hot air.
- Slope of roof should be against the wind direction to protect the animals from hot wind.
- Shady trees must be planted all around the covered area and within the open area.

b. Hot Humid Zone

The loose housing system is best suited systems of animal housing and needs very little modification. In this zone the basic problem of animal housing is to removal of the hot and used air from the microclimate of the animals.

c. High Rainfall Area

In this zone, the rain fall is very high and frequent and in addition to covered area, a 15 cm raised covered platform should be constructed in open area where the animals may take rest during heavy and long rains. The roof of raised platform and the covered area should have enough slope and extended margins (eaves) to facilitate the quick disposal of large volume of rain water and to protect the animals from shower coming with wind. Efficient drainage system should be provided for timely disposal of water.

d. Temperate Zone

This zone consists of cold and snow falling areas of Himalayan hills. In this zone the covered area should be completely covered having limited air flow, windows should be small and situated at a higher level. If possible, the floor should be fitted with wooden planks and rubber sheets. The ridge- ventilation should be provided with controlled openings. The animals will be permitted to go in open area only in sunny part of the day.

Sanitation

The word sanitation (derived from Latin word -Sanitas meaning health) refers to the establishment of favourable environmental conditions which is essential for the maintenance of the health of a subject. Intensive livestock production systems has confined them in a comparatively smaller space where they passes faeces and urine and other wastes like feed waste, wash water, soiled bedding etc are also generated there. These livestock farm waste may be in solid, semisolid or liquid form and has very good manurial value (average daily dung & urine production has been shown in Table 5.2.1 while their nutritive value in Table 5.2.2). The precise quantity of excreta voided from an animal depends on factors like species, size, age, sex, diet composition, production level and existing animal husbandry practices. If these wastes remain in the shed for longer duration of time, owing to their nutrient contents, acts as a breeding ground for microbes and disease spreading insects. Further, their decomposition produces foul smell. Hence, the environmental condition in the shed get vitiated and increasing fly and insect population not only causes annoyance to the animals but also spread certain infectious and contagious diseases. By the passage of time, the place becomes so much unhygienic that it will be impossible to keep animals there. Therefore, to conserve the health of the animals these materials needs to be disposed of properly as frequently as possible.

Table 5.2.1 : Average daily dung & urine production from different livestock species.

Species	Quantity excreted/animal/day	
	Dung (kg)	Urine (kg)
1. Cattle	23.5	9.0
2. Buffalo	32.5	–
3. Sheep	1.13	0.6
4. Pig	2.7	1.5
5. Horse	16.1	3.6
6. Poultry	0.04	–

Source : www.inseda.org

Disposal of Manure

In loose housing system, solid and liquid manures can be collected separately or both of them can be flushed out with water using pressure hose. Depending on various factors (size of farm, species, labour availability, end use, climate, level of mechanization etc) either of the two methods can be employed in their removal.

Table 5.2.2 : Average nutritive value of excreta of farm animals (in % of the original matter).

Species	Water (dung)	Nitrogen (N_2)		Phosphorus (P_2O_5)		Potash (K_2O)		Calcium (CaO)	
		Dung	Urine	Dung	Urine	Dung	Urine	Dung	Urine
1. Cattle	82.4	0.30	1.21	0.18	0.01	0.18	1.35	0.36	0.01
2. Buffalo	81.1	0.26	0.62	0.18	–	0.17	1.61	0.46	–
3. Sheep	61.9	0.70	1.47	0.51	0.05	0.29	1.96	0.46	0.16
4. Pig	80.7	0.59	0.38	0.46	0.10	0.43	0.99	0.09	–
5. Horse	76.8	0.47	1.29	0.30	0.01	0.30	1.39	0.17	0.45
6. Poultry	57.0	1.46	–	1.17	–	0.62	–	–	–

Source : Kolay(2007) & www.inseda.org

a. **Solid manure:** Solid manures can be removed by collecting them in basket, using shovel and wheel barrows, using bullock cart, tractor trolleys. Bullock/tractor operated dung cleaner may also be used to collect them. From stables and byres solid form of manure can be removed at least twice in a day while from piggeries once in a day. The collected manures can be stored properly in pits for decomposition and destruction of fly eggs and larvae. The manure pit should not be located near to the animal habitations and human dwellings, milking parlour and a minimum distance 10 m should maintained between a manure pit and wells, tanks etc. It should not be in the direction of prevailing wind. Amount of manure produced on farm decides number and size of pits required. A simple pit can be constructed by digging out in a suitable elevated place 4×3 square meter pit with 2 m depth with a roof made from locally available materials. Allnutt's manure pit with roof is ideal in our condition. It is a double receptacle pit with wall on three sides lay down on impervious platform having a gutter running across the open side. In order to prevent the larvae from climbing up ledge is provided few inches from top in the inner surface of side walls and partition wall.

b. **Liquid manure:** In loose system of housing a shallow 30-40 cm wide gutter is constructed along the long axis of the shed where covered and open area meets. Through this gutter all liquid manure runs out of the shed. Outside the shed, liquids from all the animal buildings are brought to meet at a point. At this point of junction the pipes should meet in an inspection chamber. From this chamber one pipe should carry the liquid to the settling tank, constructed of cement concrete with a perforated portion from top to bottom. Liquids pass through these perforations and retain the solids on the inlet side. The outlet pipe leads to the liquid storage tank made up of RCC and closed air tight with a manhole at top. The outlet pipe should be bent down to dip into the liquid, thereby preventing the backflow of gases from the storage tank. The size of the storage tank will depend upon the size of the farm and the number of animals kept and one adult cattle requires 0.3-0.4 m^2 of storage tank space.

c. **Liquid and solid manure:** At the places where facilities of ample supply of water with pressure are available, both solid and liquid manures can be flushed out of the shed. However, this method of manure disposal requires sufficiently wide elaborate construction of network of underground drains. The liquid manure can be directly led into the field for fodder production or can be supplied to bio-gas plants as slurry.

The excreta of the diseased animals should be disposed properly to check the spread of disease. For disposing excreta of diseased animals methods like burning, burial, chemical, drying/spreading may be used.

Disposal of Carcass

When animals die their carcasses should be disposed in hygienic manner. Special care should be taken to dispose the carcasses of animals died due to anthrax, foot and mouth disease, rinderpest swine fever, PPR, leptospirosis, tuberculosis, haemorrhagic septicaemia, black quarter, calf scour, glanders, Q fever, new castle disease, fowl cholera, bacillary white diarrhoea etc. The carcass should not be dragged on the ground and materials like feed refuge, bedding, excreta, ropes etc should also be disposed of properly. Carcass can be disposed either by cremation or burial method.

a. **Cremation method:** In this method carcass along with the other infective materials are burnt to ash by using wood or coal or gas or electricity. It is the best method when sufficient fuel is available. However, it is costly and half burnt carcasses create nuisance to the public health. Cremation method can be carried out by specially constructed furnaces, cross trench or double pit incinerator and flame gun.

b. **Burial method:** In this method carcasses are buried in a deep pit of appropriate size and highest part of the carcass remains at least 1.5 m below the earth surface. After placing the carcass in the pit skin should be sliced and kerosene or phenol should be poured and further sufficient quantity of quick lime or bleaching powder should be spread over the carcass then covered with earth. Twigs of thorny plants should be spread on the grave to discourage the animals from getting the part of carcass out.

Carcasses of animals dying due to non-infectious or non-contagious diseases can be utilized for manufacturing various useful products like meat, bone and blood meals, hide, animal fat, glue etc. In some countries the place where this conversion takes place is known as knackeries.

Disinfection & Commonly Used Disinfectants

Disinfection refers to the conversion of a place or item from a potentially infective state into one which is free from infection by using physical (wind, sunlight, heat, and electricity) or chemical (solid, liquid and gas) agents. The process involves destruction of pathogenic microorganisms and their spores and agent is known as disinfectant. Sometimes surface antimicrobials agents designated as disinfectants are used otherwise as sterilizing agents, sanitizers or antiseptics. Sterilizing agents are used to produce complete destruction of all forms of life. The most commonly used sterilizing agents are heat, chemicals and irradiation. Sanitizers combine cleaning and disinfection functions and are the agents that reduces the number of bacterial contamination to a safer level as per public health requirements. Substance which is used to remove or correct offensive odours are known as deodorant. An Antiseptic is a substance or agent that inhibit or retards the growth of microorganisms on living tissue. Antiseptics are usually the weakest and least toxic of the surface antimicrobials. Germicide is an agent which kills all microorganisms however in some instances excludes destruction of bacterial spores. Factors like desired objectives, composition and concentration of chemicals used, time, environments in which procedures are to be conducted differentiate agents into disinfectants, sanitizers, antiseptics and sterilizing category.

a. **Physical agents:** Fresh air and wind are germicidal (drying and oxidizing effect), however their action is slow. Direct sunlight is a disinfectant due to its ultra-violet range of wave lengths (most effective 2800 & 2400 A). Sunlight has an oxidizing and drying effect. It may not be a powerful disinfectant (U V rays little capacity to penetrate glass or roof materials etc) but it is an important cleanser. Use of dry heat, hot water/steam, boiling of items is other effective methods of disinfection.

b. **Chemical agents:** There are many disinfectants which vary in their composition and concentration. An appropriate and effective disinfectant should be effective-for longer duration of time, when diluted with water and at room temperature, must be free from strong and objectionable odours, should not be destructive to the materials, must not be toxic and irritant when inhaled and should be cheaper. Chemical compounds act as disinfectants due to their oxidizing, reducing power, corrosive and coagulant effect on the albumen of bacteria and parasites. Comprehensive cleaning of the item prior to the application of disinfectant is prerequisite for an effective disinfection process. Further, efficacy of the process depends on the type of the target organisms, their requirements for multiplication and their resistance, concentration of disinfectant and the duration of contact with surfaces, surrounding temperature, presence of organic matter etc. Contemporary disinfectants are intricate combinations of chemicals, soaps, and detergents etc which improve penetration of the active ingredients. Brief descriptions of common disinfectants used (historically or currently) in animal productions are given below.

Chemical agents	Description
1. Sodium carbonate (washing soda)	It is more effective as a cleanser than as a disinfectant. Its power increases when used in a hot solution. It is used as disinfectant @ 4% solution in FMD outbreaks.
2. Sodium hydroxide (caustic soda)	It is more powerful detergent and disinfectant. A 2% solution can be used to disinfect farm buildings after the outbreaks of contagious diseases like FMD, swine fever etc. A 5% solution destroys spores. It is highly corrosive in nature and during application protective clothing and goggles should be used.
3. Formaldehyde (HCHO), Formalin	Formaldehyde is bactericidal, virucidal and fungicidal in nature and it is a very good disinfectant in both liquid and gaseous for. Formalin is a 40 per cent solution of formaldehyde gas. It can be used as liquid

Contd.

	for disinfecting floors, stalls etc @ 5% strength. Fumigation with formaldehyde is effective against most viruses and all bacteria including their spores. About 15-20 ml formalin should be mixed with 15g of potassium permanganate for each cubic meter of air space. It has strong, irritant odour, corrosive in nature and has fibrolytic properties.
4. Alcohols	They are generally bactericidal in nature and only ethyl alcohol, benzyl alcohol, ethylene and propylene glycol has practical application. To increase its potency used in combination with other disinfectants.
5. Organic acids	Like acetic, citric, lactic, formic and propionic acids have bactericidal and mild viricidal properties and are sometimes used in meat and poultry packing plants and in calf and pig barns. Acetic acid @ 2% can significantly reduce levels of FMD virus on contaminated surfaces.
6. Inorganic acids	Generally sulfuric acid and hydrochloric acid is used in disease control. However, their use is limited. They are effective against FMD virus but are highly toxic, irritating to the skin and eyes and highly corrosive to metals.
7. Phenolic compounds (Phenol or carbolic acid, cresol, lysol etc)	They are effective against both Gram-negative and Gram-positive bacteria, yeasts, fungi and some viruses (enveloped and naked lipophilic) but not effective against bacterial spores. Phenols have high dilution coefficient (small change in concentration results in greater difference in killing rate), strong odour, irritating, toxic in nature and presence of organic matter reduces efficacy. Carbolic acid coagulates bacterial protein and a 5% solution can be used to disinfect metallic objects, clothings. Cresol is effective against spores and Lysol is three times more potent than phenol. Cresol/Lysol in 2-3% solution in hot water destroys a variety of organism.
7. Quaternary ammonium compounds (quats)	In appropriate dilutions, QACs are effective, less toxic, non-corrosive, without noticeable odour, tasteless and biodegradable disinfectants. These are cationic neutral detergents. Even in the presence of hard water and/or moderate amounts of organic material, they have a broad spectrum of antibacterial, antifungal, antiviral and sporicidal activities. QACs lack efficacy

Contd.

	against *Mycobacterium tuberculosis.* Can be used from 1:1000 to 1:20000 concentrations.
8. Iodine and iodine-based disinfectants	Aqueous iodine (Lugol's solution) or alcoholic iodine solutions (tinctures of iodine) are commonly used as antiseptics. Iodine has strong odour, stains, less soluble in water and is costly. Iodophors are disinfectants formed by combinations of iodine with various carrier compounds. Iodophors are used for general disinfection and cleaning, bovine teat dips, and surgical scrubs. Hard water and the presence of large amounts of organic material reduce the activity of iodophors. Commonly used iodophors are Betadine, Wokhadine etc.
9. Chlorhexidine	Chlorhexidine and its analogs are commonly used at concentrations below 4% as skin cleaners, teat dips and antiseptics; they are also used for cold sterilization of surgical instruments and for disinfecting equipment, barns and buildings. The low toxicity of Chlorhexidine makes it useful in combination with other disinfectants.
10. Chlorine or hypochlorite disinfectants	It has bleaching and germicidal properties and is commonly used in disinfection, sanitizing and water purification. These compounds are inexpensive, readily available, convenient to use, effective against vegetative bacteria, virus, fungi and posses minimal environmental hazards. However, they are corrosive in nature, has strong odour and efficacy reduces due to presence of organic matters. Sodium hypochlorite and calcium hypochlorite (chlorinated lime, bleaching powder or chloride of lime, Calx Chlorinata) has broad spectrums of antibacterial and antiviral action and are compatible with most detergents. Bleaching powder possesses powerful oxidizing and deodorant properties and are germicidal due to the liberation of nascent chlorine and nascent oxygen gas. It contains 30 per cent available chlorine. Its use should be restricted in milking byre as it may taint the milk. A 10 per cent, mixture of the powder with water may destroy anthrax. It is frequently sprinkled in barnyards and on piles of manure.
11. Calcium oxide (Quicklime, CaO)	When mixed with water, calcium oxide (quicklime) becomes lime wash, which has biocidal effects. Quicklime is spread on the animal discharges, manure present on floor,

Contd.

	wall and ground to act as deodorant and disinfectant. It has also been used to retard putrefaction of buried carcasses after depopulation.
12. Slaked lime (Calcium hydrate, Ca(OH)2)	Lime, as quicklime or milk of lime, is particularly suitable for use in dairies and cowsheds.
13. Perchloride of Mercury (Mercuric Chloride, Corrosive Sublimate,$HgCl_2$)	It is a very powerful bactericide even in very dilute solution because the free mercury ions coming into contact with the protoplasm of the bacteria, unite with it to form the insoluble albuminate of mercury. The ionisation, however, may be increased by the addition of a small amount of an acid. A 1-1000 to 1-2000 are strengths more commonly used.

Chapter - 6

Dairy Bovine Production

Rana Ranjeet Singh

Origin & Distribution of Cattle

All types of domestic cattle belongs to class Mammalia, subclass Ungulata, order Artiodactyla, suborder Ruminantia, family Bovidae, subfamily Bovinae, tribe Bovini, which includes genus *Bos*. The aurochs or urus (*Bos primigenius*) is the probable ancestor of most of the present day *Bos taurus* and *Bos indicus* cattle. The humped *Bos primigenius namadicus* were the forerunner of today's zebu cattle while humpless *Bos primigenius primigenius* gave rise to modern European cattle. Both category of cattle possess the same number of chromosomes (2N=60). Other related animals in the Bovini tribe include the gaur cattle (*Bos gaurus*), banteng (*Bos banteng*), kouprey (*Bos sauveli*) of South and East Asia, bison (*Bison bison*) of North America, European bison (*Bison bonasus*) and the yak (*Poephagus mutus*) of central Asia. According to Payne & Hodges (1997), domestication of aurochs (*Bos primigenius*) took place first probably on the lower mountain slopes in the region of western Asia known as the 'fertile crescent'. They suggested that the first domestic cattle were humpless (*Bos taurus*) Hamitic Longhorn type and were domesticated around possibly 9000 BP (years before the present i.e. 1950). About 2000 years later the humpless (*Bos taurus*) shorthorn type evolved or was developed in the region and these two types of humpless cattle are believed to the ancestor of *Bos taurus*.

Humped (*Bos indicus*) zebu cattle were domesticated in the northwest arid areas of the Fertile Crescent around *c.* 5000 BP.

Most of the cattle indigenous to tropics belong to zebu species. The most distinguishing external features of *Bos indicus* or zebu cattle, which separate it from European type of cattle is the presence of hump. Hence, the term humped cattle is frequently used as synonym to zebu cattle. Hump may be present either in cervico-thoracic region (posterior part of the neck i.e. above 6th or 7th cervical to 4th or 5th thoracic vertebra) or in thoracic region (on the shoulder above 1st to 9th thoracic vertebra). Factors like breed, sex and age of the animal affects shape and size of hump. Other common traits possessed by most of the zebu cattle are-presence of a long face, upright horns, drooping ears, narrow body, sloping rump, long legs, prominent brisket and dewlap (particularly in males) and thin & loosely attached skin. Mature body weight vary considerably among the different breeds and it may be as low as 200 kg to above 400 kg. The coat colour varies from white to grey and black. They have comparatively lower basal metabolic rate, better heat tolerance and diseases resistance (tick borne diseases) than the taurus cattle. They are late maturing, poor potential for milk production and have strong maternal instinct. In the United States they are called as Brahman cattle.

There are more than 800 breeds of cattle distributed all over the world. In India there are 34 breeds of cattle, divided into milch, dual and draught purpose breed. Around 80 per cent of the total cattle population belongs non-descript or *deshi* type. Different breeds of cattle have been discussed in chapter 3 while cattle population, their distribution around the world and among different states of the India has been discussed in detail in chapter 1 of this book.

Origin & Distribution of Buffalo

Domestic buffalo or water buffalo *Bubalus bubalis,* belongs to the class Mammalia, subclass Ungulata, order Artiodactyla, suborder Ruminantia, family Bovidae, subfamily Bovinae, tribe Bovini, which includes group *Bubalina* and *Syncerina*. Syncerina includes only species *Syncerus caffer* (the African buffalo). Bubalina (the Asian buffalo) includes three species: *Bubalus depressicornis* or Anoa which lives in Indonesia, *Bubalus mindorensis* which lives in the Philippines and *Bubalus bubalis* which descended from *Bubalus arnee* or wild Indian buffalo. The domestication of this species occurred about 5000 years ago (Borghese, 2005).

Buffaloes are massive animal with strong legs and large hooves. The colour of hair may be black, dun, creamy yellow, dark, light grey or white while colour of the skin varies from jet black to light pink. Body of young animals remains covered with good coat of soft hairs which becomes coarser and sparser as the animal grows. They lack dewlap, hump and tuft of hair at the preputial opening. Horns are generally broad, flat and nearly rectangular in shape at the base and has prominent ridges across the long axis. Horns of buffaloes are comparatively longer than cattle. Buffaloes are perfectly suited to the tropical climate and muddy terrain, but they exhibit signs of heat stress when exposed to direct solar radiation. Buffaloes are better converter of poor-quality roughage into quality animal protein as they are reported to have around 5 per cent higher digestibility of crude fiber than high-yielding cows and about 4-5 per cent higher efficiency of utilization of metabolic energy for milk production. Buffaloes are mainly kept for milk, meat and draught purposes.

The domestic buffaloes may be classified into two broad categories i.e. the swamp buffalo and the riverine (river/water) buffalo. These two groups of animals differ in their morphology, utility and genetics.

The Swamp buffaloes are usually dark grey in colour however black, black and white, or even all white animals may also be found. They are short stocky animals with short face, wide muzzle, short thin legs. Mature body weight is comparatively lesser (ranges between 300 and 600 kg) than the water buffaloes. They have long, gently curved horns. They are less adaptable to heat stress in comparison with riverine buffalo and are found only in swampy and marshy areas in hot climates. They are the working beast of the rice fields of South East Asia (Philippines, Malaysia, Singapore, southern China and Indonesia etc). They are also reared for meat production but seldom for their milk. In Philippines it is known as Carabao and in Malaysia as Kerban. The swamp buffalo has 48 chromosomes, of which 19 pairs are metacentric. The largest concentration of swamp buffalo is found in rice growing countries of South East Asia (Philippines, Malaysia, Singapore, southern China and Indonesia etc).

Majority (about 90%) of the total milk produced by the buffaloes in the world comes from water buffaloes. River buffaloes are generally black, large in size with long curled horns and are mainly found in India, Pakistan and in some other countries of Western Asia. The mature weight is approximately 1100 kg for males and 550 kg for females. They have higher milk production capacity than the swamp buffaloes. The river buffaloes are better adapted to a large range of environmental conditions. In contrast to the swamp buffaloes they prefer to enter clear water to get relief from heat stress. Water buffaloes are primarily used for milk production however they are also used for meat production and draught purposes. The River buffalo has 50 chromosomes of which 5 pairs are submetacentric, while 20 are acrocentric. The river and swamp buffalo are inter-fertile and produces progeny with 49 chromosomes. Majority of the water buffaloes are confined to the areas where animal husbandry is poorly developed and badly organized.

There are more than 70 breeds of buffalo in the world. India is considered as the home tract of some of the best buffalo breeds. Around 60 per cent of the total buffalo population belongs non-descript or *deshi* type and remaining 40% is classified into 17 breeds as per ICAR. Most of the buffaloes in India belong to riverine type though swamp type is also found in eastern parts of the country. Different breeds of buffalo have been discussed in chapter 3 while buffalo population, their distribution around the world and among different states of the India has been discussed in detail in chapter 1 of this book.

Important Traits of Cattle & Buffalo

Milk production: The milk yield in a lactation period is known as lactation yield. The lactation yield in Indian breeds is very low in comparison to exotic breeds. Milk yield increases over lactations and attains maximum yield between fourth to sixth lactation. Milk production continues to increase after parturition and reaches peak production within 4-8 weeks after calving. This yield is known as peak yield which can be held for few weeks after which milk production decreases till the end of lactation. The maintenance of peak yield for long period is known as persistency, slow decrease in dairy milk yield after reaching peak yield is necessary. The maintenance of peak yield for more time is important for better milk production. Some of the economic traits of Indian breeds of cattle & buffalo have been shown in Table 6.1.1 & 6.1.2.

The length of milk production period after calving is known as lactation period. Lactation records are normally expressed on a 305 day basis because milk yields from varying length are not comparable. For comparison of milk yield of different breeds and animals the milk yield should be converted into fat corrected milk using the following formula.

4% FCM = 0.4 total milk + 15 total fat

Table 6.1.1 : Some of the economic traits of Indian cattle breeds.

Breeds/Traits	Sahiwal	Red Sindhi	Tharparkar	Gir	Hariana
1. Age at first calving (months)	41 (38-48)	44 (39-49)	43 (39-53)	53 (44-61)	51 (41-49)
2. First calving interval(days)	473 (418-473)	496 (436-562)	447 (418-474)	511 (456-541)	497 (434-523)
3. First service period(days)	177 (136-189)	154 (152-158)	160 (130-183)	176 (135-259)	214 (136-303)
4. First lactation milk yield (kg)	1907 (1597-2125)	1476 (1312-1694)	1877 (1326-2139)	1498 (1125-1859)	916 (730-1170)
5. First lactation length (days)	314 (228-330)	305 (284-345)	300 (268-317)	293 (230-394)	267 (257-315)

Source : ICAR (2002)

Reproduction efficiency: It determines number of calves produced and hence the life-time-milk production as we know milk is obtained from a cow or buffalo after calving. Traits like age at first calving, service period, calving interval etc can be used to evaluate breeding efficiency.

The age of the animal at first calving is very important for high life time production. Higher age at first calving will result in lesser life time production due to less no of calvings and below optimum age at first calving will lead to birth of weak calf, difficulty in calving and less milk production in first lactation.

Period between date of calving and date of successful conception is known as service period. The optimum service period helps the animal to recover from the stress of calving and also to get back the reproductive organs to normal. For cattle the optimum service period is 60-90 days. Too short service period may lead to poor persistency of milk production whereas long service period will result in lesser life time production.

Inter calving period is the period between two successive calvings. It is more, profitable to have one calf yearly in cattle and at least one calf for every 15 months in buffaloes.

Dry period is the period from the date cow stops milk production to the next calving. A minimum of 2 - 2½ months of dry period should be allowed. Too low dry period may result in decreased milk production in the next lactation.

Table 6.1.2 : Some of the economic traits of Indian buffalo breeds.

Breeds/Traits	Murrah (PAU, Ludhiana)	Jaffarabadi (GAV, Junagarh)	Nili-Ravi (CIRB, Nabha)	Bhadawari (IGFRI, Jhansi)	Surti (MPUAT, Rajasthan)
1. Age at first calving (months)	44.4	64.2	39.9	48.6	53.2
2. First lactation milk yield (kg)	2000	1763	1932	711	1618
3. Total lactation yield (kg)	2032	2097	2153	781	1745
4. Lactation length (days)	347	384	341	272	311
5. Fat%	7.66	7.85	7.10	7.2 to 13	8.10
6. Dry period (days)	170	189.7	150	297	234
7. Service period (days)	197	189	170	179	207
8. Calving interval (days)	511	533	481	478	510

Source: Sethi (2000)

Apart from the above mentioned traits, body weight, type and conformation, efficiency of feed utilization and disease resistance are the other yardstick for selecting dairy animals.

Care and Management of Different Categories of Cattle & Buffaloes

Care and Management of the Newborn Calf

1. Immediately after the birth, any membrane or mucous adhering to the mouth, nostrils, eyes and ears of the newborn calf should be carefully removed to facilitate normal breathing.
2. Usually a cow will lick her calf dry. However, under certain situations if a cow doesn't do this or not allowed to do so (e.g. in case of Johne's disease), make the calf dry by rubbing with clean, dry towels or bedding over the head and ribs, this will not only dry off the calf but also enhances thermoregulation.
3. Make sure that calves should begin active breathing or respiration within 30 seconds, especially after difficult birth. This can be done by tickling the inside of the nose with straw,

finger, or other instrument or pinch the nasal septum (Immediately after thorax has emerged from cow) or use a 20 gauge 1.5" needle and place in acupuncture site in center of nose (nasal rostrum) or by pouring cold water on the calf's head, which causes the grasping reflex in the calf or artificial respiration should be induced by pressing and relaxing alternatively the chest wall or if the young one is smaller in size hold it by the hind limbs, lift it with the head down and swing it. Care is taken so that the young one is not slipped out of hands.

4. The navel cord of the calf is tied 2.5 cm away from its body and cut about one centimeter below the ligature and dip the navel in 7% tincture of iodine. Re-apply antiseptics to the stump for 2-3 days.

5. Record the birth weight of calves.

6. A normal healthy calf gets up within 30 minutes time. It is very essential that a calf should receive colostrum within a few hours after its birth. Colostrum is the secretion from mammary gland of mammals during first few days after parturition. Milk produced on second and third day after calving is called as transition milk. Colostrum contains higher levels of immunoglobulins (IgG) fat, solids, protein, minerals and vitamins needed to provide the calf with good nutrition right at the start of life. It also acts as a laxative and has a disinfectant action on the bowels and it clears the meconium. Colostrum is calf's sole source of immunity to common herd pathogens in the first few hours following birth as nature of the bovine placenta prevents transfer of immunoglobulins from the dam to the calf during gestation.

7. Feed ample amounts of colostrum as soon as possible within the first hour after birth. Use a nipple bottle if necessary. Provide a second feeding within 12 hours of birth. The intestine of the newborn calf begins to lose its ability to absorb the large antibody molecules in the colostrum within 1 hour after birth (called 'closure' of the gut). By the 9 hours after birth there is a 50% reduction in ability to absorb antibodies. Gut closure is complete by 24 hours of age. Also during this time, normal diges-tive enzymes in the abomasum and small intestine do not function or function with limited activity, allowing immunoglobulins to reach the small intestine without being

degraded. Further, colostrum contains enzyme inhibitors that allow immunoglobulins to escape intestinal degradation.

8. The quantity of first colostrum feeding should be around 4 to 5 percent of the calf's body weight and during first 24 hours after birth a calf should receive colostrum @ 12 to 15 percent of its body weight to guarantee adequate absorption of antibodies by the digestive tract. Usually the colostrum obtained in second milking after calving contains about 60 to 70 percent of the antibodies present in the first colostrum milk. Colostrum feeding should be continued for the first 3-4 days of life. The calf needs 3-4 liters of colostrum daily in 3 equal feeds.

9. Good-quality colostrum looks thick and creamy. Colostrum which is thin and watery should not be fed to calves. Always use colostrum obtained from a healthy cow.

10. Wash the udder and teats well before collecting colostrum. Filtering colostrum may help to remove some debris.

11. Colostrum may be stored at room temperature for 1 to 2 days, refrigeration for up to 7 days and freezing for up to 1 year. Frozen colostrum should be thawed using warm, not excessively hot, water baths, or by using a microwave at low or medium power and removing the thawed colostrum frequently before it overheats.

12. In the absence of colostrum, a mixture of 2 eggs in 30 ml castor oil is good. Another colostrums substitute can be prepared by mixing an egg, 275ml warm water, 5ml castor oil, 525 ml warm whole milk, 10,000 IU of vitamin A and 80 mg of Aureomycin and mixture can be given three times a day for four days. Injectable immunoglobulin products produced from purified bovine blood may be useful to increase circulating antibody levels in calves with failure of passive transfer that are older than 24 hours of age (gut closure has already occurred).

13. Weaker calves need assistance. Hence, a weak calf must be assisted to its feet and be held so it can nurse or it may be fed colostrums using a clean nipple bottle. For too week calf stomach tube can be used to feed the colostrum.

14. If a calf shows signs of anaemia, an injection of 150 mg of iron dextrin solution may be given within a few hours of birth. An injection of vitamins A, D, and E is also recommended.

15. Meconium is passed within two hours of first suckling. If meconium is not passed an enema with 1 teaspoon of sodabicarb in one liter of lukewarm water may be given.
16. Marking of calf with appropriate method of identification is done before removal from dam.
17. Calves may be kept in individual calf boxes.

System of Calf Rearing

There are two system of calf rearing viz. natural suckling and weaning. Natural suckling is the most common practice of calf rearing followed by rural dairy owners of India. Farmers allow the calf to suck milk to initiate milk let down and allow sucking liberally after completion of milk or they assign one udder solely for the calf. Female calves are given more opportunity to suck the milk from the dam than the male calves. To limit the suckling sometimes anti-suckling devices are used in case of males. In natural suckling, if dam dies during lactation, calf rearing becomes very difficult. There is possibility of transmission of disease from dam to the calf. Calf may consume more milk or less milk subsequently leading weak calf or more incidences of diarrhoea. This method also delays the early development of the calf into a ruminant.

In weaning system, calf is removed from the mother immediately after birth if weaning is to be practiced at birth. Weaning has many advantages like dam's milk production can be recorded properly for future selection and progeny testing, calf feeding is made scientific, calf is not required for letting down of milk and weaning prevents teat injuries. In many cases calf is being kept for first 12 hours with dam and then it is separated for weaning.

Feeding and Management Practices for Growing Calves

Feeding of calf from 0-3 months of age: After completion of colostrum feeding period calves are usually fed warm milk (preferably at body temperature) @1/10 of the body weight twice a day in two equal feedings. Milk is deficient in Fe, Cu, Mn, Zn and Mg. Hence, supplementation of these minerals is needed. Weaker calves may be fed three times a day by dividing the milk or milk replacer into three equal amounts. Calves are taught to drink milk from a pail or nipple pail. Milk is poured into a pail placed on a comfortable height for the calf to drink. The calf is taken to the pail and its nose brought in contact with the milk afterward feeder should dip his fingers in the milk and calf is allowed to suck it. Then slowly he should lower his fingers in the milk. Once the calf takes one or two mouthfuls fingers should be removed. This procedure may be repeated patiently each time when calf stops drinking milk and raises its head. When there is time and labour constraint weaned calves may be fed from nipple pails. It is a pail with one or more rubber nipples or bottles with nipples from which calves suck milk. If young calves are kept in group, they should be tied up during feeding time and at least up to half an hour after feeding to prevent them from licking and sucking each other's protruding body parts. After milk feeding, mouth of calf is wiped with clean towel and some salt or mineral mixture is rubbed in the mouth of the calf. When calves are fed milk they should show one kg weight gain per 1.39 kg dry mater in milk. Clean water should be available to them round the clock. Nutrient requirements of pre-ruminant cattle and buffalo calves have been given in Table 6.2.1.

Table 6.2.1 : Nutrient requirements of pre-ruminant cattle and buffalo calves.

Age (days)	Body weight (kg)	Daily gain(g)	DCP (g)	TDN (kg)	ME (Mcal)	Ca (g)	P (g)	Vit. A (mg)	Vit.D (IU)
0-15	25	200	80	0.40	1.5	2.5	1.5	1.5	200
15-30	30	300	90	0.50	1.7	3.0	2.0	1.5	250
31-60	40	300	125	0.80	2.4	3.5	2.5	1.7	250
60-90	50	350	150	1.00	3.6	4.0	3.8	2.0	360

Source : Ranjhan (1991).

Rearing of calf solely on milk up to three months of age gives best result but it is a costly affair. Hence, some methods have been evolved to economize cost of calf rearing by replacing some quantity of milk with good quality milk replacer ration, concentrate mixture (calf starter) and roughage. There are many ways in which calves can be raised, some of them are discussed Table 6.2.2 & 6.2.3.

Table 6.2.2 : Milk feeding schedule.

Body weight	Age (days)	Colostrum	Milk
Upto 30 kg	0-4	1/10th of body weight	----
Upto 30 kg	5-30	----	1/10th of body weight
31-60 kg	31-90	----	1/20th of body weight

Buffalo milk is high in fat content hence buffalo calves should be fed two litres of colostrum/milk up to 30 kg of body weight and afterward 0.5 kg of additional milk is fed for every 5 kg increase in body weight. Milk, skimmed milk and calf starter is provided as mentioned in table above. Good quality green fodder should be provided after two weeks of age.

Table 6.2.3 : Milk feeding schedule with calf starter.

Age (days)	Body wt.	Colostrum (kg)	Milk (kg)	Skimmed milk (kg)	Calf starter (g)
0-5	Upto 30 kg	1/10th of body wt	----	----	----
6-30	Upto 30 kg	----	1/10th of body wt	----	----
31-60	25-30 kg	----	1/15th of body wt	1/25th of body wt	125
61-90	31-70 kg	----	1/25th of body wt	1/15th of body wt	250

Milk replacer: Milk replacer is a constituted feed for dairy calf. Milk replacer should contain highly digestible protein that will clot in the abomasum, lactose, fat, minerals and vitamins. Milk replacers are usually made up of ingredients such as skim milk powder (60–75 percent), vegetable or animal fat (15–25 percent), butter milk powder, whey protein (5–10 percent), soy lecithin (1–2 percent) and vitamin-mineral premix. Milk based replacers are costly, hence some protein in the milk is replaced by vegetable based products. Plant protein sources are less digestible as compared to milk protein sources. A milk replacer should contain at least 20 to 22 percent crude protein when only dairy products are used as sources of protein while it should contain higher crude protein (22-24%) when plant protein sources are used. Crude fat content should be around 10 to 20 percent. Fat from animal sources are better than plant sources. Calf should be gradually shifted from milk to milk replacer diet. Generally calves at an early age do not consume much milk replacer. Hence, it is mixed with one litre of milk daily to increase the acceptability. Further, it is diluted by clean water to the extent of 1:8. The composition of a milk replacer has been given in Table 6.2.4.

Table 6.2.4 : Composition of a milk replacer.

Ingredients	Parts
Wheat	10.0
Fish meal	12.0
Linseed meal	40.0
Milk	13.0
Coconut oil	7.0
Citric acid	1.5
Molasses	10.0
Linseed oil	3.0
Mineral mixture	3.0
Butyric acid	0.3
Antibiotic mixture	0.3
Rovimix (A, B_2, D_3)	0.015

Source : Arora (1979).

Calf starter : Calf starter is a special type of concentrate mixture which contains nutrients required by the preruminant calf. Feeding of calf starters should be started as early as about 3 days of age but not later than 10 to 12 days of age. A good quality and palatable calf starter should contain approximately 75 to 80 percent total digestible nutrients (TDN) and around 16 to 20 percent crude protein (CP). A calf may be taught to eat starter by rubbing on its tongue and muzzle or putting a small amount in milk pail after each feeding. A variety of calf starters are available and the ingredients can be changed as per the availability of the feeds in different regions of the country. One of the examples of calf starter has been mentioned Table 6.2.5.

Table 6.2.5 : Composition of a calf starter.

Ingredients	Quantity
Crushed maize	35 kg
Crushed barley	15 kg
Groundnut cake	30 kg
Wheat bran	10 kg
Fish meal	07 kg
Mineral mixture	02 kg
Common salt	01 kg
Antibiotic	100 gm
Vitablend (A, B_2, D_3)	15 gm

Source : Mishra & Singh (1993).

Roughage feeding: Feeding good quality leguminous fodder (Berseem, Lucerne, Cow pea etc) or hay can be started from the 2 week of age on free choice basis. This practice will help in early development of rumen function. Legume hay is better than cereal hay. Calves are generally not allowed to graze from the pasture and fed silage before 3 months of age because high moisture content of these feeds reduce feed intake and growth of the animal. Growing calves should have access to fresh water all the time.

After the completion of three months of age feeding of milk is discontinued. The rumen is almost developed by this time and calf can digest fibrous feeds. Starter feed can be gradually replaced by standard ration and good quality fodder and hay. Nutrient requirements for growing cattle and buffalo are presented in Table 6.3.1 & 6.3.2.

Table 6.3.1 : Nutrient requirements for growing cattle calf.

Live Weight (kg)	Growth rate (g)	D M (kg)	DCP (g)	TDN (kg)	ME (Mcal)	Ca (g)	P (g)	Carotene (mg)	Vit. A (IU)
70	550	2.1	255	1.4	5.0	8	6	7	3
	600	2.3	260	1.5	5.4	8	6	7	3
80	550	2.4	260	1.5	5.4	9	7	7	3
	600	2.5	270	1.6	5.8	9	7	8	3
100	550	2.8	270	2.1	7.6	10	8	10	3
	600	3.0	280	2.2	7.9	10	8	10	3
120	550	3.2	280	2.3	8.6	14	11	11	4
	600	3.7	290	2.4	8.6	14	11	11	4
140	550	3.7	290	2.5	9.0	14	11	16	6
	600	4.2	300	2.6	9.4	14	11	16	6
160	550	4.2	300	2.7	9.7	15	12	16	6
	600	4.7	310	2.8	10.1	15	12	16	6
180	550	4.7	310	2.9	10.4	15	12	18	8
	600	5.2	320	3.0	10.8	15	12	18	8
200	550	5.2	320	3.1	11.2	16	13	22	9
	600	5.7	330	3.2	11.5	16	13	22	9
220	550	5.7	330	3.3	11.9	18	14	24	10
	600	6.2	340	3.4	12.2	18	14	24	10
240	550	6.2	340	3.5	12.6	20	16	28	12
	600	6.7	350	3.6	13.0	20	16	28	12
260	550	6.7	350	3.7	13.3	20	16	30	13
	600	7.2	360	3.8	13.7	20	16	30	13

Source : ICAR (2002).

General management practices: General management practices like identification, dehorning, castration, removal of extra teats, handling and control etc has been discussed in detail in chapter two of this book.

Table 6.3.2 : Nutrient requirements for growing buffalo calves.

Live Weight (kg)	Growth rate (g)	D M (kg)	DCP (g)	TDN (kg)	ME (Mcal)	Ca (g)	P (g)	Carotene (mg)	Vit. A (IU)
70	450	2.0	156	1.3	4.7	8	6	7	3
	550	2.1	168	1.5	5.4	8	6	7	3
80	450	2.2	177	1.4	5.0	9	7	7	3
	550	2.3	183	1.6	5.7	9	7	8	3
100	450	2.7	198	1.7	6.1	10	8	10	3
	550	2.8	210	1.8	6.5	10	8	10	3
120	450	3.1	279	1.9	6.8	14	11	11	4
	550	3.2	291	2.1	7.6	14	11	11	4
140	450	3.6	300	2.2	7.9	14	11	16	6
	550	3.7	316	2.2	7.9	14	11	16	6
160	450	4.1	264	2.5	9.0	15	12	16	6
	550	4.2	276	2.6	9.4	15	12	16	6
180	450	4.6	288	2.7	9.9	15	12	18	8
	550	4.7	300	2.8	10.1	15	12	18	8
200	450	5.1	312	2.9	10.4	16	13	22	9
	550	5.2	329	3.1	11.2	16	13	22	9
220	450	5.6	345	3.1	11.2	18	14	24	10
	550	5.2	357	3.2	11.5	18	14	24	10
240	450	6.1	375	3.3	11.9	20	16	28	12
	550	5.2	390	3.4	12.2	20	16	28	12
260	450	6.2	410	3.5	12.6	20	16	30	13
	550	6.5	420	3.6	12.8	20	16	30	13

Source : ICAR (2002).

Housing: Provision of good housing is very important during early life of calf. Inappropriate housing together with unhygienic condition inside the house results in heavy calf mortality. Under Indian condition, calves are generally reared in groups either in ordinary stall barns or often in the same house together with adult animal. At organized farms specially designed houses solely for calves are built. It is better to keep calves in individual pens for a minimum one month but if possible should be kept up to three months of age. Calves can be reared individually in small pens of 1.0×1.5 m size or in stalls in single/double rows with partitions for individual calf or in a barn with individual calf hutches up to one months of age. Individual rearing of calves will reduce the incidence of naval sucking and spread of diseases among the calves. From one month onward till three months of age calves can be reared in small groups in single row or double row stalls or larger hutches. Irrespective of the types of houses used for rearing calves it should be clean, dry, well ventilated and comfortable during all seasons. Older heifer calves from six months of age till breeding age is housed separately from suckling calves. Generally, male calves are not retained after six months of age. Housing of different categories of animals have been discussed somewhere else in this book.

Recording of body weight: Body weight of growing animals must be recorded at many stages i) within eight hours postpartum, ii) weekly up to 4 months of age, iii) twice in a month up to 6 months, iv) once in a month after 6 months of age.

Prevention and control of diseases in calf : The best way to reduce calf mortality is to carry out good husbandry practices. Some common ailments may be treated by the farmer, but once the calf is seriously ill or large numbers of calves are ill, services of specialist veterinarian should be sought. Neonatal diarrhoea or scour is the most common disease of calves, characterized by a dull appearance of the eyes, listlessness, diarrhoea and sometimes an abnormal rise in temperature and respiration rate. Faeces of scoured calves are often a chalky-white colour and have a very offensive smell and often calves die within a very short time. Scouring may be due to mechanical causes (hair balls in their abomasums), nutritive causes (rapid drinking of milk or cold milk feeding) or infective agents. Scour of nutritive origin can be cured by starving the calf for 24 hours and should be fed small quantities of milk diluted with a little lukewarm water. Scours caused by infective agents can be treated by the antibiotics.

Pneumonia is the second most common health problem for young calves. Symptoms are- rapid rise in temperature, increase in respiration rate and the calf 'gasps' for breath. Generally it is caused by bacteria and viruses but environment and management practices under which calf is reared plays very important role in the occurrence of the disease. Moist air, drafts, exposure to wet or chilling cold conditions, and sudden temperature change are predisposing factors for pneumonia. For treatment of affected calves early diagnosis, use of suitable antimicrobial agents and proper supportive therapy is necessary.

Table 6.4 : Common infectious agents causing scours in calves.

Infectious Agent	Age of affected calves
Bacteria	
Escherichia coli (*E. coli*)	Usually less than 5 days.
Salmonella (*S. typhimurium*)	Usually 2 weeks.
Clostridium perfringens type C	Usually 5 to 10 days but can occur up to 1-2 months.
Viruses	
Rotavirus	Usually between 5 to 15 days.
Coronavirus	Few days to several weeks.
Bovine virus diarrhea (BVD)	Any age but more common in 6-24 months age group.
Protozoal Parasites	
Coccidiosis (*Eimeria*)	Usually from 15 days to 6 months.
Cryptosporidium	Usually from 5 to 30 days.

Septicaemia often occurs during the first week of life and is defined as the presence of bacteria and their toxins in the bloodstream. There are a number of causative agents but important among them are *Escherichia coli* (septicaemia colibacillosis), *streptococci, Mycoplasma bovis, Klebsiella etc.* The infective agents may enter through mouth or the open navel of the newborn calf. Affected calf shows signs of fever, loss of appetite, clay coloured diarrhoea which becomes watery and blood stained, become depressed, weak, enter a state of toxic shock, and may die within 12 hours of onset of clinical signs. In umbilical infection umbilicus will be swollen. Early diagnosis and treatment with suitable antibiotics, supportive therapy (fluids or anti-inflammatory drugs) can be given. Calves surviving the initial septicemia may experience complications such as painful septic arthritis (joint ill) that is difficult to treat.

The young calf is mostly susceptible to attack by internal parasites. Calves suffer from roundworms, threadworms, hookworms, lungworms, tapeworms, coccidia and other internal parasites. Under tropical and subtropical climate condition diarrohea caused by *Ascaris vitulorum* is most common. Symptoms of parasite attack are -a 'falling back in condition', diarrhoea, a dry rough coat, a pot belly and a swelling under the jaw, weight loss, mild-to-severe anaemia, stunted growth and, in the case of lung worms, coughing and difficulty breathing. Internal parasites are best controlled by good calf-house hygiene, proper rotational grazing and a proper deworming schedule.

External parasites like ring worm, ticks, lice and mange can be checked by periodic spraying or dusting of calves and pens with proper insecticides.

Care and Management of Heifers

Heifer is a female bovine from one year of age up to first calving. Sometimes this term used to denote females until their second calving. Heifers are the future cows of the herd and productive and reproductive performance of cows depends on the care and management at their heifer stage. Rearing period of heifers can be divided into two stages, viz. from weaning to first service and from first service to calving. The aim in heifers rearing should be to attain the maximum growth and development with concurrent earliest sexual maturity of the animal, simultaneously with least cost of production. Growth rate (gain in body weight or increase in height at weight) basically depends on genetic merit of the animal and is influenced by feeding regime and prevailing environmental conditions. Further, growth rate varies considerably among the different breeds of a species as per the breed and adult body weight. Hence, nutrient requirements for growth will vary according to the target growth rate for a breed. Under feeding of growing animals will result in delayed maturity. Nutrient requirements of energy, protein and minerals (Ca & P) are high in faster growing animals.

Heifers can be reared by In-door or Out-door (gazing system) system but former method is expensive. In Out-door system they should be rotationally grazed around paddocks containing quality forage. Even if under grazing system land availability is limited; one grazing field should not be gazed for more than 5 days by the heifers. Concentrate and minerals should be provided when quality fodder is not available. Grazing land or paddock must have provision of shade for heifers and supply of clean drinking water. Under In-door system they should have access to shade, forage, clean water, minerals etc. The shade should be orientated east west, be approximately 3 m in height and be constructed from suitable materials. Around 2.5 to 3 square meter of covered floor space should be provided per animal. Heifers raised indoor can be fed cheap roughages and/or crop by-products (straw, stover, etc) but they necessitate additional high-quality forage and concentrate supplements.

Routine periodic weighing and comparison with a normal growth curve for the breed can help identifying poor growers' in the herd and necessary corrective measures can be adopted at an early stage. Heifer should be groomed regularly. They should get opportunity for exercise. Heifers should be treated with kindness and should be trained to be lead with halter from an early age. About 2 weeks before calving, heifer should be introduced to the milking herd routine so that she becomes accustomed to the milking parlour & feeding schedule of milch herd before calving. Heifers should be dewormed at every 4-6 months interval. The weight and age of the heifers should be the basis for giving first service.

Reproduction Management of Dairy Animals

Puberty may be defined as the time at which oestrus first occurs, being accompanied by ovulation. Average age at puberty is one of the most important components of the herd net reproduction because of its relation to the number of calves obtained each year and the feed intake up to calving. Timing of puberty is highly variable and dependent on a number of genetic and environmental factors such as breed, nutrition, live weight gain, season of birth and presence of male etc. Age at puberty in *Bos indicus* cattle in the tropics and subtropics range between 16 and 40 months. *Bos indicus* cattle reach puberty later than *Bos taurus* x *Bos indicus* crossbreeds or purebred taurine cattle. Temperate taurine breeds of dairy cattle reach puberty at 30-40% of their adult body weight, compared with 45-55% for beef cattle. Zebu heifers raised traditionally attained puberty at higher percentage of adult weight. Age at first calving for indigenous cattle ranges between 38-50 months and it is much later in buffaloes as compared to about 30 months in exotic breed of cattle. This delayed sexual activity is due to under feeding because of scarcity of feed and fodder, heavy parasitic load, improper housing, heat stress and breed itself. Age at first calving and mature body weight of some Indian cattle & buffalo breeds have been shown in Table 6.5.1.

Recommended minimum age at breeding is about 14-15 months in most of the temperate milch breed of cattle and minimum body weight ranges between 263-272 kg for Jersey to 374-397 kg for Holstein and Brown Swiss.

Oestrous cycle: Cattle and buffalo are polyestrous species which breed throughout the year. However, mating is limited during oestrus (behavioural manifestation of sexual receptivity) which coincides with the time of ovulation. Length of oestrous cycle in cattle & buffalo is around 21 days. Duration of oestrus is about 18 (12-30) hours in cattle and 21 (17-24) hours in buffalo. Ovulation in cattle takes place around 30 hours after onset of oestrus or 10-12 hours after the end of standing heat while in buffalo it occurs about 32 hours after onset of oestrus or 15-18 hours after the end of oestrus. Efficient and accurate detection of oestrus are essential in dairy herd using artificial insemination because inaccurate heat detection will ultimately result in lower conception rate in the herd. A good record keeping system will give an idea when a cow is expected to come in heat based on the fact that the average cow comes in heat every 21 days. A cow can be observed at least twice daily (once in the early morning and once in the late evening) for signs of heat between days 17 to 25 after the last oestrous cycle. Further, more frequent daily observation will reduce the number of missed heats. The cardinal signs of oestrus are exhibited more efficiently in loose housing system than in stanchion barn. Cows should not be observed for signs of heat at the times of feeding or other routine work is being done. Some of the heat detection methods have been discussed below.

Table 6.5.1 : Age at first calving and mature body weight of some Indian cattle & buffalo breeds.

Breeds	Weight at maturity(kg)	Age at first calving(months)
1. Sahiwal	301 - 544	37.4 - 48..8
2. Red Sindhi	317 - 454	39.0 - 50.9
3. Rathi	295 - 386	40.0 - 52.0
4. Kankrej	430 - 650	45.0 - 47.0
5. Gir	319 - 568	43.3 - 61.5
6. Ongole	363 - 591	36.0 - 54.0
7. Hariana	287 - 499	41.0 - 60.0
8. Tharparkar	293 - 544	37.5 - 53.0
9. Murrah	446-567	40.5 - 55.1
10. Nili-Ravi	454-567	37.9 - 43.0
11. Surti	319-413	47.2 - 55.4
12. Bhadawari	346-467	46.0 - 51.4
13. Mehsana	335-567	44.2 - 54.5

Source: Taneja (1999).

Methods of Oestrus Detection in Cattle & Buffalo

1. Behavioural Signs of Oestrus

- Restlessness, bellowing, excessive walking, frequent urination, drop in feed intake, slight swelling, reddening of the vulva, pale white or opaque mucus discharge from the vulva, smelling other cows and attempts to mount other cows all these signs indicate that cow is coming into heat.
- Standing when mounted by other cows (best indication of heat in the dairy cow), an increase of swelling and reddening of the vulva, a thin stream of clear mucus discharge from the vulva indicates the cow is in standing heat.

- Dairy cows stay in standing heat for an average of 15 to 18 hours and a bloody discharge from the vulva indicates the heat period is over. Record the date and observe the cow for signs of heat again in 14 to 18 days. In some dairy animals the signs of heat are very feeble. Watching for the next heat on the basis of the bloody discharge from the vulva will help in detecting heat in these cows.

2. Heat Detection Aids/Devices

- **Heat patch:** It is a plastic heat detecting patch glued to the top of the sacrum between the tail head and hipbones of the cow by the branding tag glue. It contains vials containing fluid and colour, when the cow is mounted by another animal (4 to 5 seconds of constant pressure) vial containing colour breaks and discolour the total fluid. The KaMaR patch (turns white to red) and Bovine Beacon patch (turns white to bright pink) is the examples of heat patches.
- **Chin-ball marking device**: It is a special halter with a cone-shaped unit having a stainless steel ball-bearing attached under the chin of the marker animal. It contains a marking fluid that is available in many colours. When the marker animal mounts and dismounts a cow in heat, the fluid leaves a visible colour on the cow's back and rump.
- **An implanted computer chip:** The computer chip is enclosed in a pressure sensitive case and is implanted beneath the skin on the tail head region of the cow. The device records when the cow is mounted which is read by a reading device.
- **Electronic heat detection systems:** This is based on the measurement of the pressure applied by a riding cow to a transmitter embedded in a tag and fixed to the tail head region of the cow supposed to come in heat.
- **Teaser animals:** 'Teasers' are any other cows, bulls or steers which are sexually aggressive as a result take an interest in an oestrus cow, inclined to mount and thus identify cows in heat. They are more effective if they are equipped with some device for marking the mounted cows. A bull (marker bull) which has been surgically modified (vasectomy, penectomy, penis deviation and a device in the prepuce to prevent erection) so that it cannot breed the cow but can be used for heat detection. Disadvantages for using a bull for heat detection is that it (vasectomised bulls) may spread disease from cow to cow,

requires extra feeding cost and all bulls can be dangerous. To overcome these disadvantages spade heifers, steers, or estrogen treated cows may be used for the purpose.

3. Physiological Changes

a. Milk indicators

- Milk progesterone, declines below 1ng/ml for 3-5 days around oestrus.
- Milk yield usually declines after the onset of oestrus because milk is withheld in the udder.
- Milk temperature increases by about 0.1^0C during oestrus.

b. Cervical mucus

- Conductivity increases during oestrus as more Na^+ is produced.
- Viscosity decreases
- Electrical resistance decreases (vaginal probe)
- Maximum ferning is observed when mucus is dried on glass slide
- Vaginal pH decreases

4. Oestrus related odours can be detected by trained dogs from vaginal mucus, urine or milk.
5. Physical activity increase which can be measured by electronic pedometers (Afimilkpedometer, DeLaval ALPRO Activity meter etc).
6. Body temperature rises by $0.5\text{-}0.8^0C$ during oestrus but drops on ovulation (radio telemetry).
7. Laparoscopic techniques.

Time of Breeding

The accurate timing of breeding cows naturally or by AI is vital as it seriously influences the conception rate. Cows should be inseminated from the middle to the last half of the standing heat period. However, it is better to breed late (up to 6 hours after standing heat) than to breed in the first half of the heat period. As a routine practice, if a cow is seen showing standing heat in the morning, it may be inseminated in the evening. If the signs of standing heat are observed in the evening, the cow may be bred next day morning.

Pregnancy Diagnosis

The precise diagnosis of pregnancy is essential in the establishment and maintenance of optimal reproductive performance in the herd. It is important for the farmer to know as early as possible that a mated cow is not pregnant so that cow can be rebred with the minimum delay. Some of the methods of pregnancy diagnosis have been described below.

1. **Non-return to oestrus:** Absence of oestrus approximately 21 days after service is widely accepted as an indicator of pregnancy. The percentage of cows not seen in oestrus at about this time is known as the non-return rate and is a well-known method of estimating pregnancy rates. Two conditions, anoestrus and some cow in a herd show signs of oestrus while they are pregnant may affect the reliability of this method.
2. **Clinical methods:** Rectal palpation of the foetus is the most popular method of pregnancy diagnosis in which reproductive organs are examined via rectum. By this method early signs of pregnancy like asymmetry of uterine horn, amniotic vesicles and fully developed corpus luteum in one ovary can be detected at about 30-35 days post service but requires expertise. At about 65 days post service pregnancy can be diagnosed accurately (enlargement and bulging of pregnant horn, foetal membrane slip, etc) by this method. In most cases pregnancy diagnosis is done at about 90 days post service, uterus is flaccid and both placentomes and foetus is palpable.

Pregnancy can be diagnosed using ultrasound techniques (B-mode RT) in cattle and buffalo as early as 12 & 30 days post mating, respectively. The process requires expensive equipments.

3. Immunologic tests
   - Early pregnancy factors can be detected in the serum of cattle within a few days after conception.
   - Progesterone levels are estimated in the milk and blood of cattle & buffalo about 22-24 days after mating/ insemination. At this time progesterone level is low in non-pregnant animals and higher in pregnant animal.
   - Pregnancy specific protein B (bPSPB) can be detected in the serum of pregnant cow by radioimmunoassay from day 24 of pregnancy until parturition.

- Estrone sulphate is the principle estrogen produced by the conceptus. It can be measured in plasma, urine or milk of cow at about 72 days post mating.

Care and Management of Pregnant Cows and Down-Calvers

Gestation is the period from conception to parturition. Gestation period in cows is little over nine months (280-285 days) whereas in buffalo it is longer (310 days) than cattle. The gestation length in river buffalo ranges from 305-320 days while in swamp buffalo it ranges between 320-340 days. Expected date of parturition can be calculated from a record indicating date of service. Gestation length may be affected by breed (beef cows have longer gestation length than dairy cows), sex of calf (longer for male calves), year of calving, parity, twinning and heifers conceiving at younger age.

Care and Management of Pregnant Cows

All pregnant animals should be looked after carefully. Proper nutrition and other management practices will guarantee better production, a healthy calf and a healthy mother. They should not be made to walk long distances, run fast, chased, frightened or allowed to fight. Usual signs of pregnancy are: cessation of oestrus, alteration in temperament; fattening tendency, easily induced fatigue, enlargement of abdomen, enlargement of mammary gland, increase in weight and 'quickening' i.e. signs of life exhibited by foetus in uterus. Pregnant animal ration should contain all nutrients in adequate quantity. In cattle, 0.5 to 1 kg extra concentrate mixture depending on the condition of the animals should be given from sixth month of pregnancy for meeting the demands of the developing foetus, building-up body reserves for the ensuing lactation. Further, heifers should be fed extra ration for growth in addition to other requirements. It is usual to add 0.45 kg starch equivalent and 0.15 kg digestible crude protein to the cows' ration during pregnancy. Feeding in preparation to calving is called as steaming up. It is common to start with 1.5 kg concentrate/day from two months before calving and increase it to up to 4 to 5 kg /day till two to three days before parturition. Steaming up increases daily milk yield lengthens the lactation period and also increases the butter fat percentage.

Care and Management of Down-Calvers

Around 1-2 weeks before expected date of calving, advance pregnant animals are separated from the herd and shifted to a clean, dry and well ventilated calving pen under intensive care. This practice will provide them quiet environment, better protection from climatic stress and lesser chances of getting infection from herd mate. Sufficient dry bedding (wheat/paddy straw) should be provided. Cow should be kept on a light, laxative diet and ample amount of clean freshwater should be available all the time. After parturition the external genitalia, flank and tail should be washed with water containing some antiseptic.

Care at Calving

Parturition is a critical time for both the dam and neonate and is particularly difficult for heifer because of her inexperience and small size of pelvis through which calf has to pass. The golden rule is not to interfere with the normal calving process. The signs of impeding calving are relaxation of muscles around the tail head, enlarged vulva and distension of the udder, sometimes leaking of milk from teats. The relaxation of muscles of tail head region is the best sign of impeding calving which occurs 24 hours before calving. The cow becomes restless, agitated and seeks for a quiet and sheltered place. Parturition or birth process is considered to occur in three stages of labour: first stage is the preparatory stage, during which pelvic ligaments loosen and the cervix dilates. During second stage expulsion of the foetus through the pelvic canal takes place and in third and final stage expulsion of the foetal membranes and initial involution of the uterus occurs.

(i) **First stage:** During this stage regular contraction of the myometrium begins at a rate usually of 12–24 contractions per hour. The cotyledonary attachments of the placenta begin to loosen and the cervix shortens and dilates. There are frequent signs of distress and cow tends to bellow, kick in the abdominal region, arches back and raises tail. This stage is terminated by breaking of water bag and appearance of calf's hooves. The length of first-stage labour is usually between 6 and 24 hours, generally shorter in older cows.

(ii) **Second stage:** This stage is characterized by onset of regular contractions of the abdominal muscles compressing the abdominal contents. The myometrial contraction frequency increases. The pressure of the foetus against the cervix and anterior vagina stimulates the release of oxytocin from the posterior pituitary, which stimulates further contractions of the myometrium. This mechanism is a typical neuro-endocrine reflex arc, known as Ferguson's reflex. Water bag appears at the vulva and the calf's front feet are soon apparent inside.

The rupture of water bag or amnion provides lubrication for calf's passage through the birth canal. The cow lies down to give birth whereas buffalo give birth in half standing position. The expulsion of head is most difficult process and if it's size is large particularly in heifers complicates the process and external intervention is required. The umbilical cord usually breaks spontaneously during expulsion of the foetus. Second stage of labour is usually completed in between 0.5 and 4 hours. In unnatural presentation or difficulty in parturition professional help should be sought.

(iii) **Third stage:** After expulsion of the foetus myometrial contractions continue, resulting in the separation and expulsion of the foetal membranes. This process may take up to 4-8 hours, in case of retention of placenta it can be removed manually after 12 to 18 hours postpartum. Cows should not be allowed to eat placenta as it may cause indigestion. Once the calf is on foot and placenta has been expelled, it is considered as a successful parturition. After the foetal membranes have been expelled, contraction of myometrial and release of oxytocin and PGF2á continues resulting into in an initial rapid rate of reduction in the size of the uterus by the day 15 postpartum. After this period the rate of involution of uterus is slow and in normal cows considered to be complete by around 23-35 days postpartum whereas in buffalo it takes around 16-39 days (average 28 days) in swamp buffalo and 15-60 days (average 45 days) in river buffalo. The rate of uterine involution may be influenced by month of calving and parity etc.

Care and Management of Milking Cows & Buffaloes

The udder of a cow is made up of four mammary glands (modified sweat gland) known as quarters. These quarters are composed of two halves i.e. right and left, demarcated by median suspensory ligament. Each half is further divided into two separate chambers by thin membrane. Hence, there is no communication among the four quarters. The rear quarters are usually larger than the front quarters and two rear quarters normally produces about 60 percent of the total milk. The udder remains attached to the lower abdominal wall by means of suspensory ligaments. The udder has alveoli that manufacture the milk. The raw material for the synthesis of milk is carried by the blood to the alveoli. The alveoli contain milk cavities and from each alveolus a tubule leads to small ducts which further lead to large milk ducts. These large milk ducts empty into a gland cistern. Each quarter has a teat that provides an outlet for the milk. Milk passes from the gland cistern through the teat cistern and then through the streak canal at the end of the teat. A circular muscle (sphincter muscle) at the end of the teat controls the flow of milk.

The let-down of milk is controlled by a combination of nervous and hormonal actions. When a cow is prepared for milking by suckling by a calf, washing and drying of the teat & udder, pressure of teat cup, nerves in the teat sends signal to the brain. In response to these stimuli the nervous system transmit a message to the posterior pituitary gland to release oxytocin which reaches to the udder in about one minute and causes contraction of the muscle surrounding alveoli and ducts. This forces the milk into the larger ducts and gland cistern. The effect of the oxytocin lasts for about 5 to 7 minutes. Milking must be completed within this time to remove most of the milk from the udder. When a cow is mistreated frightened, excited or disturbed before or during milking hormone adrenalin is released which counteracts the effect of oxytocin and stops the milk let down. This process will result in incomplete milking. The udder produces milk all the time but when pressure from produced milk increases, the rate of production slows down. Hence, regular milking is important to maintain high production. High producing cows will stop increasing the amount of milk in the udder after about 8 to 10 hours. However, a low producing cow takes about 16 to 20 hours. About 80 percent of the milk is removed from the udder at each milking. If too much milk is left in the udder, the pressure builds up quicker. This causes the cow to dry up sooner.

The freshly calved cows should be shifted to the milking herd once the colostrum period is over. The management and feeding of lactating animal is very important as the level of production and profitability from dairy farming depends largely on the amount of care and judicious feeding provided during this period. The milking cows should preferably be housed in loose houses. Details of housing requirements have been discussed somewhere else in the book. Milking cows should be divided into different groups depending on their level of production. When housed in groups, the number of cows in each paddock should be less than fifty. There should be provision to protect the animal from heat stress especially in Northern Indian conditions by suitable means.

Feeding of dairy animals constitutes around 70 per cent of the cost of production. The most important objectives in the feeding of dairy animals are-to allow maintenance and growth to mature body weight, provide nutrients for the production of a calf and support optimum quality and quantity of milk. The feed and fodder requirements of dairy cows and buffaloes are calculated based on their dry matter (DM) content. Requirement of dry matter depends on body weight and production status of the animal. The dry matter requirement of cattle varies between 2.0 to 2.5 per cent of their body weights while this requirement for buffalo and crossbred cattle varies between 2.5-3.0 percent of their body weight. Two-thirds of the total requirement of DM of the milking animal should be met through a mixture of cereal and leguminous green fodders and dry roughages. The remaining one third of DM should be met through concentrate mixture. As a thumb rule the producing cows (having body weight of 400 kg, producing milk @ 6kg/day and maintained on good quality fodder) should be fed concentrate mixture at the rate of 1 kg for every 2.5 kg of milk produced and buffaloes at the rate of 1 kg for every 2 kg of milk produced over and above the maintenance requirement of the animal. Daily nutrient requirements for maintenance, pregnancy and lactation for cattle and buffalo have been indicated in Table 6.6.1, 6.6.2 & 6.6.3.

Table 6.6.1 : Daily nutrient requirements for maintenance.

Body weight (kg)	Dry feed (kg)	DCP (g)	ME (Mcal)	TDN (kg)	Ca(g)	P(g)	Carotene (mg)	Vitamin A(100IU)
200	3.5	150	6.0	1.7	8	7	21	9
250	4.0	170	7.2	2.0	10	9	26	11
300	4.5	200	8.4	2.4	12	10	32	13
350	5.0	230	9.4	2.7	14	11	37	15
400	5.5	250	10.8	3.0	17	13	42	17
450	6.0	280	12.4	3.4	18	14	48	19
500	6.5	300	13.2	3.7	20	15	53	21
550	7.0	330	14.4	4.0	21	16	53	23
600	7.5	350	15.5	4.2	22	17	64	26
650	8.0	370	16.2	4.5	23	18	69	28
700	8.5	390	17.3	4.8	25	19	74	30
750	9.0	410	18.0	5.0	26	20	79	32
800	9.5	430	19.1	5.3	27	21	85	34

Source : Ranjhan (1991).

During the first and second lactation, in order to allow growth of lactating cows add about 20 and 10 per cent of the maintenance allowance, respectively.

Table 6.6.2 : Daily nutrient requirements for maintenance and pregnancy (last two months of gestation).

Body weight (kg)	Dry feed (kg)	DCP (g)	ME (Mcal)	TDN (kg)	Ca(g)	P(g)	Carotene (mg)	Vitamin A(100IU)
250	4.9	270	10.8	3.0	14	12	51	21
300	5.6	290	12.4	3.4	16	14	56	25
350	6.4	320	13.2	3.7	21	16	67	27
400	7.2	350	14.1	4.0	23	18	76	30
450	7.9	400	15.9	4.4	26	20	86	34
500	8.6	430	17.3	4.8	29	22	95	38
550	9.3	465	18.8	5.2	31	24	105	42
600	10.0	500	20.2	5.6	34	26	114	46
650	10.6	530	21.6	6.0	36	28	124	50
700	11.3	550	22.7	6.3	39	30	133	53
750	12.0	600	24.2	6.7	42	32	143	57
800	12.6	630	25.6	7.1	44	34	152	61

Source : Ranjhan (1991).

Some worker has suggested the concept of Challenge or Lead feeding for high yielding animals, in which higher levels of concentrate is fed to challenge the cow to achieve her maximum potential of milk production. About 3 days postpartum, amount of concentrate is increased by 0.45 to 0.90 kg per day for 2 to 3 weeks postpartum and at this time, the cow should be fed about 0.90 kg of concentrate per 45 kg of body weight. The level of concentrate feeding is continued till milk production levels off. At peak production, level of concentrate feeding is held constant. When milk production decreases, the level of concentrate feeding is also decreased. For every 1.36 kg reduction in milk yield 0.45 kg of concentrate is reduced. However, this practice is labour intensive, needs good record keeping and daily milk weighing.

Table 6.6.3 : Daily nutrient requirements for milk production (nutrients required per kg of milk fat)

Percent milk production	DCP (g)	ME (Mcal)	TDN (kg)	Ca(g)	P(g)
3.0	40	0.97	0.270	2.5	1.8
4.0	45	1.13	0.315	2.7	2.0
5.0	51	1.28	0.370	2.9	2.2
6.0	57	1.36	0.410	3.1	2.4
7.0	63	1.54	0.460	3.3	2.6
8.0	69	1.80	0.510	3.5	2.8
9.0	75	2.06	0.500	3.7	3.0
10.0	81	2.16	0.600	3.9	3.2
11.0	85	2.34	0.650	4.1	3.4

Source : Ranjhan (1991).

Methods of Milking

Hand and machine milking are two methods of milking. Hand milking is the most common methods among the farmers and machine milking is practiced at organized government, private farms and military dairy farms.

1. **Hand milking:** In hand milking, the upper part of the teat is closed tightly by the hand to prevent milk back flow of milk into the udder cistern at the same time pressure is applied over the milk filled lower portion of the teat. Increased pressure inside the teat stretches the sphincter muscle and the teat is forced to open. Cows are milked from left side. Milking is done either by full/whole hand or by stripping. Methods of hand milking have been shown in figure 6.1.1. Full hand milking is done by grasping the teat with all five fingers and pressing it against the palm. The teat is compressed and relaxed in quick succession, thus milk comes out much quicker than stripping. Knuckling is a faulty method in which milker bends his thumb against the teat during the milking. This method should be avoided to prevent damage to the teat tissues. Stripping is done by firmly holding the teat between thumb and fore-finger and drawing it down to the length of the teat and at the same time pressing it to cause the milk to flow down in a stream.

2. **Machine milking:** Modern milking machines employ alternating negative and atmospheric pressure through a double-chambered teat cup (one inside of the liner & another between the metal shell and the outside of the liner) assembly. A vacuum is pulled in both chambers. The vacuum inside of the liner is continuous while the vacuum in between the metal shell and the outside of the liner (pulsation chamber) alternates between atmospheric pressure and the vacuum. When the teat cup is applied to the teat, the end of the inside chamber is filled by the teat. During the open or milk phase the vacuum applied inside the liner is constant and keeps a constant negative pressure at the end of the teat. Hence, due to the high pressure difference between inside and outside the teat milk pushes open the sphincter muscle and enters into milking machine. The vacuum applied to pulsation chamber maintains the liner from collapsing under the vacuum. During the massage or rest phase, the vacuum inside pulsation chamber is briefly off thus air enters pulsation chamber and instantly reaches atmospheric pressure which collapses rubber inflation around the teat. This aids blood and lymph to flow out of the teat.

Figure 6.1.1 : Hand milking and dipping.

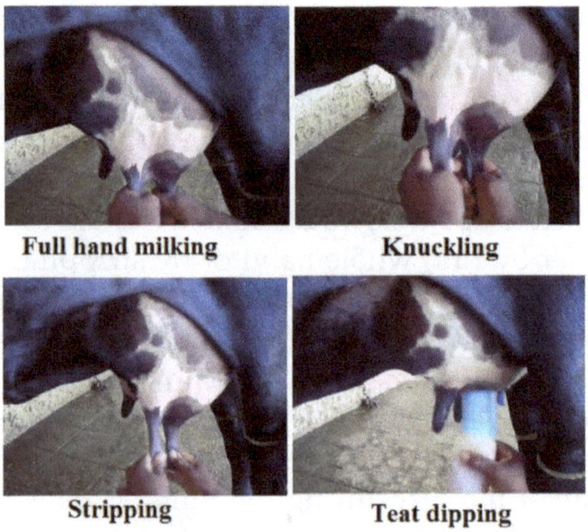

The basic layout of the three main types of milking machines (Bucket, Direct-to-can and Pipeline) is same. Bucket type milking machine consists of a lidded bucket, pulsator and teat-cup assembly or cluster. It requires manual attachment to a vacuum supply when it is moved from cow to cow during milking. Milk is poured from the buckets into milk cans positioned in the milk room or in the cowshed. The system is mechanically simple with relatively low investment, maintenance costs compared with other types. In direct-to-can milking, milk is directly drawn from the udders to the milk cans via a specially designed lid which connects the milk can to the vacuum supply. Manual lifting, carrying and pouring of milk is eliminated which results in improved milking efficiency. When full, the milk can is replaced with an empty one and milk cooling may begin by inserting an in-can cooler into the cans of milk. In pipeline milking, the advantages are that the milk is transported under vacuum from udder to dairy for cooling and storage, cleaning and disinfection of the milking equipment is done in-situ with very little manual involvement. In addition, devices can be inserted into the milking pipeline to reveal clinical signs of mastitis, indicate the milk yield from each cow and automatic removal of the cluster when milk flow end. Components of a milking machine are (Figure 6.1.2.)

1. **Milking Cluster:** It consists of four teat cup assemblies, a claw, a long milk tube and long pulse tube. Each teat cup consists of an outer shell, an inner rubber liner, short milk and short pulse tubes. The teat cup shells are usually made up of cylindrical stainless steel tube. The rubber liner is a flexible rubber sleeve having a mouthpiece, and when assembled in the shell under tension, forms an annular space (pulsation chamber) between the liner and shell. This pulsation chamber is connected to the pulsator through a nipple on the side of the shell via the claw. Claw is a component of the milking unit, which connects the short pulse tubes and the short milk tubes from the four teat cups to the long pulse tube(s) and long milk tube, respectively.

Figure 6.1.2 : Milking machine.

The claw is made of stainless steel or a combination of plastics and stainless steel and usually weighs about 0.5 kg and a milking cluster weigh about 2.5 kg. The weight of a milking cluster is important as lesser weight results in incomplete and heavier cluster weight results in falling off of milking units during milking.

2. **Pulsator:** It is a device (valve) which alternates vacuum and atmospheric air between the rubber liner & teat cup shell. The valve is activated either by a pneumatic (vacuum) or an electrical signal from a pulsator controller. Pulsator mostly pulsate the four liners of a cluster simultaneously.

Pulsation rate is the number of times per minute the milk phase and the massage phase occur (two phases equal constitute one cycle). A rate of 60 pulsations per minute, means the teat cup opens 60 times and closes 60 times per minute. Pulsation rates are generally set between 45 and 65 cycles per minute. The nursing calf sucks and swallows 80-120 times per min.

Pulsation ratio is the proportion of one cycle that the milking machine is in the milking phase versus the massage phase. Conventional ratios are 50:50, 60:40, and 70:30. The longer the vacuum is applied to the chamber, longer the teat chamber liner, or inflation, is held open, allowing the pressure differential caused by vacuum to open the teat sphincter.

3. **The Vacuum Pump:** The function of the vacuum pump is to extract air from the pipeline system and in the majority of milking machines it is a rotary exhauster driven from either an electric motor or a stationary internal combustion engine, using pulleys and V-belts. For milking cows vacuum levels of 50 (14.8 in Hg or 375 mmHg) kilopascals (kPa) and 44 kPa (13.0 inHg = 330 mmHg) are the most commonly used levels. Vacuum gauge indicates abnormal levels and fluctuations in vacuum. Vacuum regulator is fixed into the main vacuum pipeline near the milking units to maintain the working vacuum level in the milking machine.

Factors Influencing Quantity and Quality of Milk

All types of milk have the same kind of constituents but in varying amount. There are various factors which are associated with variation in both quantity and quality of milk. In order of the degree of variation, generally milk fat shows the greatest daily variation followed by protein, ash and lactose. Some of the factors affecting both quantity and quality of milk are as follows.

A. Inheritance

(i) **Species:** The milk yield varies from species to species. Temperate breeds of dairy cows yields higher than that of indigenous buffalo and cattle breeds. Further, milk of each species has a characteristic composition. The average chemical composition of milk of different species is given in Table 6.7.1 & 6.7.2.

Table 6.7.1 : Chemical composition of milk of different species.

Species	Water (%)	Fat (%)	Protein (%)	Lactose (%)	Ash (%)
1. Ass	90.0	1.3	1.7	6.5	0.5
2. Camel	86.5	3.1	4.0	5.6	0.8
3. Goat	86.5	4.5	3.5	4.7	0.8
4. Human	87.7	3.6	1.8	6.8	0.1
5. Mare	89.1	1.6	2.7	6.1	0.5
6. Sow	89.6	4.8	1.3	3.4	0.9
7. Whale	70.1	19.6	9.5	-	1.0
8. Elephant	67.8	19.6	3.1	8.8	0.7

Source : De (1977).

(ii) **Breed:** The breed of the animal has a distinct relation to the quantity of milk produced and especially to the fat content. Within a breed variation in milk yield is far greater than that of the fat percentage.

Table 6.7.2 : Chemical composition of cow and buffalo milk.

Milk	Fat (%)	Protein (%)	Lactose (%)	Total Solids (%)
1. European cow	3.90	3.47	4.75	12.82
2. Zebu cow	4.97	3.18	4.59	13.45
3. Crossbred cow	4.00	3.46	4.88	12.83
4. Buffalo	7.64	4.36	4.83	17.96

Source : FAO (1977) & Mudgal *et al.* (2003).

Average chemical composition of milk of some of the cattle & buffalo breeds has been given in Table 6.7.3 & 6.7.4.

Table 6.7.3 : Chemical composition of cow milk of different breeds.

Breed	Water (%)	Fat (%)	Protein (%)	Lactose (%)	Ash (%)
1. Sahiwal	86.42	4.55	3.33	5.04	0.66
2. Red Sindhi	86.07	4.90	3.42	4.91	0.70
3. Gir	86.44	4.73	3.32	4.85	0.66
4. Tharparkar	86.58	4.55	3.36	4.83	0.68
5. Holstein	87.74	3.40	3.22	4.87	0.68
6. Ayrshire	87.10	4.00	3.58	4.67	0.68
7. Brown Swiss	86.59	4.01	3.61	5.04	0.73
8. Jersey	85.09	5.37	3.92	4.93	0.71
9. Guernsey	85.39	4.95	3.91	4.93	0.74

Sources : De (1977).

(iii) **Individual:** Strains and individual animal of a breed differ in total milk production. Each animal tends to yield milk of composition that is characteristic of an individual. Normally, cows will not produce daily milk yield more than 8-10 percent of their body weight in contrast in goats this is around 20 percent or more of their body weight. Hence, large cow will produce more milk.

Table 6.7.4 : Composition of buffalo milk from different countries (in %).

Breed	Water	Fat	Protein	Lactose	SNF	Minerals
1. Bulgarian	82.56	7.50	4.33	4.80	9.94	0.81
2. Carabao	78.46	10.35	5.98	4.32	11.19	0.84
3. Chinese	76.80	12.60	6.04	3.70	10.60	0.86
4. Egyptian	82.09	7.96	4.16	4.86	9.95	0.78
5. Hungarian	83.79	7.22	3.65	4.56	8.99	0.78
6. Italian	81.94	7.85	4.28	4.97	10.21	0.75
7. Murrah	82.76	7.38	3.6	5.48	9.86	0.78

Source : Rao and Nagarcenkar (1977).

B. Physiological Status

(i) **Age:** Lactose and protein content of milk decreases with age while fat content remains same up to fourth lactation thereafter it decreases with age. In extreme old age milk yield may be as low as 50 percent of maximum mature capacity.

(ii) **Variation of different quarters of udder:** Normally the forequarters produce 40% and rear quarters 60% of the total milk produced. The fat% of the milk from different quarters varies less and also independent of the variation in milk yield. The orders in which the quarters are milked affect both the amount of milk and fat percentage.

(iii) **Effect of oestrus:** In general oestrus is accompanied by a decrease in milk yield and increase in fat percentage.

(iv) **Stage of lactation:** The stage of lactation affects both quantity and quality of milk produced. Colostrum is very different from milk of established lactation in its composition and general properties. The change from colostrum to milk takes place within a few days after calving. As lactation progresses milk yield increases till peak production has reached thereafter it decreases.

Table 6.7.5 : Composition of colostrum (%) of different species.

Species	Water	Fat	Total protein	Lactose	Ash
1. Cow	77.50	3.60	14.30	3.10	1.50
2. Buffalo	68.00	15.00	13.60	3.10	-
3. Ewe	58.80	17.70	20.10	2.20	1.00
4. Doe	81.20	8.20	5.70	3.40	0.90
5. Sow	69.80	7.20	18.80	2.40	0.60

Source : Long (1961) & Bhat (2010).

(v) **Gestation period:** There is decline in milk production in later months (fifth) of gestation. It may be due to the hormones produced by the ovaries which exert an inhibiting effect on the secretion of the lactogenic hormones produced by the pituitary gland.

(vi) **Milk yield:** There is a tendency for increased yield to be accompanied by a lower fat percentage and vice versa.

C. Environmental Factor

(i) **Feed:** Both quality and quantity of milk may be affected by the feed. Inadequate feed supply can be single most important factor which can limit the milk production. Different types of feed offered to the animals results in change in composition of milk. Feeding of ration low in energy and protein content leads to lower SNF in the milk particularly the protein. High concentrate, low roughage diet, finally ground hay, fish oils are known to reduce fat content of the milk. Change from stall feeding to grazing results in increased SNF in the milk.

(iii) **Effect of season:** The percentage of Fat and solid-not-fat show variations round the year. The intensity of the effect of the season exerted through weather conditions varies with the locality. Effect of weather on milk production in detail has been discussed somewhere else in this book.

D. Procedure of Milking

(i) **Frequency of milking:** Milk synthesis is a continuous process and as milk accumulates in alveoli and storage area of udder exerts back pressure and gradually inhibits milk secretion. Hence, more frequent removal of milk is necessary. It has been observed that milking cows three times a day increases milk production 10-25% over twice a day milking. Further, milking four times a day increases milk production by 5-15% over thrice a day milking.

(ii) **Variation from day to day:** Daily variation in milk yield may be due to fear, excitement, oestrus, disease, under feeding, incomplete milking and other factors.

(iii) **Portion of milking:** With the progress of milking, each successive portion of milk has an increased fat content. Foremilk may be as low as 1 percent milk fat while last drawn milk may be as high as 8-15 percent.

(iv) **Irregularity in milking:** Frequent changes in milking time and interval influences both quality and quantity of milk.

E. Miscellaneous Factors

(i) **Fear/Excitement:** Both quantity and quality of milk is liable to temporary fluctuation during period of excitement.

(ii) **Efficiency of milker:** May affect fat and SNF percent of milk.

(iii) **Administration of Drugs and hormones:** Certain drugs may affect transient change in fat percent, injection or feeding of hormones may lead to change in milk yield and composition.

(iv) **Diseases:** Mastitis, digestive disturbance and other diseases are responsible for a decrease in milk yield and affect quality of milk in various ways. Any disease which is severe to produce an increase in the body temperature or reduce the appetite for feed is accompanied by a decline in milk yield. Reduction in milk yield and change in milk composition due to mastitis depends upon severity of attack.

Dry Cow Management

The day from which the cow stops giving milk to the day it calves is called as dry period. A normal dry cow does not give milk but it should be in advance pregnancy. The significance of dry period is to give an opportunity to the cow to recuperate its body condition lost during the lactation which results in higher milk production in the succeeding lactation. Conditioning for the dry period is done during the last few weeks of lactation. Cows should not be too fat or too thin at the end of the lactation. Weight is controlled by changing the concentrate to roughage ratio i.e. higher amount of concentrate to thin cows and lower amount of concentrate to the fat cows. An ideal length of dry period is around 60 days. However, for indigenous cattle and buffaloes it is much higher. The high producing crossbred cows and purebred exotic cows need to be dried by using a suitable method. There are three methods of dry off the cow viz. abrupt cessation of milking, intermittent milking and incomplete milking. In abrupt cessation method, the milking of cows is stopped all of a sudden. The build up pressure of milk left in the udder causes regression of milk secretory cells and helps in drying off. This method is suitable for low to medium milk producing cows. Before drying off, the concentrate or grain is not fed for 2 or 3 days then water and forage intake is reduced for 1 or 2 days. After the feed has been reduced for the recommended time, milking process is stopped. In intermittent milking method, the cows are milked once every second or third day till milk production completely ceases. In incomplete milking, the cows are milked gradually to smaller and smaller quantities of milk for last few days. This method is preferred for high yielding cows because high producing cows need to have their production level reduced before drying off. Routine treatment for mastitis at drying off is recommended. Udder is observed for any abnormal swelling for 2-3 weeks after drying off. Dry cow is separated from milch animal herd and allowed to get plenty of exercise.

Clean Milk Production

Clean milk is defined as the milk drawn from the udder of healthy animals, which is collected in clean dry milking pails and free from extraneous matters like dust, dirt, flies, hay, manure, extraneous water and drug residues and contains relatively small numbers of bacteria none of which are pathogenic to humans. Freshly drawn milk from udder is seldom sterile as udder itself harbour bacteria (non pathogenic in uninfected udder) and contamination occurs during production, procurement and transit which can adversely affect its quality. The quality of milk can be adjudged from the details given in Table 6.7.6.

Table 6.7.6 : Bacteriological standards for milk.

Grade	Methylene blue reduction test (time in hours)	One hour resazurin test (Resazurin disc number)	Total plate count of milk (standard plate count per millilitre)
Very good	5 and above	-	Not exceeding 0.2 million
Good	3 to 4	4 or higher	Between 0.2 to 1 million
Fair	1 to 2	3.5 to 1	1 to 5 million
Poor	0.5	0.5 to 0	Over 5 million

Raw milk and pasteurized milk is considered satisfactory when coliforms are absent in 1:100 and 1:10 dilution, respectively. The standard plate count of pasteurized milk shouldn't exceed 30,000 per millilitre.

Factors Altering Quality of Milk

The quality of milk is highly influenced or altered by various factors like off flavour and odour, microbial content / contaminants, somatic cells in milk, antibiotic and drug residues, residues of plant protection agents etc. Milk may acquire off-flavours and / or odour from various sources, these are feeds and weeds flavours (wild onions and garlic, strong flavoured feed stuffs like lucerne silage or such other silages etc), cow-barny flavour (develops due to poorly ventilated and or unclean and inadequately cleaned milking barn), rancid flavour (due to presence free fatty acids), oxidized flavor (due to exposure of milk to sunlight, contact of milk with oxidizing agents such as rust, copper and chlorine), fly sprays, medications, strong flavoured disinfectants etc.

There are certain bacteria capable of multiplying in unprocessed milk and then surviving pasteurization and / or refrigeration process. The primary source of these bacteria is the environment which includes air, dust, dusty feeds, dusty barn, dirty equipment, operators, diseased animals, contaminated udder etc.

There are three types of somatic cells typically found in milk: i) Epithelial, ii) Macrophages and iii) Polymorphonuclear leucocytes

(PMNs). Milk from non-infected glands contain predominantly epithelial cells and macrophages while milk from infected glands (udders), generally contains high concentrations of PMN's with little increase in other cell types. Consequently somatic cell count is an important indicator of udder health.

It has been observed that trend of use of various pesticides, herbicides, insecticides, growth hormones etc for higher fodder productivity, antibiotics hormonal preparations and feed additives to enhance animal productivity has increased. These substances enter the food chain through animals and poses danger to human, animal and ecosystems health.

Advantages of Clean Milk Production

- It enhances better keeping quality and commercial value.
- It is safe for human consumption as it is free from disease producing micro-organisms.
- The products prepared from clean milk are of very high quality.

Measures to Produce Quality Milk

1. Milkers and their milking procedures

- Milkers should be free from contagious diseases like Cholera, Typhoid, Diphtheria and Tuberculosis etc.
- Milkers should be clean and avoid sneezing, coughing, smoking as well as spitting and chewing of tobacco just before and after milking.
- Milkers should thoroughly clean their hands and arms before milking. Finger nails should be kept trimmed. Preferably liquid soap hand wash be used and clean towel to dry the hands.
- Milkers should avoid wrong milking practice like knuckling and incomplete milking.
- Pre-dipping is very effective step in mastitis control. Pre-dip should cover 75% of teat surface and should stay on the teat for minimum of 20-30 seconds.
- If calf suckling is followed, at an appropriate time calf should be separated and tied.
- Drying of the teats with individual paper or towel is the most important step. Drying does more to lower SCC and reduce

clinical mastitis than other steps. The teat and teat end must be wiped clean and dry.

- Drawing of milk should be started 45-90 seconds after stripping. If the timing is delayed, there will be longer milking times and less milk harvested and may lead to incomplete milking resulting in favourable condition for bacteria to grow in udder.
- A minimum of 75% of the teat must be covered with dip. So as to replace the milk film with a layer of germicide after milking, thereby preventing more growth of bacteria and reducing chances of environmental mastitis.

2. Milking animals and their environment

a) Health of the animals

- Animals should be free from diseases like Tuberculosis, Brucellosis Salmonellosis, Shigellosis, Entero-pathogenic *E. coli.*, Streptococcus and Staphylococcus and Pseudo-cowpox, Louping ill (Tick borne encephalitis), FMD etc.
- Periodical testing should be done for T.B., J.D. and Brucellosis etc.
- Feeding of feedstuffs giving off flavour should be avoided.
- Feeds contaminated with aflatoxins, phytotoxins, heavy metals and radioactive substance should be avoided.

b) Cleaning of animals

- Grooming, brushing and washing should carry out perfectly to remove dirt, dust, dung, loose hairs and such extraneous material from hind quarters, abdomen, udder and tail of the animal.
- Udder and teats should be washed with warm non-staining antiseptic solution.
- Waterless cleaning of teats by pre-dipping and wiping with disposable paper napkin is the most modern practice to be followed.

c) Environment

I. Animal houses

- Keep house as dry as possible by providing adequate ventilation, light and cleanliness by removing dust, dirt, mud, dung, urine etc.
- Long axis of the house is preferred in such a direction that maximum benefit of sunlight can be obtained.
- Disturbance by flies, mosquitoes, rats, cockroaches, birds etc should be minimized.
- In pakka construction periodical white washing is advocated and floor should not be slippery.
- Floor swiping should be done before one hour and there after washing may be done to restrict dust particles to rise and contaminate milk.
- Clipping of udders (mostly in exotic and crossbred cows) is probably one of the most important steps in producing quality milk. Flaming the hairs from udders 3-4 times in a year is a new approach and takes 75% less time.

II. Milking equipment or utensils and their functions

- Microbial contamination of milk with dust at the time of milking can be minimized by using small topped container not the buckets.
- Milking pails should have smooth surfaces (stainless steel is preferable), so that it would be easy to clean, wash and dry using permitted sterilizing disinfectants.
- No traces of detergents are left in the pails which may become extraneous material while using for milking.
- Straining should be done to remove sediments and other foreign materials. If cloth is used it should be washed and dried daily otherwise it will spoil the milk by contributing bacteria from its surface.
- Utensils should be washed the immediately after disposal of milk.

Care and Management of Bull

Careful selection of an appropriate bull is extremely essential as bull is said to be half of the herd, i.e. half the inherited characteristics of all the calves are obtained from him. Good health and proper growth of calf starts and gets accelerated in the neonatal life. The general management and feeding of young bulls should be similar to that of heifers. Every effort should be made to feed them sufficiently so that they can be brought into service at an early age. Faulty management practices of young animals are often responsible for retarded growth and late maturity of the animal.

Feeding: A well balanced ration in relation to energy, protein, minerals and vitamins should be given to the bulls. Before puberty the amount of feed provided depends on growth rate and should be fed with growth promotion ration. Once the maturity is reached, principles of feeding are to maintain the condition neither too fatty nor too lean. Six to eight hours of grazing on a quality pasture is sufficient to maintain health and fertility. Under stall fed condition they should be fed 15-20 kg of green fodder, 6-8 kg of straw and 2-2.5 kg of concentrate. Bulls should be provided liberal amount of clean drinking water.

Housing: Bulls are housed individually. The bull sheds are constructed towards one end of the farm. Detail of the bull shed has been discussed somewhere else in this book.

Health: Bulls should be regularly tested for brucellosis, trichomoniasis, campilobacteriosis, leptospirosis, tuberculosis and Jhone's disease. They should also be checked for febrile conditions and foot and mouth disease.

Management: Generally most of the bulls are aggressive. Hence, a copper or brass ring of 5-6 cm diameter should be inserted in the nose of the bulls when they are about 8-12 months of age by punching a hole in the septum with a special pair of pliers or some other sharp instrument. This ring is replaced by heavier and stronger ring (6-8 cm diameter) at the age of two years. Sometimes farmers punch a hole in the septum of the nose of their bulls and pass a rope through it with a knot at one end that is suitably large to prevent the rope slipping through the hole. Bull should be handled by giving them grooming, washing and exercise. As a routine practice bulls should be groomed daily in the morning. The sheath and soiled abdominal portion should be cleaned. The bull should be washed daily except in severe cold condition in winter. Daily exercise is most important for bulls as it helps in keep them in trim and good health. If bulls are managed indoors without access to a paddock they should be given walk for 15-30 minutes, either in the early morning or in the evening.

Selection of Bulls: Following points should be considered while selecting bulls for breeding purpose.

- Production record of ancestors and closely related animals
- The anatomic and physical characters of bull
- Health
- Fertility
- Economy
- Age
- Production record of offspring

Evaluation of Breeding Soundness in Males

1. General physical condition
   - The bull should be healthy and free from apparent diseases.
   - It should not be pot-bellied as it may create problem during semen collection.
   - The locomotion should be normal and hoof should not be over grown.
   - Anterior portion (shoulder) of the bull should be massive and well developed.
   - Bull should have masculine voice and aggressive.
   - The bulls should be allowed to mount before final selection so that physical defects like spondylosis traumatic reticulitis could be detected.
2. Clinical examination of reproductive system
   - Inspection of testicles and epididymis: The scrotum should be symmetrical, normal size and contour wrinkle. The testicles should be of same size or a difference between two testicles should be within 10 percent. It should be mobile within scrotum. No adhesion or inflammation should be present. The consistency of testicle is normally turgid and elastic. In normal bulls tail of epididymis feels filled and spongy. Small and firm tail indicates poor semen production from the testicle.

- Inspection of penis and prepuce: Penis and prepuce should be examined for adhesion and tumor etc. Movement of penis should be normal.
- Inspection of accessory sex organs: Accessory sex glands particularly seminal vesicles should be checked for inflammatory changes, if any.

3. Mating behaviour of bull should be observed.

Service Management : The active breeding life of bulls varies between 5 to 16 years. In the temperate countries young bulls can be first used as early as 10-12 months of age, but in the tropics bulls sexually mature at an older age. In well-managed herds they can be of used at 18 months of age, but zebu bulls can be used after 2 years and attain maximum breeding powers considerably at further later age. Under standard feeding and management practices bulls attain a body weight of 250-300 kg at about two years of age and by this time they should be trained for semen collection. However, actual semen collection should be started at about 30 months of age. Buffalo young bulls can be used once or perhaps twice a week, but older bulls may be used up to five times a week. Mature bulls can serve up to 100 cows a year, but usually a bull is allowed to serve only 50 or 60 cows. The acceptable ratio of breeding females to breeding males is likely to be somewhat lower in the tropics than in the temperate climate as heat stress affects the fertility and vigour of the bull.

Care and Management of Bullocks

In India, cattle and buffalo are the major sources of draught animals. Generally male cattle and buffalo are used for draught purposes but in few eastern and southern states heifers are also used for draught purposes. As per CARTMAN (Livestock and Environment for Sustainable Development), bullocks constitute 88%, cows less than 2% and buffaloes only 9% of the work animal population. These animals are used for pulling agricultural implements, transport carts, lifting water, cane and seed crushers etc. Majority of farmers in India follow mixed farming system in which bullock fits well. There are no major physical differences between draft animals of the same species, even though certain breeds are more renowned for their draft capabilities. Usually, a thick and muscular neck indicates heavy draught capacity, while a slender and sinewy neck denotes speed. A tight skin is generally considered to denote great sensitivity and fiery temper.

Population: In the year of 2003, work animal population comprised of 54319 thousand of cattle and 5834 thousand buffalo. Annual growth rate of these animals were negative during 1997-2003. Highest work cattle population was observed in Madhya Pradesh and work buffalo population was highest in Uttar Pradesh. There are very good draught purpose breeds in India which have been discussed somewhere else in this book.

Work performance: The amount of draught power which can be delivered by an animal depends on the species, breed, size, sex, body weight, health, nutrition, environment, training for work, terrain condition, type of harness and implements used. In general, the most important criterion for draft effort is live body weight which indicates potential of an animal to exert a force or a pull which can utilized through working implements. Power developed by various animals in terms of Horse power (HP) is bullock- 0.74; buffalo- 0.75 and cow- 0.45. Bullock could develop draught equivalent to 1/5 to 1/6 of their body weight. Routaray (1985) observed that the average power out-put decreased with the increase in working time for all the draught conditions and similar was the trend in crossbred and local breed of bullocks also because of fall in working speed time. At higher draught values although the power output increased yet the bullocks could work only for shorter duration in a continuous operation. Roy *et al.* (1972) reported that the average efficiency of crossbred bullocks were higher than the Hariana while ploughing both wet and dry land and average area covered was also higher in crossbred cattle. This was attributed to bigger size and higher speed of crossbred under both the conditions. Upadhyaya and Madan (1982) reported that the performance of crossbred bullocks could be comparable to Hariana bullocks under hot summer condition, if crossbreds put to heavy loads for a short duration of time. Under different climatic conditions Hariana bullocks travelled faster than crossbred bullocks during first hour but during succeeding hours of walking and carting exercise, Hariana travelled at lesser speed while crossbred travelled almost at the same speed as in first hour of work. The reduction in the speed was marked in Hariana while pulling the load but crossbred travelled at almost similar speed. Many studies have been conducted on the draught ability of bullocks during different seasons, using local yokes/harnesses in continued working conditions of 6-8 hours; performance of bullocks has been summarized in Table 6.8.1.

Table 6.8.1 : Draught ability of animals.

Animal	Draught equivalent to % of body weight		
	Summer	Winter	Duration of work (hours)
1. Bullock	12	14	7
2. Male buffaloes	10	12	6

Source : Srivastava (1989).

Feeding and Management: Bullock of indigenous, crossbred cattle and buffalo males can be put to work when they are about 3-4, 2-3 and 2-2.5 years of age, respectively. Bull calves intended to be used as bullocks should be castrated when they are between 1-2 years of age. Untrained bullocks are yoked with old and experienced bullock and the yoke is connected with fairly heavy wooden log with the help of a rope or iron chain and then pair of bullocks are driven in the field. While training, owner talks and pats the animal and at the end of exercise they are hand fed. After a few days of empty training when bullocks become easy to be led, it is introduced to plough or cart yoke again with a trained bullock. Bullocks are normally trained in 3-4 weeks period for most of the agricultural operations but it is advantageous to take light work in the first year of training. On an average buffalo working life is 20 years. The work has been categorized into two types for bullocks- i) normal work where bullocks work for 2-4 hours of a farm duty of pulling cart without extra load or thrashing etc. ii) heavy work where bullocks work for a minimum of 8 hours of ploughing or pulling a loaded cart. The nutrient requirements for both categories are given in Table 6.8.2.

Table 6.8.2 : Daily nutrient requirements for working bullocks and buffaloes.

Live weight	Normal work				Heavy work			
(kg)	DM (kg)	DCP (g)	TDN (kg)	ME (Mcal)	DM (kg)	DCP (g)	TDN (kg)	ME (Mcal)
200	4.00	0.24	2.0	7.2	5.0	0.25	2.7	9.5
300	5.80	0.33	3.1	11.4	7.0	0.42	4.0	14.4
400	7.60	0.45	4.0	14.4	9.8	0.57	4.8	17.3
500	9.40	0.56	4.9	18.0	11.2	0.71	6.4	23.0
600	11.2	0.66	5.8	20.8	13.0	0.82	8.0	18.8

Source : Sen *et al.* (1978).

Use of bullocks should be avoided in direct sunlight during hotter part of day particularly in summer season and should be used in morning and late afternoon. A proper work and rest should be followed and during rest period feed and water should be offered. Besides the common diseases of bovine, bullocks are prone to luxation of patella, yoke gall and horn injuries etc. Grooming the legs and quarters of bullock with wisp made up of straw before and after heavy work is desirable. Extra attention must be paid towards the condition of hoof of the animal. Renewal of shoes in bullock working in field should be done once in 2 or 3 months, while those on road work may require change of shoes every month.

There are various instruments like loading car (sledge type, winch type, suspended weight type and hydraulic type), tread mill, dynamometers (spring, hydraulic and load cell type), abney level which can be used for measuring draught ability of the animal. Further, sensors have been developed for immediate measurements of heart rate, respiration rate, body temperature, stepping rate, pull, angle of pull and speed of working under field condition. For measuring and quantifying fatigue, animal fatigue score card developed by Upadhyay & Madan (1985) can be used (table 6.8.3). It is based on visual and physiological symptoms. In this method, for oxen and buffaloes maximum 40 points have been assigned and an animal is considered to be fatigued at 40% of the maximum score points.

Table 6.8.3 : Fatigue score card for draught animals.

Parameters	Fatigue score					Total score
	1	2	3	4	5	
1. Respirationrate /min	Ro +15	Ro + 30	Ro + 45	Ro + 60	Ro +75	5
2. Heart rate/min	Ho + 10	Ho + 20	Ho + 30	Ho + 40	Ho + 50	5
3. Rectal temp.(°C)	To + 0.5	To + 1.0	To + 1.5	To + 2.0	To + 2.5	5
4. Frothing	First signs	Starts to dribble	Continuous dribbling	Froth on upper lips	Full mouth frothing	5
5. Leg coordination	Stride uneven	Occasional dragging	Leg movement not coordinated frequent foot dragging	Complete lossof coordination in all legs	Unable to move	5
6. Excitement	Composed	Disturbed	Nostrils dilated, bad temper	Prominent eyeball movement	Furious and tying to stop	5
7. Inhibition offorward movement	Brisk	Free movement	Slow walking	Very slow	Stop walking	5
8. Tongue protrusion	Mouth closed	Occasional opening ofmouth	Frequent appearance of tongue	Continuous appearance of tongue	Tongue fully out	5

Note : Ro, Ho and To are respiration rate, heart beat and body temperature values at rest.

Table 6.8.4 : Work rest schedule for animals.

Bullocks:	2 h work + 1 h rest + 2 h work + 1 h rest + 2 h work3 h work + 1 h rest + 3 h work4 h work + 2 h rest + 3 h work
Buffaloes:	4 h work in the morning + 3 h rest + 4 h work in the evening2 h work + 1 h rest + 2 h work + 1 h rest + 2 h work + 1 h rest + 2 h work in continuation.

Source : Singh (1999).

Healthcare Practices of Dairy Animals

Health is the state of body structure and function which on assessment of larger number of similar individuals are observed to be identical. As per, WHO health is a state of complete physical, physiological and mental well being of an individual. Disease may be defined as deviation from physical, physiological and mental state of affairs. Health and disease are subjected to a complex of contributory factors associated with the animal (resistance to disease), organism itself (virulence of organism) and with the environment (opportunity potential of organism and stress) in which it must manage to exist.

Factors affecting health of livestock: There are number of factors which affect health of an animal. Various metabolic dysfunctions, mechanical factors, infective agents, qualitative and quantitative deficiency of feed, faulty management practices, improper cleaning and sanitation of facilities and equipment, inadequate ventilation, climatic stress, industrial waste, non-observation of proper quarantine period, any type of toxicity, dietary mistakes, unhygienic & polluted drinking water supply, lack of preventive measures like scheduled vaccination against infectious or contagious diseases, lack of control of disease carriers, like flies, birds, and rodents, keeping sick animals with healthy animals, unlimited access to the farm visitors and ageing is apparently associated with decline of health.

There are numbers of ways in which infectious/contagious diseases are transmitted from diseased to healthy animals. Some of them are mentioned below.

a. Direct contact with affected animal

b. Indirectly, it can transmitted by

- Contaminated water, feedstuffs, infected premises / stockyard.
- Human being may convey infective material on their hands, clothing and boots etc to the surroundings of susceptible animals.

- Other animals like stray dogs may act as an important source in spreading diseases like Rabies, skin infections and worm infestations etc.
- Birds, flies, pests, insects, lice and ticks may act as passive carriers of infection.
- Exchange of breeding stock may lead to spared of disease.
- Introduction of a potential carrier animal (animal carrying, infective agent but not showing signs of disease but intermittently shed infective agents through its excretions) in a new area.

Steps to be adopted in controlling spread of contagious diseases: following measures can be undertaken in controlling spread of contagious diseases.

- Isolation
- Immunization
- Intervention with the life cycle of infective agent

Herd should be inspected at least twice daily (in the morning and evening) to detect unhealthy animals. Sings of ill health has been already discussed somewhere else in this book. Animal showing some signs of disease must be isolated from the apparently healthy animals without loss of time. Animals must be grouped into three categories- 1) affected 2) suspected and 3) healthy. Affected animals should be kept in separate buildings with separate arrangements including separate bedding, buckets, grooming tools, handling tools and animal attendants. These animal attendants are not permitted to go to the healthy animal shed but if it isn't possible then he must go to the healthy animals first and then to the sick animals. He should wear separate dress, gumboot while working with isolated herd and must be disinfected when leaving isolation unit. Affected animals should be treated as per the nature of diseases. Period of isolation must extend beyond the recovery period till the infection has passed.

Immunization/vaccination may be practiced when the incidence of disease is very high or reservoirs of infection are so extensive that control by above measures is impracticable. It would be impossible to eradicate diseases completely caused by organisms which can live outside the animal body for longer period and still remain pathogenic *viz.* Anthrax, H.S., B.Q., F.M.D etc. Vaccination is a process of artificially inducing active immunity in animals against specific infectious diseases by introducing vaccines into their systems.

Table 6.9.1 : Vaccination schedule for dairy animals.

Disease	Vaccine	Dose(ml)	Schedule
1. Foot and mouth disease	FMD vaccine Raksha FMD vaccine	5 m 13 ml	First: 4 months of age Booster: 2-4 weeks laterRevaccination: twice in a year(September & March)
2. Haemorrhagic septicaemia	Alum ppt. H.S. VaccineH.S. vaccine	5 ml 3 ml	First: 4 months of age Revaccination: Annually before monsoon
3. Black quarter	Alum ppt. B.Q. vaccine	5 ml	First: 6 months of age Revaccination: Annually before monsoon
4. Brucellosis	Brucella cotton - 19 strain vaccine	5 ml	Calfhood (at 4-8 months of age) vaccination, in case herd is suspected of infection.
5. Anthrax	Spore vaccine	1-2 ml	First: 6 months of age Revaccination: Annually before monsoon. In endemic areas only.
6. Theileriosis	Rakshavac-T-vaccine	3 ml	First: over 3 months of age Revaccination: Annually.In endemic areas.

Note : Add dose, route & repeat vaccination may vary for different manufacturers.

Along with immunization, strict isolation and slaughter of affected animals and those in close vicinity may still be necessary in the event of outbreak of highly contagious diseases, which are liable to spread in the area. Carcasses must be disposed off either by burial or by burning. Adequate disinfection of premises, equipments etc should be carried out.

Interference with the life cycle of the causal agent or with course of disease includes procedures such as disinfections and disinfestations, treatment of sick animals and measures aimed at the elimination of carrier animals and intermediate hosts if such exist.

Deworming: Internal parasites can be controlled by periodical deworming and maintaining proper hygienic condition in the shed. For the effective control of these parasites all the animals of a herd should be dewormed at a time after fasting for 24 hours. It is difficult to control them on a farm where animals go for grazing. Young animals should be dewormed at monthly interval up to the age of six months. Adult animals should be dewormed twice (before monsoon and after monsoon) in a year. Piperazine salts can be used to control round worm infestation in the calf and first can be given at 10 days of age. Now-a-days many broad spectrum anthelmintic drugs are available which can be used.

Antiparasitic drugs have been used successfully to control parasitic diseases in animals over the years, as they are safe, cheap and effective against a broad spectrum of parasites but due to the emergence of drug resistance in many target parasites and issues of residues in the food chain and environment have increased. Hence, control methods of parasites through vaccines could be a better option in near future.

General Care and Management of Sick Animals

For early recovery from diseases sick animals need proper care. Following steps can be undertaken for management of sick animals.

- Sick animal should be separated from rest of the animal and kept in comfortable and sanitary environment.
- Adequate protection from extreme climatic conditions viz. cold, heat and rains should be provided.
- Provide proper medicinal treatment as per the type of disease.
- Well-balanced ration should be provided including green fodder as per the requirement of animal. The diet should be easily digestible, palatable and nutritious.
- Ample amount of fresh and clean drinking water should be provided.
- If possible, light grooming should be practiced.
- In urgent situations, prompt life saving measures (as mentioned below) should be followed
- The animals requiring emergency treatments should be transported with least or no discomfort.
- In case of bleeding, stop haemorrhage by pressure or tourniquet.
- In case of fracture, temporarily immobilize the fractured part using sticks/splints.
- In case of respiratory distress, then clean air passage, remove the mucous debris obstructing the passage, for relief in respiration.
- In snake bite, a tourniquet may be applied above the site of bite and in scorpion and wasp bite ice packs may be placed at the site.
- Commence suitable emergency treatments as per the symptoms shown by the animal.

Common diseases of dairy animals and their control measures: The disease conditions normally affecting cattle and buffaloes, their causes, modes of transmission, symptoms and their prevention and control measures are given in the Table 6.9.2.

Table 6.9.2 : Common diseases of dairy animals and their control measures.

Name of disease	Causative agent	Mode of transmission	Main symptoms	Prevention & control
Foot and mouth disease (FMD, Aphthous fever)	RNA virus(Strains-A,O,C, SAT1,SAT2, SAT3 & ASIA-1).	Direct and indirect contact with infected animals and their secretions (saliva, blood, urine, faeces, milk and semen).	Salivation, sores on feet, tongue and inside of mouth, stamping of feet, lameness, off-feed, drop in milk production	Segregation and other sanitary measures,pre seasonal vaccination with polyvalentvaccine.
Haemorrhagic septicemia	*Pasteurella multocida Type-I*	Ingestion of contaminated feed, water and contact with infectedanimals.	High fever, salivation and difficulties in swallowing, painful, hot swellings on throat, neck and dewlap, swollen tongue and laboured breathing.	Segregation, avoidance of infected pasture,feed and watersources, pre-monsoonvaccination and adequatesanitation.
Brucellosis (contagious abortion, Bang's disease)	*Brucella abortus*	Feed, pasture, water, milk contaminated with an aborted fetus, fetal membranes and uterine fluid and discharges.	Incidence of abortions during 7th to 9th month of pregnancy, full-time still-births, retained placenta etc. in the herd. Hygromas on the knees, stifles, hock, between the nuchal ligament and the primary thoracic spine.	Elimination of carriers from herd, calf hood vaccination at 6 months of age.

Contd.

Black quarter (Black leg)	*Clostridium chauvoei*	Water and food contaminated with blood and excretions.	Lameness, crepitating swellings often on the hips and shoulder, high temperature & death in 3 days.	Avoidance to the contaminated pasture in endemic areas, proper disposal of dead body, sanitation & vaccination.
Anthrax	*Bacillus anthracis*	Water and feed contaminated with blood and excretions or by wound infection.	History of sudden death, high fever, swallowing, breathing difficulties, swelling over throat & neck.	Proper sanitation, quarantine, proper disposal of dead body and annual vaccinationbefore rainy season.
Mastitis	Bacteria, fungi, yeasts, injury, improper milking etc.	Infective agents from dirty floor, milker's hands, cow's body etc. enter into udder through injuries on udder and teats.	Discomfort during milking, udder swollen, hot and painful in acute cases, milk is whey like with milk clots or even blood clots and increase in temperature.	proper dry hand milking, washing or wiping of udder and teats with mild antiseptic before and after milking, proper hygiene in the sheds etc.
Tuberculosis	*Mycobacterium bovis*	Direct and indirect contact with infected animals and their secretions or excretions. Bacteria enter system by ingestion or inhalation.	Weakness and loss of appetite, emaciation, swelling of superficial body lymph nodes, chronic intermittent hacking cough and difficult breathing	Segregation and other sanitary measures

Contd.

Johne's disease (Bovinepara tuberculosis)	*Mycobacterium paratu berculosis*	Contaminated food and water	Gradual and chronic weight loss and emaciation, Non responsive diarrhoea with watery fluid faeces, submandi bular edema, debility and death	Segregation and other sanitary measures
Milk fever or parturient paresis	Due to acute fall in calcium level possibly due to draining of the Ca during early lactation.	Occurs generally during the early stages of lactation	Loss of appetite, constipation, general depression, animal lies with its head resting on the chest wall and the nose pointing towards the flank, temperature sub-normal, animal may develop nervousness and die in 6-24 hours if unattended.	Provision of mineral supplements to high yielders during late pregnancy and early lactation. To prevent further secretion of milk - stop milking.
Retention of placenta	As a consequence of abortion,difficult parturition etc.	–	A portion of membrane hangs out from the vulva, chocolate coloured discharge with foul smell, milk yield goes down.	Clean the hind quarters of cow with warm water and take care that the hanging part of the membrane does not get pulled out. Seek veterinary aid.
Bloat	Accumulation of gas / foam in rumen	Greedy feeding on lush green fodders, obstruction in oesoghagus	Greatly distended abdomen especially on the left side.	Care in feeding green fodders, feeding after wilting, feeding dry fodders with green fodders.

Raising of Buffalo for Meat Production

Buffalo meat is also known as Carabeef or buffen. Buffalo meat is produced primarily in Asia. World buffalo meat production and leading countries in buffalo meat production have been shown in Table 6.9.3. Uttar Pradesh, Andhra Pradesh, Maharashtra, Punjab and Kerala are the main buffalo meat producing states in India. In the year 2011-12, India has exported 9, 85,491.29 MT of buffalo meat to the world for the worth of Rs. 13,725.25 crores. Major export destinations were Vietnam Social Republic, Malaysia, Egypt, Arab Republic, Saudi Arabia and Jordan (APEDA, 2013).

Until few decades ago buffaloes were mainly considered as milk and/or work animals. Since this animal was not considered as potential meat producer therefore only underfed surplus young ones, old animals at the end of their productive career and culled animals (low producing or injured) were used for meat production. Many studies conducted in different countries have confirmed that if water buffaloes are reared and fed for slaughter at 12-18 months of age give a very good meat with optimum live weight. Though, quality and quantity of buffalo meat depends on many factors like type of animal, breed, age, feeding intensity, management system and environmental conditions. Further, the importance of buffalo as work animal is constantly decreasing. Hence, scientists are giving more attention to exploit its meat producing capacity. This new avenue together with work and milk production will play more important role in the economic development of world's poor areas and difficult habitats. Many workers in India gave resolute effort to develop appropriate feeding regime for male buffalo calves. Generally, feed formulations were based on NRC (1976) requirements or Morrison's standard for beef cattle, growth rate, feed efficiency and carcass evaluation was done at predetermined slaughter weight (about 300-350 kg). A daily gain of around 500 g and about 50-55 per cent of dressing yield was obtained.

As per an estimate, every year in India around 10 million male calves are lost due to some economic reasons. If these losses are minimized and these male calves are reared for meat production, will not only increase the income but will also increase the availability of meat for domestic consumption and export market. We are citing an example of how these animals of lesser economic importance can be utilized in nutritional and economic security of the farmers. In a village demonstration farm with Hind Agro Industries Limited, male calves are purchased from the farmers at the age of eight to ten months and are then quarantined for 15 days and during this period vaccinations and deworming are practiced. Subsequently, they are brought to the main farm and are fed on a high protein/high-energy diet in order to gain an additional weight of 120 kg in four months. They are raised organically and are provided 2.5 - 3.0 kg of pellets; 3 kg of spent brewer's grains and 2 kg whole sugar cane/sorghum per day, this diet provides 20% crude protein and 65% TDN with a dry matter intake of 4.0 to 5.0 kg/animal/day. Antibiotics, hormones and growth promoters are never fed to the animals. The Murrah calves grow at the rate of 900 to 1000 g/head/day with a feed conversion ratio of 5.0:1. Meat from such animals is tender, lean juicy and goes to the wet market (Ranjhan, 2007).

Table 6.9.3 : Leading buffalo meat producing countries of the world.

Countries	Production (in million tonnes)			% of world buffalo meat production
	2006	2008	2010	
1. India	1.38	1.44	1.49	42.56
2. Pakistan	0.67	0.71	0.75	21.49
3. Egypt	0.30	0.38	0.40	11.38
4. China	0.29	0.31	0.31	8.87
5. Nepal	0.14	0.15	0.16	4.63
World + (Total)	3.14	3.34	3.50	

Source : FAO (2010)

Labour Management

Human labour is an important input in farm production irrespective of type of technology and level of mechanization used on the farm. Human labour is employed for cleaning, disinfection, offering feed, milking other routine and/or periodic dairy farm practices. The farm labour force in India has been classified into following categories.

i) **Farmer's own labour:** Indian farmer performs multidimensional role as he is the owner, manager and labourer. Due to the personal interest this is best type of labour though it is unpaid labour.

ii) **Family labour:** Family labour consists of various kind of farm work done by the farmers spouse and his children. This is the unpaid labour force on a small farm.

iii) **Hired labour:** When family labourers are not sufficient or not available for performing farm operations then hired labourer is engaged. Hired labour is classified as under

a) **Casual labourer:** They are engaged as and when they are required.

b) **Permanent labourer or attached labourer:** They are hired for a year or more. They are paid wages in cash or kind or both. Wages are fixed at the time of employment and provided with addition facilities like housing, clothing etc.

c) **Contract labourer:** They are group of workers assigned with specific task to perform on a specific payment.

Further, they could be skilled, semi-skilled and unskilled labourer. Skilled labourers are those who have gone through specialized training to carry out certain specific job. A semi-skilled labourer is those who have not gone through any training but has gained the experience by working under some experienced worker. Unskilled labours are those who are required to perform non-specific day to day work.

Dairy farming is a labour intensive enterprise as compared to crop production and other allied enterprises. Some routine activities like milking, heat detection, care of advanced pregnant animals and calf etc requires experienced and devoted workers. The dairy farmer who hires labourer must be skilful in managing people. On a large dairy farm half of the workers should have earlier experience of working at a dairy farm and 2-3 workers should be trained for the specialized jobs at the dairy farm. In addition to these, at large dairy farms technically qualified persons (farm manager, veterinarian) related with dairying will have to be hired.

Farm labour efficiency: Farm labour efficiency refers to the amount of work performed by a worker on the farm per unit of time or it is the amount of output per unit of labour. The labour efficiency of a dairy farm is measured by considering amount of milk produced per labourer or number of cows maintained by per labourer. Labour efficiency (returns per labour per day) of a farm can be estimated by subtracting cost of all input except labour from gross income, divided by number of workers in days.

Farm Record and Record Keeping

A record is an accounting of events or data. Actually, records do very little except to remind one of an event. Records must be processed or summarized in order to provide information. Good records are needed to serve as the basis for sound management decisions. The dairy farmer must have records for evaluation, planning, coordination and execution of farm activities. Further, periodic analysis of the business is necessary to make progress that will result in increased net income. Recording can be done most easily if animals have some form of identification. Thus, animal recording and identification are inseparable. Good farm records should contain information of essential parameters, must be easy and simple to understand and should have systematically arranged continuous (no missing data) information.

Advantage/ Utility of Keeping Farm Records

- Assist in farm planning and farm business analysis.
- Can provide data to scientist for research purpose.
- Can be used for performance evaluation.
- Help in scientific management of available resources.
- Can assist in enhancing production/income of farm.
- Provide data to extension workers for giving feedback information from farmers to scientist and vice-versa.
- Provides data to breeder for selection and breeding programme.
- Provides data to policy makers for future planning.
- Helps in reducing the cost of production.
- To adapt easily to government policies- rules for price fixation, getting subsidies etc.
- To furnish required information for getting credit.
- To know about productivity of farm in short duration of time.

Constraints in keeping farm records in India: There are many reasons due to which Indian dairy farmers don't maintain farm record. Some of them are

- Low level of literacy of farmers.
- Farmer plays role of owner, manager and laborers.
- Subsistence type of farming (producing for own consumption not for business purpose).
- Fear of taxation.

- Non availability of farm record book, lack of knowledge of record keeping etc.
- Small herd size.
- Mainly involve in crop production, less emphasis on livestock rearing.

Records can be maintained in many ways (in books with permanent leaves, loose leaves, books, envelops and computer etc). Following kinds of records can be maintained at a livestock farm:

- History card and pedigree sheet.
- Production & reproduction records of individual cows and of the herd.
- Feed use records.
- Cattle breeding register.
- Calving register.
- Calf register.
- Growth record of young animals.
- Herd health register.
- Cow identification records.
- Daily livestock register.
- Financial records (entire farm and enterprise) etc.

Selection and Culling of Dairy Animals

Desirable dairy animals are selected for breeding and production purposes based on their physical appearance, health, milk production and pedigree record. Physical appearance i.e. conformation & dairy character is known as type and good dairy type cow produces more milk. Health records include health of the individual animal as well as the health of the herd and history of immunisation. Milk production records include past performances which can be evaluated from cow's own past performance or the possible production of the off spring of the cow or production records of the daughters of the young bulls. Pedigree i.e. record of the ancestors, provides most important information for selection.

Culling is the process of elimination of undesirable animals from the herd. Animals are culled from a herd due to various reasons like poor production, reproduction performance, very poor condition, stunted growth, suffering from incurable diseases, positive animals for serious infectious diseases (Tuberculosis, Johnes disease, Brucellosis etc), loss of one or more quarters/ teats, undesirable breed characters, disabled animals, very old animals, male animals or other animals surplus on the farm or not useful in the farm and calves born with congenital defects etc. Animals are culled once or twice a year. In an organized herd culling rate (total culled/total cows ×100) should be less than 20 per cent, and animals culled due to diseases, production and other reasons should be less than 5, 10 and 5 per cent, respectively.

Preparation of Animal for Show

Showing dairy animals provide an opportunity to the farmer to exhibit their best animal, brings fame to the farmer and popularizes best breed as it act as an advertising media. It provides a platform where comparison among superior animals within and between breeds can be made. It further make possible to exchange experiences among the breeders, chance to learn about better feeding and management practices. It provides an opportunity to sell the limited number of animals. It helps in building a better competitive spirit among the participants.

The process of getting animal ready for the show is known as fitting. It enhances the features of the animal and includes training and grooming of the show animal.

1. **Selection of show animal:** Animals having true breed characteristics, better dairy conformation, proper growth & development as per the age of the animal, good temperament and no physical defects should be selected. While selecting young animals for different age group category show (6 months, 6 months to 1 year, above 1 year) best animal approaching towards upper age limit should be selected. Recently calved animals should be preferred over cows in their advance lactation.

2. **Feeding:** Similar type of feed mixture should be provided to the animal during fitting period and at the show. Feeding should be done in such a manner that the animal should be in good condition at the time of show and they should not be too thin or fat. Linseed meal is fed to the show animal as it enhances the gloss of hair and hide quality.

3. **Training:** First make the animal familiar to the use of halter then begin training to lead by walking the animal to walk

slowly with short steps and to hold its head in clockwise motion initially for short periods and gradually move up to longer periods of time. The animal should be led by the exhibitor from the left side of the animal at its head, holding the halter with the right hand, and close enough to have secure control of the animal. Further, animal must be taught to set-up or pose by giving it a signs by putting pressure on the point of its shoulder with the hand. While posing the front legs of the animal should be squarely placed and the hind leg nearest to the judge should be positioned slightly further back than other rear leg.

4. **Care of hooves:** A too long and bad shaped hoof affects the appearance and the way animal walks. Hence, hooves should be trimmed at least two weeks before the show.

5. **Care of horns:** There is no bias between horned and dehorned animal till the head illustrates true breed character of the animal. If horns are present, its surface should be made smooth and symmetrical with the help of wood file and scrapper etc.

6. **Clipping:** Generally head, neck, tail, and in some cases the belly area should be clipped. The entire head region should be clipped by beginning at the nose then neck to the withers, and down to the point of the shoulder. Tail is clipped from a point just above the switch to the rump. Enough hair is left on the tail setting so that rump looks like long and level. Clipping of udders of cows or springing heifers is also a desirable practice. The belly of a cow should be clipped to show the milk veins. Clipping of belly in case of young heifers is not recommended. Animals should be clipped not more than 3 or 4 days before showing.

7. **Grooming:** Animals should be groomed once daily for a month which will make hair glossy and hide loose & pliable. Currycomb should be used to remove excess dirt on the legs and flanks only and a soft bristled brush for other body parts. After brushing, smooth the hair lightly with a flannel cloth moistened with oil. A light mineral or vegetable oil is best.

8. **Washing:** Generally showman wash the animal frequently when fitting is started. Animal should be washed with mild soap thoroughly. Be sure to wash under the flank. Ears should

be cleaned with a towel and water or rubbing alcohol, to remove wax build-up and dirt. Animal should be tied in the sun for about an hour after washing.

9. **Blanketing:** Washing and blanketing an animal should be started as soon as she is put in the fitting quarters. Blanketing speeds up the shedding of long hair and makes the hide soft and pliable. It also protects the new coat of hair grown under the blanket. Once blanketing has been started it should be continued till the show is over. Blankets made of burlap are inexpensive and satisfactory. An old woolen or cotton blanket can also be used. When weather is extremely hot, blanket should be removed during the daytime.

10. **Showmanship:** Following guidelines should be followed during show

    - Before entering the show ring, the exhibitor and the animal should look their best.
    - The exhibitor should be dressed in a clean white shirt and pants.
    - The animal should be clean, dry and fitted for the show.
    - Use a halter that fits properly and correctly when placed on the animal. A leather halter with leather lead strap is best.
    - External applications of cosmetics that affect only appearance may be used (hoof polishes and false switches).
    - The maximum permissible length of naturally growing hair anywhere on the topline should not exceed 1 inch.
    - When entering the show ring, keep the head of the animal held up, move the animal at slow and steady pace for impressive style, attractive carriage and graceful walk.
    - Enough space should be allowed between animal and exhibitor so that the judge can clearly view his or her animal from the side.
    - It is important to keep eye contact with the judge at all times, watching for signals from the judge and respond rapidly to requests from the judge and officials.
    - Pose the animal with feet placed squarely and the hind leg nearest to the judge slightly behind the other one. When

posing cows in milk, the hind leg nearest the judge should be slightly ahead of the other one.

- Be natural and avoid over showing, undue fussing and maneuvering.
- Keep showing until the entire class has been placed and the judge has given his or her remarks.
- Criticizing/challenging/threatening/interfering with the judge, show management or other exhibitors while in the show ring or other conduct should not be done.

Judging of Dairy Animals

Judging of dairy animals is a process of comparing dairy cows with an ideal dairy type. This is done with the help of Dairy Cow Unified Score Card developed by Purebred Dairy Cattle Association (PDCA). This score card describes general traits of a dairy cow and is same for all dairy cattle breeds. It also includes breed characteristics and shows body parts of a dairy cow. Some points have been assigned to each divisions of the score card and in an ideal cow it would add up to 100. These divisions on the score card are- frame, dairy strength, rear feet and legs, and udder.

Procedure

- An animal or group of animals are viewed first from a distance of 20 to 30 feet (6 to 9 m).
- The animals should be standing on the level ground or with the front feet slightly higher than the rear.
- First impressions are usually more accurate so make written notes on each animal as they are viewed on the first inspection.
- Head, chest and front legs should be observed from front view.
- Barrel, loin, hips, pin bones, tail head, udder and legs are observed from the rear view.
- Topline, rump, barrel, heart girth, shoulders, udder, teats, flank, legs and neck is observed from side view.
- Use Dairy Cow Unified Score Card to compare each animal.
- If there is an easy top or bottom animal or top and bottom pair in the class, make note of it.

- During judging animals are usually walked slowly around the judge in a clockwise direction. Style, carriage, straightness of legs and topline, udder attachment, and blending of parts are observed while they are walking.
- After observing the animals from a distance, go for a closer inspection and observe the shape of the withers, quality of hair, mellowness of hide, texture of udder and development of the mammary veins.
- A class of four animals is compared by pairs. That is, the top pair, middle pair, and bottom pair and when a pair is close on score, decision can be made on the mammary system score. If the mammary system score is equal in the pair, dairy character score can be used to determine the placing. In a close pair, when a cow is dry and the other milking, milking cow is ranked above the dry cow.
- Look out individual animals for any defects as described in the Dairy Cow Unified Score Card. The influence of the defect on the placing ranges from disqualification to slight discrimination depending on the degree of seriousness. For example, permanent lameness will put a cow at the bottom of the class regardless of other traits she may have.

DAIRY COW UNIFIED SCORECARD

	MAJOR TRAIT DESCRIPTORS	Perfect Score
1.	**Frame:** The skeletal parts of the cow, with the exception of rear feet and legs. Listed in priority order, the descriptions of the traits to be considered are as follows	**15**
a.	**Rump (5 points):** Should be long and wide throughout. Pin bones should be slightly lower than hip bones with adequate width between the pins. Thurls should be wide apart. Vulva should be nearly vertical and the anus should not be recessed. Tail head should set slightly above and neatly between pin bones with freedom from coarseness.	
b.	**Front End (5 points):** Adequate constitution with front legs straight, wide apart and squarely placed. Shoulder blades and elbows set firmly against the chest wall. The crops should have adequate fullness blending into the shoulders.	
c.	**Back/Loin (2 points):** Back should be straight and strong, with loin broad, strong and nearly level.	
d.	**Stature (2 points):** Height including length in the leg bones with a long bone pattern throughout the body structure. Height at withers and hips should be relatively proportionate. Age and breed stature recommendations are to be considered.	

Contd.

e.	**Breed Characteristics (1 point):** Exhibiting overall style and balance. Head should be feminine, clean-cut, slightly dished with broad muzzle, large open nostrils and strong jaw.	
2.	**Dairy Strength:** A combination of dairyness and strength that supports sustained production and longevity. Major consideration is given to general openness and angularity while maintaining strength, width of chest, spring of fore rib and substance of bone without coarseness. Body condition should be appropriate for stage of lactation. Listed in priority order, the descriptions of the traits to be considered are as follows:	**25**
a.	**Ribs (8 points):** Wide apart. Rib bones wide, flat, deep and slanted towards the rear. Well sprung, expressing fullness and extending outside the point of elbows.	
b.	**Chest (6 points):** Deep and wide floor showing capacity for vital organs, with well-sprung fore ribs.	
c.	**Barrel (4 points):** Long, with adequate depth and width, increasing toward the rear with a deep flank.	
d.	**Thighs (2 points):** Lean, incurving to flat and wide apart from the rear.	
e.	**Neck (2 points):** Long, lean, and blending smoothly into shoulders; clean-cut throat, dewlap and brisket	
f.	**Withers (2 points):** Sharp with chine prominent.	
g.	**Skin (1 point):** Thin, loose and pliable.	
3.	**Rear Feet and Legs:** Feet and rear legs are evaluated. Evidence of mobility is given major consideration. Listed in priority order, the descriptions of the traits to be considered are as follows:	**20**
a.	**Movement (5 points):** The use of feet and rear legs, including length and direction of step. When walking naturally, the stride should be long and fluid with the rear feet nearly replacing the front feet.	
b.	**Rear Legs-Side View (3 points):** Moderate set (angle) to the hock.	
c.	**Rear Legs-Rear View (3 points):** Straight, wide apart with feet squarely placed.	
d.	**Feet (3 points):** Steep angle and deep heel with short, well-rounded closed toes.	
e.	**Thurl Position (2 points):** Near central placement between the hip and pin bones.	
f.	**Hocks (2 points):** Adequate flexibility with freedom from swelling.	
g.	**Bone (1 point):** Flat and clean with adequate substance.	
h.	**Pasterns (1 point):** Short and strong with some flexibility, having a moderate, upright angle.	
4.	**Udder:** The udder traits are evaluated. Major consideration is given to the traits that contribute to high milk yield and a long productive life. Listed in priority order, the descriptions of the traits to be considered are as follows:	**40**

Contd.

a.	**Udder Depth (10 points):** Moderate depth relative to the hock with adequate capacity and clearance. Consideration is given to lactation number and age.
b.	**Rear Udder (9 points):** Wide and high, firmly attached with uniform width from top to bottom and slightly rounded to udder floor.
c.	**Teat Placement (5 points):** Squarely placed under each quarter, plumb and properly spaced.
d.	**Udder Cleft (5 points):** Evidence of a strong suspensory ligament indicatedby clearly defined halving.
e.	**Fore Udder (5 points):** Firmly attached with moderate length and ample capacity.
f.	**Teats (3 points):** Cylindrical shape; uniform size with medium length and diameter; neither short nor long is desirable.
g.	**Udder Balance and Texture (3 points):** Udder floor level as viewed from the side. Quarters evenly balanced; soft, pliable, and well collapsed after milking. (In the Holstein breed, an equal emphasis is placed on fore and rear udder (7 points each). All other traits are the same as listed above.)

Source : Purebred Dairy Cattle Association (Copyrighted) 1943. Revised and copyrighted 1957, 1971, 1982, 1994 and 2009.

Breed Characteristics: Except for differences in colour, size, and head character, all breeds are judged on the same standards as outlined in the Unified Score Card. If any animal is registered by one of the dairy breed associations, no discrimination against colour or colour pattern is to be made.

Ayrshire: Strong and robust, showing constitution and vigor, symmetry, style and balance throughout, and characterized by strongly attached, evenly balanced, well-shaped udder.

- Head- clean cut, proportionate to body; broad muzzle with large, open nostrils; strong jaw; large, bright eyes; forehead, broad and moderately dished; bridge of nose straight; ears medium size and alertly carried.
- Colour- light to deep cherry red, mahogany, brown, or a combination of any of these colours with white, or white alone, distinctive red and white markings preferred.
- Size- a mature cow in milk should weigh at least 1200 lbs.

Brown Swiss: Strong and vigorous, but not coarse. Adequate size with dairy quality. Frailness undesirable.

- Head- clean cut, proportionate to body; broad muzzle with large, open nostrils; strong jaw; large, bright eyes; forehead, broad and slightly dished; bridge of nose straight; ears medium size and alertly carried.

- Colour- body and switch solid brown varying from very light to dark; muzzle has black nose encircled by a white ring; tongue and hooves are dark brown to black.
- Size- a mature cow in milk should weigh at least 1400 lbs.

Guernsey: Strength and balance, with quality and character desired.

- Head- clean cut, proportionate to body; broad muzzle with large, open nostrils; strong jaw; large, bright eyes; forehead, broad and slightly dished; bridge of nose straight; ears medium size and alertly carried.
- Colour- shade of fawn and white markings throughout clearly defined.
- Size- a mature cow in milk should weigh 1200-1300 lbs.; Guernsey does not discriminate for lack of size.

Red & White: Rugged, feminine qualities in an alert cow possessing adequate size and vigour.

- Head- clean cut, proportionate to body; broad muzzle with large, open nostrils; strong jaw; large, bright eyes; forehead, broad and slightly dished; bridge of nose straight; ears medium size and alertly carried.
- Colour- must be clearly defined red and white; black-red and brindle is strictly prohibited.
- Size- a mature cow in milk should weigh at least 1400 lbs. and be well balanced.

Holstein: Rugged, feminine qualities in an alert cow possessing Holstein size and vigour.

- Head- clean cut, proportionate to body; broad muzzle with large, open nostrils; strong jaw; large, bright eyes; forehead, broad and moderately dished; bridge of nose straight; ears medium size and alertly carried.
- Colour- black and white or red and white markings clearly defined
- Size- a mature cow in milk should weigh at least 1400 lbs.

- Udder- equal emphasis is placed on fore and rear udder (7 points each), all other traits are the same as listed on the PDCA scorecard.

Jersey: Sharpness with strength indicating productive efficiency.

- Head- proportionate to stature showing refinement and well chiselled bone structure. Face slightly dished with dark eyes that are well set.
- Colour- some shade of fawn with or without white markings; muzzle is black encircled by a light colour ring; switch may be either black or white.
- Size- a mature cow in milk should weigh at least 1000 lbs.

Milking Shorthorn: Strong and vigorous, but not coarse.

- Head- clean cut, proportionate to body; broad muzzle with large, open nostrils; strong jaw; large, bright eyes; forehead, broad and slightly dished; bridge of nose straight; ears medium size and alertly carried.
- Colour- red or white or any combination (no black markings allowed).
- Size- a mature cow in milk should weigh 1400 lbs.

Factors to be Evaluated

The degree of discrimination assigned to each defect is related to its function and heredity. The evaluation of the defect shall be determined by the breeder, the classifier or judge, based on the guide for discrimination and disqualifications given below.

- **Horns:** No discrimination for horns.
- **Eyes:**
  - Blindness in one eye: Slight discrimination.
  - Cross or bulging eyes: Slight discrimination.
  - Evidence of blindness: Slight to serious discrimination.
  - Total blindness: Disqualification.
- **Wry Face:** Slight to serious discrimination.
- **Cropped Ears:** Slight discrimination.
- **Parrot Jaw:** Slight to serious discrimination.

- **Shoulders-**Winged: Slight to serious discrimination.
- **Capped Hip:** No discrimination unless affects mobility
- **Tail Setting:** Wry tail or other abnormal tail settings: Slight to serious discrimination.
- **Legs and Feet**
  - Lameness- apparently permanent and interfering with normal function: Disqualification.
  - Lameness- apparently temporary and not affecting normal function: Slight discrimination.
  - Evidence of crampy hind legs: Serious discrimination.
  - Evidence of fluid in hocks: Slight discrimination.
  - Weak pastern: Slight to serious discrimination.
  - Toe out: Slight discrimination.
- **Udder**
  - Lack of defined halving: Slight to serious discrimination.
  - Udder definitely broken away in attachment: Serious discrimination.
  - A weak udder attachment: Slight to serious discrimination.
  - Blind quarter: Disqualification.
  - One or more light quarters, hard spots in udder, obstruction in teat (spider): Slight to serious discrimination.
  - Side leak: Slight discrimination.
  - Abnormal milk (bloody, clotted, watery): Possible discrimination.
- **Lack of Adequate Size**
  - Slight to serious discrimination. (Note: Guernsey does not discriminate for lack of size.)
- **Evidence of Sharp Practice**
  - Animals showing signs of having been tampered with to conceal faults in conformation and to misrepresent the animal's soundness: Disqualification.

- Uncalved heifers showing evidence of having been milked: Slight to serious discrimination.

- **Temporary or Minor Injuries:** Blemishes or injuries of a temporary character not affecting animal's usefulness: Slight to serious discrimination.
- **Over conditioned:** Slight to serious discrimination.
- **Freemartin Heifers:** Disqualification.

Economics of Dairy Farming

India is the largest milk producing country in the world and there is consistent rise in India's share in world milk production over the years. In contrast to the developed countries, in India milch animals are maintained mainly by the landless, marginal and small farmers in smaller herd size across the length and breadth of the country. Approximately 70 million rural households are engaged in milk production. These animals play multidimensional role and acts as a tool in achieving nutritional and economic security. In the recent years, this sector has emerged as an important source of rural employment and income in the country.

Successful and economic dairy business depends on the comprehensive knowledge and experience in the various facets of dairy farming like economic use of resources, feeding, breeding, general management of the herd and efficient disposal of products, etc. Further, to find out the means to reduce the cost and increase the profit, to fix the sale price of milk, to cull the unproductive cows and to encourage the farmers to practice scientific dairy farming, economic analysis of the business is essential. For the purpose of understanding of the various livestock projects of different species students may refer NABARD's model bankable projects (*http://www.nabard.org/modelbankprojects/modelbankprojects.asp#*).

Chapter - 7

Goat Production

Rana Ranjeet Singh

Origin and Domestication of Goat

Goat is a cloven-hoofed, hollow-horned ruminant which is comparatively smaller in size and belongs to class Mammalia, order Artiodactyla, family Bovidae and genus either *Capra* (2N=60) or *Hemitragus/tahrs* (2N=48). The distinction between these two genera was previously based on horn structure but the distinction has been confirmed genetically. The domestic goat (*Capra hircus/Capra aegagrus hircus*) belongs to the genus *Capra* and has possibly developed from five wild ancestors. On the basis of available information, it is most likely that bezoar is the main ancestor of most domestic goats. However, there is slight likelihood that Markhor has been involved in the parentage of some of the Indian goat breeds.

Evolutionary biology indicates that domestication of goats started during Neolithic age and it was perhaps the first ruminant and first animal after dog to be domesticated by the man around 9000-7000 BC. Domestication was associated with three of the oldest civilizations namely- (i) the Nile in northeast Africa, (ii) the Tigris-Euphrates in west Asia, and (iii) the Indus in the Indian subcontinent. Archaeologists indicate that the goat was first domesticated in the "Fertile Crescent" (landmass which extends between the Black and Caspian seas along the coast of Palestine and forms arc like a quarter moon towards the

Persian Gulf) of the eastern Mediterranean. Since their domestication, goats remained associated with variety of functions in various aspects of human life including food, employment, economics, folklore and religion.

Phylogenetic Classification of Goat

Kingdom:	Animalia
Phylum:	Chordata
Class:	Mammalia
Order:	Artiodactyla
Sub-order:	Ruminantia
Family:	Bovidae
Sub-family:	Caprinae
Tribe:	Caprini
Genus:	• *Capra* (true goats)
	• *Hemitragus* (true goats)
	• *Ovis* (Sheep)
	• *Ammotragus* (Goat-like sheep)
	• *Pseudois* (Sheep-like goats)
Species:	*a) Capra hircus* , including the Bezoar (*c.h.aegagrus*)
	b) Capra ibex (Ibexes)
	c) Capra caucasica (Caucasian tur)
	d) Capra pyrenaica (Spanish ibexes)
	e) Capra falconeri (Markhor)

Source : Devendra & McLeroy (1990), Payne & Wilson (1999) and Solaiman (2010).

Characteristics of Goat

Goats are well adapted to a wide variety of ecological zones ranging from hot and cold dry areas to humid tropical zones. Due to their narrow mouth parts they selectively feed on highly nutritive plants parts thereby causes little damage to the plant structure. Goats have ability to replenish lost water rapidly and may drink up to 45% of body weight without haemolysis. Goats sweat and pant less than sheep. Goats have often angular body with long legs in the arid regions and short legs in the humid regions. Goats are multi-purpose animals and are reared for meat, milk, skin and hair. In developed countries, goats are valued mainly for their milk, fiber, and meat; while in the developing countries, they are valued mainly for meat, followed by milk, fibre, and skins. The external features which separate the goat from sheep are the presence of beard, body remains covered with hair, has upright tail. Bucks have odoriferous tail glands. Further, they lack face glands, lachrymal pits in the skull, foot glands at least in the hind feet and the heteronymous type of horns.

Goat Population and Distribution

In the year of 2010, total world population of goats was about 909.85 million. Goats are distributed over all types of ecology and are more abundant in the tropics, dry zones and developing countries. The largest goat population is in Asia (537.77 million) followed by Africa (312 million). Collectively, these two continents accounted for about 93% of the world's total goat population (Table 7.1). Distribution of goats among different countries of the world and Indian states has been discussed in detail in the chapter one of this book.

Table: 7.1 : Global distribution of goat population (in million).

Region/Year	2006	2008	2010	% of world population (2010)
Oceania	4.20	3.62	4.92	0.54
Europe	17.89	17.82	17.12	1.88
Americas	38.07	37.51	37.60	4.13
Africa	284.98	301.22	312.45	34.34
Asia	499.14	516.92	537.77	59.11
World	844.27	877.09	909.85	

Source : FAOSTAT (2010).

Advantages of Goat Farming

Goats provide products and services important for man throughout the world. Goat raising is one of the important livestock enterprise especially in rural areas of the country. It is known as 'poor man`s cow' in India and as 'wet nurse' of infants in Europe. Followings are the advantages of goat farming in India

1. The goat is a versatile animal.
2. Ability to reproduce at shorter age.
3. Higher chances for twinning or triplets in kidding.
4. They have tremendous ability to thrive on sparse vegetation unsuitable for feeding of other livestock species.
5. They have high adaptability to varying environmental conditions.

6. Goats can be raised profitably with low investments under intensive and extensive system.
7. Goat rearing requires low investment hence suitable for landless labourers, industrial workers and marginal farmers.
8. Lesser land and capital requirement.
9. Lesser amount of feed and concentrate is required.
10. No elaborate shelter is required.
11. Unique ability to digest lower quality feedstuffs.
12. It is a quickly maturing animal with a high fertility.
13. Goat milk is cheap, wholesome, nutritious and easily digestible (small fat globules).
14. Goat milk (alkaline) is recommended for the patients suffering from dyspepsia, peptic ulcer, allergic eczema and pyloric stenosis and is preferred over cow milk in jaundice, biliary disorders, acidosis and insomnia.
15. Goat milk is suitable for preparation of various milk products.
16. Goat raising gives more production per unit of investment.
17. The goat is a small animal hence can be managed by women and children easily.
18. Well established market for goat meat is available.
19. Goat meat has no religious taboo and is relished by all sections of the society.
20. Hairs of goats are used for rope, rugs and hide is used for leather products.
21. Pashmina is used in the production of fabrics.

Disadvantages of Goat Farming

1. More vulnerable to predators and thieves.
2. Small value often makes formal credit systems uneconomical.
3. Small value makes formal insurance systems difficult to manage.
4. Prone to broncho-pneumonia.
5. Vulnerable to internal parasites.
6. Often accused for ecological degradation & desertification.

Scope of Goat Farming in India

Under livestock production systems, the importance of a livestock species increases in relation to its adaptability, scope for increasing productivity, capacity to make socio-economic contributions and competence to fill market opportunities. In India, goats are important part of rural economy particularly in the arid, semi-arid and mountainous regions. India is a rich repository of indigenous goat germplasm of varying production potential. Chevon is one of the most preferred meats in India. Further, under the influence of drivers of development change, there is increase in the demand for goat and sheep meats in domestic as well as international markets (developing counties). However, the productivity of goats under the existing traditional production system is very low as they are maintained under the extensive system on natural vegetation on degraded common grazing lands and tree lopping. Additionally, adoption of improved management practices by farmers' is very low. Recently, goat production system in the country has been slowly moving towards more intensive system of management for commercial production. Hence, in order to exploit their full production potential and to cater the demand of the market it is imperative to rear them under intensive and semi-intensive system using improved technologies for commercial production.

Goat Production Systems

Goat production systems are based on pattern of agriculture, availability and use of land. It can be broadly classified into (i) extensive, (ii) intensive and (iii) semi-intensive system.

i) **Extensive system:** In this system, goats are allowed to graze over large areas which are not suitable for other agricultural use. This is perhaps the most popular system of goat production. Goats are taken out for grazing by women and children in the morning and brought back in the afternoon. Hence, a very low level of unpaid family labour is the main input. However, for managing larger flocks generally some hired labour is engaged. This system requires accessibility to ample marginal land, availability of various browse plant species and characterized by low rainfall. Nomadism and transhumance are examples of extensive production systems.

ii) **Intensive system:** In this system, goats are fed in confinement with limited opportunity for grazing. This system of goat rearing is followed in the areas where land is mostly used for perennial cropping and grazing land is available during inter-cropping periods. Cut and carry system of top feeds is also practiced in this system. The major advantages under this system are- it allows efficient conversion of feed resources, better control of goats, lesser deterioration to the environment and efficient use of family labour.

iii) **Semi-intensive system:** In this system, goats are gazed for few hours daily and also supplied additional roughage and concentrate in the stall throughout the year. Goats are allowed to graze/browse on crop residues after harvesting. This practice results in increased fertility of soil, increased crop yield, good growth rate and easier management of the flock. This system is suitable for small to medium flock size and is favoured where labour is limited for intensive production.

Tethering is a modified form of this system. Tied to a peg or fence line and tied to a ring on a wire between two pegs are the two common methods of tethering. In the former method, 2-3 goats are tied together by a rope usually 3-5 meters in length and another end of the rope is tied to the tree/post/peg. Length of the rope represents the available grazing area and by shifting the peg or changing the post/tree new grazing area is provided. This method generally provides a circular area for grazing. In later method, a rope/wire is tied between two pegs and a ring is inserted to this rope. A sufficiently long neck rope is fastened to this ring which slides over the long rope between two pegs. Goat tied in this manner can graze either side of the wire in parallel line and thus provides a rectangular area for grazing.

Further, in some countries, goats are integrated with plantation crops (rubber, oil palm, coconuts) as extensive, intensive or semi-intensive system.

Breeds of Goats

There are about 570 goat breeds in the world (Devendra and Liang, 2012). These breeds are usually classified based on various parameters like their geographical origin, their region & productivity, their height at the withers, shape and size of the horns in conjunction with production & geographical considerations, primary product of goats & their distribution, their morphological characteristics and hair & coat characteristics. In India, there are 23 breeds of goat. According to the production type, they have been classified into dual purpose (Jhakrana, Beetal, Jamunapari, Barbari and Surti), meat purpose (Bengal, Ganjam, Malabari, Osmanabad, Sangamneri, Kutchi, Zalawadi, Gohilwadi,

Mehsana, Sirohi, Konkal Kanyal, Berari and Kannai Adu) and fibre purpose (Gaddi, Marwari, Changthangi and Chegu) breeds. Further, on the basis of body size these breeds have been classified into large, medium and small type and about 80 per cent of the breeds are medium in size. These breeds have been discussed in detail in chapter 3 of this book.

Housing of Goats

Goats do not require elaborate housing designs and materials. However, there must be provision of ample light, good ventilation and drainage facility. Moreover, it must provide goats a dry, comfortable, safe, place free from worms and must protect them from adverse environmental conditions. Different types of housing and various space requirements have been discussed in detail in chapter 5 of this book.

Feeds and Feeding of Goats

Goats usually accept a wide variety of feeds however at times their feed preferences may differ considerably among different individual animals. They can travel long distances in search of their food and usually refuse wet feeds and soiled forages. Goats have relatively lesser rumen volume to the body weight than other grazing animals. Further, they are essentially browsers who prefer to feed mainly on the leaves, flowers, fruits & twigs of shrubs. Their narrow mobile muzzle and prehensile tongue help them to select more nutritive plants parts even in the event of their short supply. However, in absence of these highly nutritive plants parts they eat coarser and less nutritive feeds. They are known to assume a bipedal stance and reaches up to 1-1.5 meters above the ground in order to get their preferred feed. Goats have the ability to differentiate among bitter, sweet, salty and sour tastes. Goats have higher tolerance to the wide variety of undesirable phyto-chemicals. Goats appear to be more efficient in digesting coarse feeds than other ruminants (sheep & cow) and can sustain water deprivation for longer time.

Feed resources: Goat feeds can be broadly divided into two categories i.e. forage/ roughage and concentrates. Roughage comprises of both green roughage (pastures, green chop, tree leaves, herbs, shrubs, climbers, vegetable wastes, pods, silages and cultivated fodders) and dry roughage (hay, pelleted forage, some by-products like straw, stovers and hulls). The ideal ratio of roughage to concentrate in their diet depends on factors like physiological status and the desired level of production. Goats managed for milk production owing to the higher nutritional requirements are usually fed higher amounts of grain than goats managed for meat production. Feed resources of goats may be classified into following categories

1. **Range land forage:** Majority of goat's feed comes from natural vegetation on common grazing lands, range land and other non-cultivable areas. Most of these areas are of poor to very poor type and further high cell wall and silica content of the vegetations from these areas lowers dry matter intake and digestibility. Moreover, most of the nutrients produced can't be utilized fully as their production is limited to rainy season and in inaccessible areas and majority of these biomass produced is being utilized by large ruminants and goats have to thrive on leftover vegetations. Grazing resources in India have been shown in Table 7.2.1. The common perennial grasses like Dub, Motha, Blue panic, Anjan, Sewan, Karad, perennial legumes such as *Dolichos lablab, Clitoria temata, Phaseolus micropteliu*m (Siratro) and *Stylosanthes* and weeds like Jharberi, Ghiabati, Punarnnava, Gokhru and Morara etc are browsed by goats.
2. **Top feeds:** Leaves of large variety of trees serves as potential feed resource for goats in the tropics. Farmers provide tree leaves together with cut grass or grazing. Usually tree leaves contains 20-40 % dry matter 4-15% DCP and 50-60% TDN and many of them are known to contain anti-nutritional factors which reduces the palatability, digestibility and at times causes metabolic disorders. Some of the common tree leaves which are fed to the goats are- Ipil-ipil, Babul, Bargad, Gular, Pipal, Pakar, Ber, Oak, Banana, Gliricida, Mulberry, Jackfruit, Cassava, Khanthal etc.
3. **Cultivated fodders:** Depending on type of soil, rainfall, availability etc a number of grasses and legumes can be grown for goats. Owing to the intensive agriculture and small holding in some areas it is imperative to rear them under stall feeding condition. Generally, goats tend to prefer less coarse grasses. The leguminous and non-leguminous fodders can be fed in the ratio of 1:1. Leguminous fodders like berseem, lucerne, cowpea and guar etc; cereal fodders like oat, maize, barley, bajra and sorghum etc; grasses like Guinea grass, Paragrass, Elephant grass, Star, Napier and Pangola grass are suitable for pasture and fodder production for goats.

4. **Post harvest feeds, agro-industrial, kitchen and food processing by-products and crop residues:** It may include potato vines, bean stalks, banana leaves & stems, cakes and meals from oil milling industry, rice, maize and wheat bran, dried brewers grain, mango, banana and potato peels, gram, arhar, wheat, rice straw, ground nut stalks. Various forms of chemical, biological or mechanical treatment can be practiced to improve the nutritive value of poor qualities crop residues.

5. **Grains and Concentrates:** Usually bulk of required nutrients of goat often come from forage but when forages are not available or are of poor quality, necessitates provision of supplements (grains or concentrates) in the diet of goats. Concentrate feeds are low in fiber but rich in energy, protein, or both. Energy concentrates contain less than 20% protein, less than 18% fiber and include all cereal grains, some root crops, flourmill and bakery by-products and many other food manufacturing by-products such as beet pulp and citrus pulp.

Protein concentrates usually contain at least 20% crude protein and includes cakes and meals from oil milling industry (groundnut, cottonseed, peanut, soybean cakes etc), alfalfa hay/meal/cubes. Urea can be excellent, cost-effective source of nitrogen and the amount of urea to be fed should not be more than 25% of the required protein in rations for milking does. Mineral supplements can be as simple as salt only or may be a complex mixture comprising of several macro and micro-minerals and vitamins, provided free choice in the form of either a loose mixture or a block.

Nutrient Requirement of Goats

1. **Water requirement:** Water is as essential to goats as to the other livestock species. Actual requirements are affected by factors like breed, climate, nature of feed, dry mater intake, ambient and drinking water temperatures, frequency of drinking, type and level of production (milk, growth, etc). They are more efficient user (low rate of water turnover per unit body weight) of available water. It is generally recommended that dry matter to water intake should be 1 to 4 (4 kg water for 1 kg feed dry matter). More water is required at higher ambient temperature and for lactating animals (an additional 2 liters of water per liter of milk produced). They should be provided with clean water *ad lib* and daily actual water intake may vary

from nothing to several litres. Goats are more sensitive to foul tasting water.

2. **Dry matter intake (DMI):** The utilisation of the available nutrients is mainly determined by the amount of dry matter intake and the digestibility of the feed intake. Dry matter intake indicates the capacity of goats to utilise feed voluntarily and it depends on breed, digestibility and environment. Digestibility of a feedstuff depends on stage of maturity of the feed, botanical composition, dietary supplements, processing and chemical treatment. Dry matter intake for dairy goats ranges between 5-6 per cent of live weight in temperate region and 4-5 per cent in tropics. Meat type goats in tropics generally consume dry matter @ 3 per cent of the live weight.

3. **Energy & protein requirements:** Feeding strategies should be able to meet energy and protein needs as per the condition of the goats. Energy requirements for different physiological stages like maintenance, pregnancy, reproduction, lactation and growth vary. Further, increased activity, quality of pasture/range land, type of terrain, distanced travelled during grazing, requires additional amount of energy above the maintenance requirements. Besides the energy requirement of lactating does, pregnant does in their last two months of gestation and young kids, good quality roughage can supply the energy requirements of other categories of goats. Hence, concentrate should be incorporated in their diets. Energy requirements of different categories of goat have been shown in Table 7.2.2.

The level of protein in the diet affects voluntary feed intake. The protein requirements are higher during growth (kids), pregnancy, lactation and mohair production etc and protein deficiency can affect voluntary intake, growth rate and age at sexual maturity even if energy in the diet is adequate. Protein requirements of different categories of goat have been shown in Table 7.2.2.

4. **Minerals & Vitamin requirements:** Minerals and vitamins are essential for the maintenance as well as proper functioning of the physiological systems of goats and their deficiencies, excesses or imbalances exert a significant influence on goat's productivity. Some of the macro-minerals like calcium, phosphorus, magnesium, sodium, potassium, sulfur and chlorides are required in a goat's diet. Ratio of calcium to

phosphorus should be adjusted in the diet of milking does. Goats also require micro-minerals like iron, copper, cobalt, manganese, zinc, iodine, selenium, molybdenum etc. Goats maintained on quality pasture will get adequate quantity of vitamins in their diets. Vitamin supplements should usually be provided in the diet of lactating and young kids.

Grazing and Feeding Systems

1. **Stall feeding or zero grazing:** In this system of goat rearing goats are confined to the stalls and are not allowed to graze. The fodder could be grown on farm or collected from communal areas. This system is generally followed when goats are reared in and around cities and for crossbreds whose nutritional demands cannot be met merely by grazing on poor quality pasture. This system protects the animals from harsh environment, thieves and predators, controlled breeding can be achieved. Main disadvantages of this system are-it are a labour intensive, inappropriate for large flocks and goats are denied the opportunity to select their own feeds.
2. **Grazing:** It is mainly dependent on the availability of the grazing resources from pastures, forests, cultivable wastelands and fallow land. Grazing resources in India has been shown in Table 7.2.1. There are many systems of grazing (kindly refer chapter 4 of this book) under which livestock is allowed to graze grassland efficiently.

Table: 7.2.1 : Grazing resources in India (2000 -2001).

Resources	Area (Million Ha.)	Percentage
1. Forests	69.41	22.70
2. Permanent pastures, grazing lands	10.90	3.60
3. Cultivable wasteland	13.66	4.50
4. Fallow land	24.99	8.10
5. Fallow land other than current fallows	10.19	3.30
6. Barren uncultivable wastelands	19.26	6.30
7. Total common property resources other than forests	54.01	17.70
Livestock units /ha of common property resource	6.07	–

Source : GOI (2006).

Table: 7.2.2 : Average energy & protein requirement of goats.

Nutrient	Requirement
A. Energy	
i. For maintenance:	131 Kcal DE/$W^{0.75}$ kg
ii. For growth:	7.8 K cal/DE g gain
iii. For pregnancy:	219 Kcal DE/$W^{0.75}$ kg
iv. For milk production	1,675 Kcal DE/kg 4% FCM
B. Protein	
i. For maintenance:	2.4 g/kg $W^{0.75}$
ii. For pregnancy	6.1 g/kg $W^{0.75}$
iii. For milk production:	58.6 g/kg 4% milk

Source : Patnayak (1992).

Care and Management of Different Categories of Goats

Care and Management of Kidding Does

Few hours before parturition doe should be shifted to kidding pen. On a goat farm either individual kidding pens can constructed or/ a movable hurdle can be used to partition a house into many kidding pens. The pen should be cleaned and disinfected properly at regular interval. The first sign of impending kidding is restlessness, getting up and down, bleat in low voice. Other signs of kidding include- udder engorgement with milk, swelling of the vulva, the belly appears shrunk and the flanks appear somewhat hollow, raised tail head and appearance of water bag. Does should be given chance to care for herself before help/ corrective measures is provided *i.e. when* there is excessive delay in the expulsion of the kid after the bursting of the water-bag or delay of more than 30 minutes between the birth of kids. After kidding hind quarters of doe is washed with disinfectant and place is cleaned, disinfected and afterbirth is disposed properly. They should be given laxative diet with salt, mineral mixture and jaggery. After two days of kidding she should be provided with normal diet.

Care and Management of Kids

1. Immediately after birth clear the mouth and nasal passages of the kid and stimulate respiration by irritating the nasal passages with straw.

2. Allow the doe to lick her kid if required assist her in doing so.
3. The navel cord should be tied, cut and dip the navel cord in 7% tincture of iodine. Re-apply antiseptics to the stump if required.
4. Check the teats of dam for supply of milk if required place them with foster mother or feed them artificially.
5. Feed colostrum as early as possibly, ideally within 2-4 hours of birth. They should receive colostrum equal to 10% of their body weight during the first 24 hours of life.
6. Mortality in kids is highest during first week of life which may be due to the malpresentation of the foetus, weaker kid, dams having malformed udders, milk fever and accidents to the dams & kids. Therefore, more attention is required during this period so that proper corrective measures can be adopted to meet the challenges.
7. Give proper attention to the weaker kid from a multiple birth.
8. Before each suckling clean the udder with KMnO4 solution.
9. Essentially, detect and isolate the kid suffering from diarrhoea or other ailment and proper corrective measures should be adopted.
10. Provide them soft bedding and clean grassy paddocks.
11. Provide them a clean, dry, draft-free environment.
12. Kids are very susceptible to cold that's why special attention is required to monitor the shed temperature and if required impose correction of ambient temperature in winter.

After completion of colostrum period, kids can be raised either with dam or artificially. Milk is the principal component of the diet of the pre-weaning kid and in the later method milk can be fed by using bottles, pails, or self-feeder units 4-5 times daily up to second week and 2-3 times daily thereafter until weaning. When milk is fed by bottle, nipples and bottles should be kept thoroughly cleaned.They can also be raised on milk replacer in order to save the saleable milk or to raise orphan kid or in case of lactation failure.

Table: 7.3.1 : Feeding schedule for kids

Age of kids(days)	Body weight (kg)	Number of feeding	Dam's or cow milk (ml)	Green feed	Creep mixture (g)
1-7	1-3	With dam	300	-	-
8-30	3-6	3	350	free choice	free choice
31-60	6-9	2	400	free choice	100-150
60-90	9-12	2	200	free choice	200-250

Source : Ranjhan (2003).

Kids can be weaned anytime up to six months of age however it is better to wean them not more than 3 months of age. Excessive milk feeding should be avoided as it may lead to improper stomach development. Kids usually start to nibble forage such as pasture grass or hay by 2 weeks and consume grain within 4 weeks of age. Higher growth rate is obtained when kids are raised on ad lib. intake of milk than the limited quantities of milk. Generally growing kids are not provided with concentrate or supplements. Supplementation of concentrate makes up for limited milk intake mainly three weeks after birth. Feeding schedule of kid and composition of creep mixtures has been given in Table 7.3.1 & 7.3.2. In order to obtain good growth rate it is better to provide them 150 g of concentrate or 200 g wheat bran during one to three months of age. Further, if leguminous fodder is available concentrate feeding is not required. Provision of concentrate feeding before weaning improves weight gain after weaning because it shapes goat well for weaning and weaning stress.

Table:7.3.2 : Composition of various creep mixtures

Ingredients	Ration-1	Ration-2	Ration-3
Maize	60	-	20
Barley/oats	-	40	45
Groundnut cake	20	22	20
Fish meal	10	10	10
Molasses	-	20	-
Wheat bran	7	5	2
Mineral mixture	2	2	2
Salt	1	1	1
Calculated DCP %	18	17	17
Calculated TDN %	78	70	75

To 100kg or more creep mixture add Aurofac@150g, vitablend/Rovimix @25g, shark liver oil @1tsf per kid/day.
Source : Ranjhan (2003).

Space for exercise is an important requirements for the kids. Kid mortality is one the most important factors for the success of goat enterprise. Factors like age, season, birth weight, hygienic condition etc affects the extent of mortality in kids. There are some common diseases like pneumonia, diarrohea, septicaemia, contagious ecthyma, goat pox, coccidiosis and internal parasites etc are responsible for higher rate of mortality in kids. In order to minimise the incidence of these diseases adequate preventive & control measures should be adopted. Other management practices like handling, restraint, identification, castration etc has been already discussed in chapter 2 of this book.

Care and Management of Fattening Stock

After weaning at three months of age, kids are provided with clean, comfortable and well ventilated housing and reared under group management system. They should have easy access to fresh and clean water. They should be dewormed and vaccinated against various diseases. Kids should be fed green leguminous fodder for their normal growth. Poor quality roughage should supplemented with concentrate mixture (DCP-12-14%, TDN-63-65%) fed @ 350-400 g daily. In our country, castrated male kids are raised for chevon production. Factors like breed (larger>shorter), sex (male>female), season, nutrition and prevailing meteorological conditions etc affects post-weaning growth. The mature body weight of goats ranges from about 10 kg for small tropical breeds to over 100 kg for the large dairy breeds (Table 7.4.1). Similarly, growth rate vary from around 50 g/day for the small tropical breeds to over 200 g/day for large European dairy breeds. Apart from breed of the animal, nutrition is the most important factor which affects growth and development of an animal. Cereal based energy feeds are fed for fatty carcass. Roughage containing 20-25 and 30-40 per cent dry matter is satisfactory for fatty and lean carcass production, respectively. Indian breeds of goats have been divided into small, medium and large category and at the age of 10-12 months of age they may achieve 20, 25 and 30 kg of live weight, respectively. The dressing percentage of goats normally varies between 44-55 per cent. Further, it has been observed that dressing percentages and muscle-to-bone ratios are better for goats raised under intensive and semi-extensive systems than the extensive systems of feeding management.

Table: 7.4.1 : Mature size and growth rate of some of the goat breeds

Breed	Sex	Body weight (kg)	Growth rate (gld)
1. Barbari	M+F	35-45	53
2. Beetal	M/F	65-85/ 45-60	–
3. Jamnapari	M/F	70-90/ 45-65	–
4. Black Bengal	M/F	14-15/ 8-13	–
5. Saanen (Switzerland)	M	90-100	222
6. Alpine (France)	M	80-90	219
7. Angora (U.S.A.)	M	50-60	165
8. Kambing/Katjang (Malaysia & Indonesia)	M	25-30	57
9. Improved Boer (South Africa)	M	100-110	291
10. Damascus (Cyprus)	M	80-90	240

Source : Dhanda *et al* (2003) and Copland (1985).

Care and Management of Breeding Goat

Bucks: Buck is half of the band hence their importance should never be underestimated in the breeding programmes. Buck from highly pedigreed does or from does with good performance records should be retained on the farm. Onset of puberty in buck is associated with the ability to release spermatozoa. A well grown buck kid may be used for breeding at the age of six months but at this age semen is less in volume with low motility score. Bucks at age of 18-24 months can be used to serve 25-30 does and thereafter when they attain full maturity may be allowed to serve 50-60 does in a breeding season. In order to achieve good reproductive efficiency optimum ratio of buck to does should be 1:40/50. Bucks should be housed separately and when required should be taken to does for breeding.

Buck should never be overfed or underfed. Generally, if good grazing area is available, they can be maintained on grazing alone and no additional supplementation is required. However, supplementary feeding of concentrate (300 g/goat/day) from 2-3 weeks prior to the mating season should be provided to the bucks (flushing). In the absence of concentrate, supplement with good legume hay should be provided in the quantity twice the recommended concentrate allowance. They should given enough exercise to keep them in thin, thrifty and sexually active condition. Annual vaccinations should be done at least 6 weeks before the start of breeding season. Periodic trimming of hooves should be done to facilitate easy walking and mating ability. In order to prevent foot rot during rainy season foot dipping should be followed at every 3-4 weeks. They should be groomed daily to keep them free from clean, free from parasites and to reduce aggressiveness. To make them free from internal parasites deworming should be practiced at 1-2 weeks prior to mating. Prior to the mating they should be tested for fertility and sexually transmitted diseases.

Does: Puberty is the age at which first ovulation takes place and most goats reach puberty at a relatively younger age. Though there are considerable differences in the age at puberty among the different breeds of goat. It may be as low as 3-4 months and as high as 18-20 months. Puberty is more intimately associated with the mature weight than the age of the animal and usually breeding in does should be delayed until the animal has attained 60-70% of its mature body weight. This intervention will result in increase in conception rate, frequency of multiple births and chances of survivability of the kids. Physical maturity is more related with growth rate hence animals with higher growth rate will attain sexual maturity at earlier age than the animals with lower growth rate. Factors like breed, plan of nutrition, time of birth and the presence of the male etc could modify the age at puberty.

Goats are seasonally (declining day length) polyestrous in temperate zone while indigenous breed of tropical zone breed throughout the year. It has been observed that when temperate breed of goats are introduced in tropical environment they gradually lose their seasonality. However, in all goats the incidence of oestrus remains higher during certain part of the year viz. in India, it has been observed to be highest during June to October. Factors like latitude, climate, day length (photoperiodicity), food availability, breed and breeding systems affects length of the sexual season. Some of the reproductive parameters of goat have been shown in Table 7.5. Goat is the most prolific among all domesticated ruminants under tropical and sub-tropical conditions and most goat breeds have average litter sizes of 1.5 or higher. These parameters vary considerably among the different breeds of the goats. Depending upon the length of the breeding season and the breed of goat, females undergo several oestrous cycles one after another during each breeding season. Goats are spontaneous ovulators and ovulation rate (number of ova released from the ovary in a particular estrous period) increases with age and peaks between 3-6 years of age and it remains higher during early part of breeding season than the later part. Other factors like body size, body weight, condition and genotype also affects ovulation rate.

Provision of rising plane of nutrition to the does in preparation for breeding is known as flushing. When does are in low plane of nutrition prior to breeding, additional supplementation (250 g of concentrate/ doe/day) for about one month has beneficial effect in bringing the does into oestrus and increases ovulation rate. They should be maintained in group management system and should be provided a clean, well drained and ventilated house. They should be vaccinated and dewormed periodically.

Methods of heat detection: Heat detection in does can be done by following methods

- Reddening and swelling of the vulva.
- Continuous bleating & restlessness.
- Frequent urination.
- Mucous discharge from the vulva.
- Flagging of the tail side to side.
- Use of teaser buck.
- Reduced appetite.
- Reduced milk yield.
- Oestrus does frequently mount other does or allow them to be mounted.
- Strong male seeking behaviour

Table 7.4.2 : Production performance of some of the Indian goat breeds

Breed	Birth weight (kg)	Weight at 12 months	Age at first kidding (kg)	Kidding interval (days)	Service period (days)[a]	Multiple birth (%) (days)[a]	Dressing (%)
1. Jamunapari	2.98±0.03	28.76±0.62	735±24	364.7+2.3	169.7+4.4	35-40	45
2. Sirohi	2.78±0.02	18.15±1.06	598.15±16.8	351.4+18.6	–	10-15	45
3. Marwari	2.3±0.04	21.23±0.09	468.43±2.73	–	–	10-15	45
4. Jakhrana	2.56±0.05	18. ±0.8	561.24±8.2	316.0+21.8	170.0+21.7	40-50	45
5. Beetal	2.8±0.01	22.5±0.65	743.5±4.8	364.7+2.3	169.7+4.4	40-50	46
6. Barbari	1.76±0.02	15.2±0.43	632.9±26.1	328.7+7.4	124.3+22.2	60-70	49
7. Mehsana	2.5-3.0	20-25	600-650	–	–	15-25	-
8. Zalawadi	2.87±0.05	21.9±0.96	500-600	–	–	50-55	-
9. Kutchi	3.12±0.05	27.82±0.37	777.5±20.4	–	–	20-25	50
10. Sangamneri	1.86±0.02	19.88±0.48	550.5±20.9	356.6+5.7		40-50	46
11. Malabari	1.62±0.02	27.25±3.17	609.9±51.1	289.7+6.5	142.6+6.7	10-20	-
12. Ganjam	2.30±0.02	16.49±0.55	42.33±2.12	376.1+6.7		5-10	-
13. Black Bengal	1.13±0.01	13.84±0.47	521.8±21.4	282.3+22.7	89.6+15.1	60-80	-
14. Gaddi	2-3	25-35	350-450	–	–	10-15	39
15. Changthangi	2.15±0.05	15.8±0.4	700-900	–	–	2-5	41
16. Chegu	2.04±0.01	15.3±0.4	627.9±10.3	279.9+5.3	–	<10	39

Source : Rai *et al.* (2005) and Acharya (1988)[a]

Table: 7.5 : Reproductive parameters of goat.

Characters	Time
1. Age at puberty	5-7 months
2. Age to breed does:	10-20 months
3. Age at first kidding	15-26 months
4. Length of oestrous cycle:	21 day intervals (18-22 days)
5. Length of heat period:	36 hours (24-48 hours)
6. Ovulation time:	Few hours after the end of standing estrous or toward the end of estrus
7. Best time to bred:	Toward the end of estrus
8. Gestation length	149 (144-150.8) days

Source : Jainudeen *et al.,* (2000), Fatet *et al.,* (2011), Solaiman (2010) and Devendra & McLeroy (1990).

Mating Systems

Depending upon the system of production, natural breeding is done by flock mating, pen mating or hand mating. In the flock mating system, fertile bucks are allowed to remain continuously all through day and night with a group of females during the breeding season. There is no need of heat detection however recording of sire, mating date is difficult without the use of marking paint. This method provides best result in terms of fertility. In order to have better results bucks are replaced on a definite period of time. In pen mating system, bucks are allowed for service with a group of females in the pen during night and stall-fed or grazed separately during the day time. This practice provides rest to the bucks and thus helps to conserve the energy. To ascertain date of mating bucks needs to be fitted with marking harness. Bucks are observed continuously for their performance and if required they are changed. In Hand mating system, female in heat is detected and brought to the designated buck of the flock for mating. Teaser bucks are employed for detecting the does in heat and in order to separate the animal from breeding flock and to record date of mating some dye mixed in grease or simple linseed oil is smeared on the brisket of the buck. The colour of the dye should be changed every 16 to 18 days. This system is practiced when exotic purebred sires are used or when services of buck is to be used over much larger flock.

Goats are generally maintained under free grazing where natural breeding is most frequently practiced and hence use of artificial insemination technique in case of goat has been generally limited. The technique involves collection of semen from the male artificially, dilution and insemination into the reproductive tract of a doe (vaginal, cervical and intrauterine) through a glass or plastic tube into a number of does. By using artificial insemination technique, it is possible to impregnate 400 does with semen from a single buck within 3 weeks, compared with 30-50 does per buck in a conventional mating system.

Methods of pregnancy diagnosis: Early detection of pregnancy in the doe is essential to identify non-pregnant females so that they can be remated immediately in successive heat or culled. Signs like non-return to estrous and techniques viz. ultrasonic scanning or sonography (6 weeks following mating), X-rays, rectal-abdominal palpation, hormonal tests, vaginal biopsies, immunological tests can be used to ascertain pregnancy in does.

Care and Management of Pregnant Does

When a doe is in good condition and good grazing is available they can be maintained on normal feeding regime during early part of gestation period (about 12 weeks) but during last 4-6 weeks before kidding extra feeding is essential. During this period does should be given sufficient amount of feed so that it can meet up the demands of fast growing foetus as well as dam can replenish the body reserves. This extra feeding results in low kid and doe mortality, higher birth weight and greater live weight gain in kids. Clean and wholesome water should be available round the clock. They are handled gently. They are placed in larger pen and transferred to kidding pen towards the end of gestation period. Soft bedding materials are provided in kidding pen.

Care and Management of Lactating Does

It has been observed that the milk yield and lactation lengths of tropical indigenous goats are lower, higher for breeds in temperate regions and in between the two when temperate breeds are introduced into tropics. Goats have comparatively flat lactation curve than cattle and buffaloes. In goats, peak milk yields are attained between 8th to 12th weeks of lactation. This indicates that goats are more persistent milker. Due to relatively larger mammary gland and volume of secretary tissues goats produces more milk per unit of live weight compared to cows and buffaloes.

Table: 7.6 : Milk yield, lactation length and butter fat (%) from dairy goats

Breed	Milk yield (kg)	Lactation length (days)	Butter Fat (%)
1. Beetal (India)	200.1	208	5.2
2. Barbari (India)	118.0	183	3.9
3. Jamnapari (India)	182.0	168	4.5
4. Malabari (India)	180.6	180	5.1
5. Dera Deen Panah (Pakistan)	204.5	130	4.9
6. Chopper (Pakistan)	75.0	105	5.2
7. Osmanabadi	180	210	–
8. Surti	150	165	–
9. Black Bengal	34	122	7.18
10. Alpine (United States)	916	276-305	3.5
11. Toggenburg (Switzerland)	605	281	–
12. Saanen (France)	527	279	3.30
13. European breeds in tropics	143-1011	106-336	–
14. European breeds in temperate zone	245-2707	180-730	–

Source : Devendra (1980), Singh and Senger (1978), Mishra (1978) and Dickinson and King (1977).

Factors like breed, strain, suckling, milking frequency, litter size, parity, health, climate and nutrition & feeding practices affects milk production from goats. They should be provided with clean and comfortable housing (conventional/loose housing) and are reared under group management system. Generally, goats are hand milked twice daily. Does should be milked completely with dry hand. Before milking, washing the udder with warm water and a sanitizer will stimulate milk letdown, as well as promote cleanliness. Hair around teat and udder should be clipped periodically. Milking does should groomed daily in order to remove dirt, parasites, increase circulation and to make them docile. Feeding most importantly affects the initial and total milk yield in lactation. Hence, availability of quality feed during lactation is crucial. A lactating doe weighing around 20-40 kg should be given 2-3kg grass/ grass with legume 1.8-2.5 kg with 300-800 gm of concentrate (16-18% CP) per day. In general, for each kg of milk produced 300-500 gm of concentrate should be given to the does.

Care and Management of Dry Does

During dry period does are least productive and no special care is required for these animals. Regular grazing for 8 to 10 hours on a good pasture is sufficient to maintain their condition. However, before breeding season they could be exposed to high plane of nutrition.

Health Care for Goats

A healthy goat can be recognized by its normal appearance, behaviour and proper functioning of its biological processes. Normally a healthy goat is an energetic animal, walk at good pace, has brightness in their eyes, good appetite, smooth and shiny coat, pink coloured visible mucus membranes (eye, mouth, nose and vulva), normal body temperature, heart beat, calm breathing, voids round and firm droppings and clear, yellow colour urine. Further, a healthy goat shows good production (growth, milk production and reproduction) performance. Normal values for most vital signs for goats has already discussed in chapter 2 of this book. Some of the diseases of goats and their control measures have been presented in Table 7.7.1 & 7.7.2.

Table: 7.7.1 : Some of the diseases of goat and their control measures

Diseases	Causative agent and clinical signs	Prevention & control
Peste des petes ruminants (PPR)	Caused by *Paramyxo-virus*, characterised by high fever, oral necrosis, catarrh, nasal discharge and diarrhoea and death. Kids are more susceptible than adults.	Chemotherapy to treat secondary bacterial infection, slaughter of infected animals and protective immunisation.
Contagious caprine pleura-pneumonia (CCPP)	Caused by *Mycoplasma mycoides var. Capri*, highly contagious disease characterized by pyrexia (41°C or greater), severe respiratory distress, high morbidity, mortality and post-mortem lesions of fibrinous pleuropneumoniae with pronounced hepatisation and pleural adhesions.	Good herd hygiene, quaran -tines, movement controls, slaughter of infected and exposed animals and treatment by arsenicals, antibiotics and protective vaccination.
Foot-and-mouth disease	Caused by *Aphtho virus*, characterised by pyrexia, salivation, foaming at mouth, vesicles in mouth and feet, lameness and agalactia in milking goats. Death of young stock may occur without clinical signs.	Provision of foot bath, chemotherapy to treat secondary bacterial infection, slaughter of infected animals and protective immunisation.

Contd.

Goat pox	Caused by *Capripox virus,* all ages of goats are susceptible but malignant form is more common type in kids, characterized by marked depression and prostration, high fever and discharges from the eye and nose, kid may die during this stage before typical pox lesions develop. Skin lesions appear on skin and on the buccal, respiratory, digestive and urogenital tract mucosa.	Good herd hygiene, quarantine and protective vaccination.
Contagious ecthyma	Caused by dermatotropic ungulate pox virus, It is a contagious and highly infectious disease of kids, characterized by the development pustular and scabby lesions on the muzzle and lips. The death may occur due to the extension of lesions into the respiratory tract.	Good herd hygiene, immediate isolation and treatment of affected kids with antiseptic ointment or astringents lotions and protective vaccination.
Brucellosis	Caused by *Brucella melitensis,* characterised by abortion about the fourth month of pregnancy. Arthritis and orchitis may occur.	Good herd hygiene, slaughter and protective vaccination.
Anthrax	Caused by *Bacillus anthracis,* characterised by high fever, staggering, excitement, depression, incoordination, trembling and difficulty breathing and sudden death.	Quarantine, treatment of asymptomatic animals, prompt disposal of dead animals, manure, bedding, or other contaminated material by cremation or deep burial method and protective vaccination.
Haemorrhagic septicaemia	Caused by *Pasteurella multocida* type I and characterised by dullness, reluctance to move, fever, salivation, and serous nasal discharge, edematous swelling in neck region.	Antimicrobial therapy (sulfonamides, tetracyclines, penicillin, and chloramphenicol) and protective immunisation.
Colibacillosis	Commonest disease of new born kids caused by *E. coli.* Two age groups appear to be susceptible (one to two days of age and 3 to 8 weeks old),	Antibiotics treatment (streptomycin, tetracyclines, neomycin, nitrofurazone etc) and appropriate supportive (fluid therapy, antihistaminic

Contd.

	peracute cases are found dead without premonitory signs. Acute cases show collapse and occasionally signs of acute meningitis manifested by a stiff gait in the early stages followed by recumbency with hyperaethesia and tetanic convulsion.	and vitamin A) therapy may be given.
Enterotoxaemia: (Over eating, pulpy kidney disease)	Caused by *Clostridium perfringens* type C & D, the course of illness in kids is short and may develop during the night, consequently may find dead kids during morning. Respirations are very rapid, saliva drips from the mouth and rise in body temperature. The animal may manifest abdominal pain.	Good herd hygiene, quarantine and protective vaccination.
Mastitis	Caused by number of pathogenic bacteria and characterised by hot, swollen, red udder, decrease in milk production, flakes or strings on the strip cup screen and blood in the milk.	Cleanliness and good hygiene, cleaning and washing of udder, regular monitoring of udder health and treatment by antibiotic injections as well as antibiotic teat infusions.
Cocciodosis	Caused by infection with *Eimeria sp.*, characterised by usually the sudden onset of severe diarrhoea with foul smelling, fluid faeces containing mucus and blood. Severe straining is characteristics, severe dehydration emaciation and complete anorexia is usual.	Good herd hygiene, immediate isolation and treatment of affected kids with antimicrobial & electrolyte therapy
Endo-parasites (nematodes, cestodes and trematodes)	*Haemonchus contortus, Ostertagia circumcincta, Trichostrongylus colubriformis, Trichuris ovis, Cooperia punctata, Oesophagostomum columbianum, Strongyloides papilosus,* Moniezia expansa, Avitelina globipunctata, *Fasciola hepatica Fasciola gigantica and Schistosoma bovis* affects goat. Symptoms	Good herd hygiene, better nutrition, rotational grazing and preventive deworming at regular interval.

Contd.

	include weakness, depression and lying down abnormally, anorexia, pale mucous membrane, bottle jaw condition, shabby hair and diarrhoea.	
Ecto-parasites (ticks, lice, mites)	Irritation to the skin, thickening and wrinkling of surrounding skin, loss of hair, appearance of red papules and general erythema of the skin, restless and tick-paralysis	Spraying or dusting or dipping or pour-on with suitable insecticides or Ivermectin injection.

Table 7.7.2 : Vaccination schedule for goats

Diseases	**Primary Vaccination**		**Dose & Route**
	First Injection	**Repeat vaccination**	
Foot & MouthDisease	At 3-4 months of age	Booster 3-4 weeks after 1st Injection and repeated twice a year (September & March).	1 ml s/c
Peste-des-Petitis Ruminants (PPR)	At 3 months of age	Every three years.	1 ml s/c
Goat Pox	At 3-4 months of age	Booster 3-4 weeks after 1st Injection and repeated annually.	1 ml s/c
Haemorrhagic Septicaemia (HS)	At 3-4 months of age	3-4 weeks after 1st Injection (2 doses at 1 month interval and repeated annually).	2 ml s/c
Enterotoxaemia	At 3-4 months of age	3-4 weeks after 1st Injection & repeated at every 6/12 months.	2 ml s/c

Dose, route and repeat vaccination time may vary for different manufacturers.

Deworming: A number of helminths considerably affect goats. Good management is the key to controlling losses from internal parasites. A regular deworming program (2-3 times in a year) should be followed with anthelmintic drugs like Albendazole, Fenbendazole, Levamisole, Tetramisole, Piperazine, Niclosamide, Oxyclozanide, Rafoxanide, Closantel, Ivermectin etc. An ideal anthelmintic should have broad spectrum activity, wide margin of safety, should be effective against both mature & immature stages of the worm and easy to administer. While practicing deworming avoid under dosing of drugs and use of same class of anthelmintics continuously. Liver tonics should be administered in animals for at least two to three days after vaccination or deworming.

Marketing and Transport of Goats

The goat keepers trade their goats to the local traders or in the nearby market whenever they are in need of money. The local traders also look for brokers to sell their animals in major markets. The goats which are collected in the local markets are taken to the local slaughter houses and some of them are sent to the terminal markets. Generally, big traders and sometimes wholesaler also buy the animals from small markets to sell them in big markets directly or through some mediators. These mediators are known as brokers or middle men in the small markets and commission agents in the big markets. In majority of the markets the business deal takes place after examination of the animals by the buyers and brokers/commission agents help them in negotiating the prices. The price of the animals can be arrived at by -buyers declare the sell price publically or through secret whispers to the sellers or directly negotiations through brokers or negotiations under the cover or hatha system or auction of the animal. Apart from these local markets goats are also sold in some of the annual fairs in the country. There are certain constraints for goat marketing in India- farmers don't follow the practice of feeding extra energy to prepare them for the slaughter purposes, livestock markets are weekly markets and are owned by private/trust/local bodies, these markets lacks even the facilities of shelter, least developed cooperative system of marketing of the goats, farmers are not well aware about market information hence are more reliant on middlemen, more involvement of middlemen reduces the profit to the producers and lack of forward agreement with butchers or big meat processing units. Goat meat is marketed in villages by slaughtering one or two animals once in a week or a group of people joining together, slaughter the animals and sharing the cost of the meat so obtained. In small towns goat meat is marketed directly to the consumers through meat shops. Transport of goats by various means has been already discussed in chapter 2 of this book.

Goat Products

1. **Milk:** In the year 2010, total milk production from goat was around 17.37 million tonnes and with a share of about 26.44% India was largest producer of goat milk in the world followed

by Bangladesh, Sudan, Pakistan and Spain (FAOSTAT, 2010). In general, goat milk is a very nutritious drink which adds a valuable protein to the diet of rural poor household. It is known as the 'poor man`s cow' in India and as 'wet nurse' of infants in Europe. The quantity and quality (chemical composition) of milk produced on factors like breed, stage and number of lactation, individual, season, environmental conditions, udder health and nutrition etc. Goat milk can be used to prepare products like evaporated/dry milk, fermented milk, cheese, yoghurt or buttermilk, ghee, butter, desserts, sweets, whey products and cosmetics. Comparative chemical composition and physical property of goat milk has been Shown in Table 7.8.1 & 7.8.2. Some of the unique properties of goat milk are

- Goat milk has greater proportion of smaller fat globule (average 3.5 μm) than the cow and buffalo milk. These smaller fat globules provide a better dispersed and naturally homogenized milk fat and thus better digestibility due to availability of greater relative surface area of fat globules for the digestive action by lipases.
- Goat milk fat has significantly higher levels of short and medium chain length fatty acids (C4:0-C14:0) than cow and human milk. This unique difference has been used for treatment of a variety of fat malabsorption syndromes in patients. Goat milk has a higher content of conjugated linoleic acid (CLA) and lower levels of cholesterol than cow milk.

Table: 7.8.1 : Average composition of nutrients in goat, sheep, cow and human milk.

Composition (%)	Goat	Sheep	Cow	Human
1. Fat	3.8	7.9	3.6	4
2. Soild-not-fat	8.9	12	9	8.9
3. Lactose	4.1	4.9	4.7	6.9
4. Protein	3.4	6.2	3.2	1.2
5. Casein	2.4	4.2	2.6	0.4
6. Albumin, globulin	0.6	1	0.6	0.7
7. Non-protein nitrogen	0.4	0.8	0.2	0.5
8. Ash	0.8	0.9	0.7	0.3
9. Calorie per 100ml	70	105	69	68

Source : Park *et al* (2007).

- Goat milk is a valuable source of amino acids and has higher amount of non-protein nitrogen.
- Goat milk proteins are smaller, softer and hence more friable curd formations of goat proteins occurs during acidification in the stomach.
- Goat milk has about 10 times more oligosaccharides than cow milk and closely resembles that of human milk which is of special interest to infant nutrition.

Table: 7.8.2 : Physical properties of goat, sheep and cow milk

Properties	Goat milks	Sheep milk	Cow milk
1. Specific gravity (density)	1.029-1.039	1.0347-1.0384	1.0231-1.0398
2. Viscosity (Cp)	2.12	2.86-3.93	2.0
3. Surface tension (Dynes/cm)	52.0	44.94-48.70	42.3-52.1
4. Refractive index	1.450±0.39	1.3492-1.3497	1.451±0.35
5. Freezing point (-^{0}C)	0.540-0.573	0.570	0.530-0.570
6. Acidity (lactic acid %)	0.14-0.23	0.22-0.25	0.15-0.18
7. pH	6.50-6.80	6.51-6.85	6.65-6.71

Source : Park *et al* (2007).

- Goat milk has higher calcium, phosphorus, potassium, magnesium, and chlorine, but lower sodium and sulfur content than cow milk.
- Goat milk is white in colour because goat converts all â carotene (yellow color) into vitamin A.
- For a human infants goat milk supplies adequate amount of Vitamin A and niacin and excesses of thiamine, riboflavin and pantothenate.
- Goat milk may have lower Reichert Meissl and higher Polenske values than the cow milk.
- Goat milk has lower saponification value and slightly lesser refractive index.

2. **Meat:** Goat meat may be referred as chevon or cabrito, which is one of the most important products from goats. Dressing percentage (carcass weight divided by the live weight times 100) is around 50 per cent. Goat meat is normally dark in colour and slightly coarser in texture. A typical feature of goat carcasses is their thin subcutaneous fat cover. Fat is yellow in

colour and is concentrated mainly around the organs (omentum, mesentery and kidneys) and not between the muscles. Goat meat on average consists of 72.3% moisture, 21% protein, 4.7% fat and 1.1% ash. There is often strong odour particularly in the meat of uncastrated male goats. "Goaty" odour of chevon is attributed to 4-methyloctanoic fatty acid (hircinoic acid). Meat has high amounts of protein, essential amino acids, fat - soluble vitamins, water-soluble vitamins (niacin, riboflavin, panthothenic acid, cyano cobalomin), iron, and phosphorus. Chevon has a biological value of about 60 and reported to be higher in thiamine and riboflavin and lower in niacin than beef, lamb and veal. Like the other species process of freezing, curing, smoking, emulsifying and cooking etc can be used to produce value added products. Leading chevon producing countries of world has been shown in Table 7.8.3.

Table: 7.8.3 : World's leading goat meat producing countries

Sr. No	Countries	Production (million tonnes)			Average meat produced/ animal (kg)	Number of animals slaughtered (million)
		2008	2009	2010	2010	2010
1	China	1.82	1.85	1.87 (36.19)	13.8	135.61
2	India	0.55	0.56	0.58 (11.33)	10.0	58.65
3	Nigeria	0.27	0.28	0.29 (5.62)	12.7	22.93
4	Pakistan	0.26	0.27	0.27 (5.31)	17.0	16.20
5	Bangladesh	0.17	0.183	0.19 (3.69)	7.0	27.30
	World + (Total)	5.00	5.13	5.17	12.4	418.07

Figure in parentheses indicate per cent of world goat meat production.

Source : FAOSTAT (2010).

3. **Skin:** Goat breeds like Maradi, improved Boer, Somali, Black Bengal, Red Sokota, Moxoto, and Sahil are recognized for their good quality skin. Goat skin or leather is an important by-product of goat and provides considerable source of income (around 5-10% of the total value of the animal). A desirable goat skin produces leather that is firm, durable and has an attractive appearance. Factors that affect quality of skin are-basic structure of the skin, damage acquired in live animal, damage incurred during slaughter and flaying; and

preservation. Pricing of skin is based on a simple grading system where it is evaluated for extent of damage caused by disease and careless handling etc. Goat leather can be used in manufacturing footwear, garments, handbags, wallets, and gloves, and drum skins.

4. **Abattoir by-products:** Apart from meat various edible (meat, blood, bone meal) and non edible (tallow, horn, hooves) abattoir by-products are also obtained from goat rearing and has some commercial value.

5. **Fiber production:** Following types of fibers are generally obtained from the goats

   a) **Ordinary goat hair:** These are multicoloured, straight, elastic, coarser, heavily modulated fibers with variable length which varies among the various breeds of goat. In certain part of world goats are occasionally clipped and these fibers are used in the manufacturing of cheap felts, interlinings, ropes, paint brushes, mulch, and mats for soaking up oil spills.

   b) **Mohair:** The word " mohair " is derived from the Arabic word " Mukhayar " meaning " best of selected fleece, "silky goatskin cloth," "cloth of bright goat hair," and "hair cloth". Mohair is strong, resilient and lustrous white fleece obtained from Angora goats. Mohair quality is determined by the fibre diameter (20-40 μm), length (50-150 mm), general appearance and contaminants such as vegetable material, dirt, medullated fibres and kemps. It is used to manufacture blankets, rugs and furnishings particularly upholstery. Good quality mohair is soft, firm, bright, twisted and solid ringlets. The lustrous characteristic of the fibre when dyed results in richly coloured fabrics, which are used for the manufacture of women's apparel. Presently there are about 2.6 million Angora goats in the world and annual mohair production is about 6.6 million kg (<0.02% of textile production of world). Leading mohair producing countries of world are South Africa, USA, Lesotho, Argentina, Turkey (Van der Westhuysen, 2005).

   c) **Cashmere/Pashmina:** Cashmere is the downy coat or winter undercoat produced by cashmere goats (produced by many breeds except Angora) predominantly in Central

Asia. Pashmina is basically fine quality cashmere grown in India by Chegu and Changthani breeds. Animals that produce cashmere have two coats-an outer longer, straight, coarse guard hair and a very fine, crimpy undercoat that is shed at the end of winter. The fine cashmere fiber is then separated by combing it out or using a commercial dehairer on the sheared fleece. The yield of cashmere fiber is in the range of 30 to 40 percent of the total fleece weight. Solid-color goats are preferred than the multicolored goats for the production of cashmere fiber. It may be white, brown, gray, or black. Cashmere fiber must be less than 19 microns in diameter (16-19 microns). Its fineness and strength allow it to be spun into very fine. Soft yams used to produce luxury knitwear. The quality of cashmere is determined by length, texture, colour and fibre diameter. Most of the world supply of cashmere fiber comes from China, Afghanistan, Iran, Outer Mongolia, Turkmenistan, Tajikistan, Uzbekistan, Pakistan, India, Australia, and New Zealand. In the plateaus of Mongolia and China about 85% of the world's cashmere is found and annual cashmere production is about 11-14 million kg. The pashmina produced by the Indian goats is either grey or white in colour. White Pashmina is preferred to grey Pashmina due to its lustre. The average Pashmina yield per year of Indian goats varies from 111 to 149 g during second to fourth year of production. The quality of Pashmina of Indian goats is finer (8-14μ diameter) but shorter in staple length.

Cashgora is a typically white, lusterless fleece with average fineness 17-23 μm, staple length of 3-9 cm and has a relatively bold (compared to cashmere) crimp produced by crossbred goats which are result of crossing of Angora goat with cashmere, meat, feral, or dairy goats.

6. **Manure:** Goat manure is a useful by-product it can be used to fertilize crops. It is better to compost the manure instead of using fresh manure. It contains higher amount of nitrogen and phosphoric acid than the manure of cattle/buffalo (Table 7.8.4).

Table: 7.8.4 : NPK composition of animal manures (percent of oven-dry weight)

Livestock	Nitrogen	Phosphorous	Potassium
1. Goat dung	2.04	0.73	0.47
2. Buffalo dung	0.75	0.20	2.00
3. Cattle dung	1.65	0.44	0.83
4. Sheep dung	1.55	0.70	0.72
5. Chicken droppings	2.87	1.28	1.95

Source : Misra & Hesse (1982).

Preparation of Animals for Show

Goat show provides an opportunity to the farmer to exhibit their best animal, brings fame to the farmer and popularizes best breed as it act as an advertising media. It provides a platform where comparison among superior animals within and between the breeds can be made. It further make possible to exchange experiences among the breeders, chance to learn about better feeding and management practices. It provides an opportunity to sell the limited number of animals. It helps in building a positive learning experience, ability to display an animal to its best advantage and better competitive spirit among the participants.

The process of getting the animal ready for the show is known as fitting. It enhances the features of the animal. Select the animals that are correct in conformation and in good condition. Animals should never be overfed fed nor under fed. Clean, fresh water must be constantly available to them. They should be given adequate exercise daily. Hooves should be trimmed one to two weeks before the show. Animals should be groomed properly before the show. Dairy goats can be clipped from a few days up to two weeks before a show. Clippers should be moved against the hair including whiskers, beard, and hair inside the ears and around the tops of the hooves. The hair on the tail is squared off below the last bone in the tail, leaving a triangular tuft of hair at the end of the tail. Meat goats are clipped at least one week to 10 days before the show and while shearing, the clippers need to run parallel to the length of the body. Hair below the knees and hocks should not be shorn and bob the hair at the end of the tail. Angora goats are not clipped and they may or may not be washed. Washing removes the natural oils in the mohair. Goats are rinsed with a mild shampoo before and after the clipping. Collars are fitted to control animal's movements in the ring. It is ideal to work with the animal ahead of time until they lead readily and respond quickly to signals. They should move forward with a slight pull on the collar and stop when you pull slightly up and back. Angora goats may be shown without a collar. The animal is shown by holding one horn and placing the other hand on the loose skin under the chin. The fleece should never be grabbed or held to control the animal. Most goats will benefit from a final touch-up cleaning with a damp cloth just prior to being shown. Further, to know more about the showmanship Washington state 4-H dairy goat fitting and showing scorecard has been given below.

Washington State 4-h Dairy Goat Fitting and Showing Scorecard

Based on Usual Order of Consideration	Points	Maximum points
Appearance of Animal		40
• Condition and thriftiness- it shows normal growth, neither too fat nor too thin.	10	
• Hair-clean and properly groomed.		
• Hoofs-trimmed and shaped to enable animal to walk and stand naturally.	10	
• Neatly disbudded-if the animal is not naturally hornless.		
• Clipped-animal neatly trimmed inside ears, around feet, udder, and tail. Clipping entire body is discouraged.	10	
• Cleanliness-as shown by a clean body as free from stains as possible, with special attention to legs, feet, tail area, nose, and ears.	10	
Appearance of Exhibitor		
• Person and clothes neat, clean, and appropriate to occasion.	10	
Showing Animal in the Ring		50
• Leading-enter, leading the animal at a normal walk around the ring in a clockwise direction, walking on the left side, holding the collar with the right hand. Exhibitor should walk as normally and inconspicuously as possible.		
• Goat should lead readily and respond quickly.		
• Lead equipment should consist of collar or small link chain properly fitted. Lead strap not permitted.		
• As judge studies the animal, the preferred method of leading is to walk alongside on the side away from the judge.		
• Lead slowly with animal's head held high enough for	10	

Contd.

impressive style, attractive carriage, and graceful walk.		
• Pose and show an animal so it is between the exhibitor and the judge as much as possible.		
• Avoid exaggerated positions, such as crossing behind the goat.		
• Stand or kneel where both judge and animal may be observed.		
• Pose animal with front feet squarely beneath and hind feet slightly spread. Where possible, face animal upgrade with her feet on a slight incline. Don't crowd other exhibitors or leave too much space when leading into a side-by-side position.		
• When judge changes placing, lead animal forward out of line, down or up to the place directed, then back through the line, finally making a U-turn to get into position.		
• To step animal ahead, use slight pull on collar. If the animal steps badly out of place, return her to position by leading her forward and making a circle back through your position in the line.		
• When judge is observing the animal, replace her as quickly and inconspicuously as possible, if she moves out of position.		
• Be natural. Over showing, undue fussing, and maneuvering are objectionable.	15	
• Show animal to best advantage, recognizing the conformation faults of your animal, trying to help overcome them.	15	
• Poise, alertness, and courteous attitude are all desired in the show ring.		
• Keep an eye on the animal and be aware of the position of the judge at all times, but do not stare at the judge.		
• Respond rapidly to requests from judges or officials, and be courteous and sportsmanlike at all times, respecting the rights of other exhibitors.		
• The best exhibitors show the animals at all times not themselves and continue exhibiting well until the entire class has been placed, and the judge has given reasons and dismissed the class.	10	

Judging Goat

Judging is an art and expertise can be achieved by repeatedly practicing it with patient over period of time. For appropriate judging it is crucial to have meticulous understanding of ideal physical features of goats of different breeds, age group and reproductive characteristics etc. The parameters of judging goats vary as per the purpose for which they are reared *i.e.* milk, meat and fiber production. The parameters of judging meat goats (structurally correct meat goats with high meat to bone ratio) have been mentioned below

Size, Condidtion and Delelopment	
• Preference is given to animals exhibiting superior growth and muscle development according to age. • Animal should have an even covering of firm flesh without excess fat. • Animal must exhibit a healthy appearance being alert. • Eyes are to be bright and clean, loose pliable skin is a must.	**16 Points**
Head and Neck	
• Head should be strong, wide and have depth to the jaw. • Bite should be aligned for optimal feed consumption being neither under/over shot. • Neck should be in proportion to the body in size and demonstrate muscling with a thick base that blends smoothly into the shoulders.	**06 Points**
Front End Assembly	
• Shoulders should be strong, wide and well muscled with an even covering of firm flesh. • Shoulder blades should be set smoothly against the chest wall and withers. • Withers should be rounded and barely defined while blending smoothly into the area of the spine, crops and ribs. • The brisket should be broad, deep and proportionally strong.	**10 Points**
Back and Rump	
• Back should be strong, wide, straight and nearly level with a smooth, even covering of firm flesh. • The loin should be long, wide and well covered. Rump should be long, broad and slightly sloping. • The hips (hooks) should be wide apart and level with the back. • The thurls should be set wide apart as the pins. Tail should set slightly above the pins, swing freely and have less than 3/8″ fat at the tail head.	**16 Points**
Hind Saddle	
• Hind saddle should be wide, deep and demonstrate smooth, firm, well defined muscling.	

Contd.

• The thigh muscle should extend well down the hind leg resulting in a short shank. • This muscling should result in a deep, wide twist when viewed from the rear and an outward curving of the thigh when viewed from the side.	**16 Points**
Legs, Pasterns and Feet	
• Hind legs should be medium in length, set wide apart and be nearly straight when viewed from the rear. • Hind legs should be nearly perpendicular from hock to pastern when viewed from the side. • Fore legs should be medium length, wide apart, straight and strong. • Bones should be clean, round and of adequate density to support weight. Pasterns should be short to medium in length and strong. • Feet should be sound and correctly support the weight of the goat. • All joints should be clean and move freely. • Animal should track wide in the fore and rear with all toes pointed forward.	**14 Points**
Heart Girth	
• Heart girth should be large with both width and depth. • Fore ribs should be well sprung. • Width must be displayed between the front legs in the chest floor. • Fullness at the elbow is desired with less than 3/8″ fat behind the elbow.	**12 Points**
Barrel	
• Barrel should have well sprung wide spaced ribs and strong support from front to rear. It should blend smoothly into a deep, strong flank.	**10 Points**
Total Points	**100**

Source : USBGA meat goat guide (US Boer Goat Association)

Chapter - 8

Sheep Production

K. N. Wadhwani

Origin and Domestication of Sheep

The domesticated sheep belongs to phylum-*Chordata*, class-*Mammalia*, order- *Artiodactyla*, family - *Bovidae*, sub-family - *Caprinae*, genus - *Ovis* and species- *aries*. They are hollow horned, cloven-hoofed ruminants of relatively smaller size. Sheep plays an important role in Indian agrarian economy as it is having multi facet utilities like meat, wool, skin, manure and to some extent milk and transport also. Marginal and sub-marginal lands of arid and semi-arid tropics are well adapted by sheep. Among the different wild species, Asiatic (*Ovis orientalis*) species of mouflon is considered to be the ancestor of present day sheep. Sheep were domesticated by man at around 9000 BC. The reference of role of sheep in the economy of mankind is found in the history of civilization of *Mesopotamia*, *Mohen-jo-Daro* and *Harappa* in Northern India.

Classification

Phylogenetic

Kingdom	:	Animalia
Phylum	:	Chordata

Class	:	Mammalia
Order	:	Artiodactyla
Sub order	:	Ruminantia
Family	:	Bovidae
Sub family	:	Caprinae
Genus	:	*Ovies*
Species	:	*aries*

Tail Types

There are four basic tail types in domestic sheep: long tailed, short tailed, fat tailed and fat rumped. All the wild species had short thin tails and from them different types have arisen since domestication. The caudal view of different tail types in sheep are: long-thin-tailed, short-thin-tailed, fat tailed, fat rumped, long-thick-tailed and root tailed.

The long thin tailed sheep are more in number than any other tail types, has about 300 breeds and are mainly found in European countries. The short-thin-tailed are commonly found in the Nepal, Tibet, India and Pakistan and are coarse wooled type. Short-thin-tailed sheep are usually quite hardy but are of a rather archaic or primitive type. In India with the exception of fat tailed Bhakarwal, all sheep can be considered thin-tailed. Bikaneri breed of sheep is having a long-thin-tailed, carpet wooled breed.

Characteristics of Sheep

Sheep is small ruminant with a high adaptability to the extreme climate, gregarious in nature and shows uniparous habit. It plays very important role in economy of arid and semi arid region of India. Though India has vast livestock population, its share in sheep population is only 6.7 percent of world total. Role of sheep rearing is established in improving economy of rural farmers. Sheep rearing of farmer gives them helping hand especially at the time of crop failure and disaster. Sheep is affectionately labelled as "mortgage lifter"/ surety against loan by the rural poor.

Though the number of sheep is only about half of the goat, they form important species of livestock in some part of the country (Himalayan region) especially in Deccan region of Peninsular. In India, sheep is primarily reared for meat production and secondarily for both fine (Northern Himalayan region) and carpet wool (North-Western Semiarid region) production. Sheep can thrive well in dry pastoral condition. Sheep can subsist on sparse vegetation whereas other species of livestock are struggling to sustain. This is because of their inherent capacity and cleft in upper lip to graze very close to the roots of herbage. The sheep contain high amount of nitrogen (golden hoofed animals), potash and phosphorus in faeces than FYM therefore penning of sheep in harvested fields (sheep folding) is common practices to improve the fertility of soil. As sheep rearing don't require large investment in buildings and equipments it opens good scope for use by small and marginal farmers.

Sheep are resistant to drought, semi-starvation, less prone to extreme weather, parasites and diseases. Their body coat is unique as it permits ventilation and also protects the skin from the hot sun, rain and abrasion. Sheep can constrict and dilate blood vessels supplying extremities for controlling heat loss. Sheep are usually raised on waste land rarely fed on quality feeds. Milk from sheep is of limited importance, they are not dairy animals.

Scope of Sheep Farming in India

India posses 71.56 million sheep as per 2007 livestock census and ranks third in the world and which was 6.67% of world population (1071 millions). In India, maximum sheep population was present in the state Andhra Pradesh followed by Rajasthan, Karnataka, Tamil Nadu and Jammu & Kashmir. During the inter censual period from 2003 to 2007, there was increase in the sheep population by about 16.41%. During 2011-12, wool production stands at the modest level of 44.3 million kg. The export earnings from different woolen products during 2010-11 were Rs.288 crores at current prices. The contribution of sheep to total meat production in the country is around 7.6 percent. Sheep skin in the form of leather and leather products is also exported.

Sheep makes a valuable contribution to the livelihood of the economically weaker sections of the society. Amongst the livestock owners the shepherds are the poorest of the lot. Realizing the importance of sheep in Agrarian economy the central government had established the Central Sheep and Wool Research Institute (CSWRI) at Avikanagar in Rajasthan. Numbers of sheep breeding farms were established during various plan periods throughout the country for evolving (i) New fine wool breeds for different agro-climatic regions capable of producing 2.5 kg. of greasy wool per annum. (ii) New mutton breeds capable of attaining 30 kg live weight at 6 months of age under intensive feeding conditions. Intensive Sheep Development Projects (ISDP's) were introduced in many of the sheep rearing districts. Setting up of wool boards in important wool producing states was envisaged and States of

Jammu & Kashmir and Karnataka have already set up these boards. Some of the states have set up wool development corporations / federations.

Advantages of Sheep Farming

- Sheep can economically convert grass into valuable meat and wool.
- Little concentrate feed is required by sheep as compared to cattle because sheep gets 80 to 90 % of their protein from forages as against the cattle.
- Sheep can survive very well on agro-forestery, agro-industrial by-products and unconventional feedstuffs like Mango seed kernel, Cassiatora seeds, Corn steep liquor, P J Pods and Babul Pods etc.
- Sheep are considered as excellent cleaners of pastures as they are capable of grazing more closely to the ground than other livestock.
- Sheep increases the fertility of soil by spreading their dung very evenly on the ground.
- Sheep are the best animals that can be raised on stubble's & weeds from the field.
- Unlike goats, sheep do not damage the trees.
- Less labour is required compared to other livestock because of strong flocking instinct.
- Three times income in a year can be obtained from sheep viz. once from sale of lambs and two times from sale of wool. There may be daily small income from milk also.
- The return on investment starts earlier and sheep manure is two times richer in K and P as compared to bovines.
- Marketing of wool is relatively easy as wool clips can be easily stored & transported.

Disadvantages of Sheep Farming

- There is lack of adequate infrastructures in production and marketing of sheep and their by-products. There is also lack of organized sheep market.
- Lack of modernization in slaughter houses.

- Predators attack sheep easily which may leads to mortality.
- Keeping sheep is not an attractive profession, especially among youngsters.
- There is heavy competition between wool and synthetic fibres.
- The price of wool varies too much from place to place, lot to lot and grade to grade.
- As a result of urbanization and industrializations there is reduction in grazing land which leads to unavailability of sufficient fodder.
- Interference of middleman in the purchase of sheep leads to less profit to the sheep keepers.
- Twins and triplets birth in Indian sheep is not common.
- Less synchronization of oestrus in sheep as compared to goats.

Comparison of Sheep and Goat

Table 8.4 : Some important differential points between sheep and goat

Sr. No.	Item	Sheep	Goat
1.	Chromosome No.	54	60
2.	Breeds in India	44	23
3.	Annual growth rate	3.28 %	2.6 %
A. Physical characters			
4.	Tail	Long, hanging and fairly broad.	Short, thin and upright sharp and little flesh.
5.	Back and wither	Round and well fleshed.	Sharp and little flesh.
6.	Thorax	Barrel shaped.	Flattened latterly.
7.	Radius (bone)	1-$1^1/_4$ time length.	Twice as long as metacarpal.
8.	General appearance	Fatty and mostly roundish type.	Taller, thinner and more angular.
9	Body cover	Wool or hair.	All Hair.
10	Flesh (Meat)	Pale, red, fine in structure.	Dark, red, coarse with goaty odour.
11	Presence of beard	Absent	Present
12	Facial gland	Present	Absent
13	Foot gland in hind foot	Present	Absent

Contd.

B. Reproductive behaviour

14	Onset of puberty	12 months	4-8 months
15	Sexual maturity	Late	Early
16	Estrous cycle	14-20 (17) days	7-24 (21) days
17	Estrus period	24-48 hrs	16-50 hrs
18	Gestation period	150 ± 3 days	150 ± 3 days
19	Time of ovulation	12-24 hrs after onset	25-30 hrs after onset

C. Feeding behaviour and digestive efficiency

20	Feeding pattern	Grazer and less selective.	Browser and more selective.
21	Browse and tree leaves	Less relished and prefers grazing.	Relished and preference.
22	Feed varieties	Lesser preference.	Greater Preference.
23	Taste sensation	Less	More
24	Recycling of urea in saliva	Lesser	Greater
25	DMI	Meat-3%Lactating-3%	Meat-3%Lactating-4-6%
26	Water intake per unit DMI	Higher	Lower
27	Digestive efficiency	Less efficient	Highly efficient
28	Retention time	Shorter	Longer
29	Moisture in faeces	More	Less
30	Urine	Less concentrated	More concentrated

Sheep Production Systems

Sheep are reared by men and women with diverse working and professional background in India. The production systems are as numerous as the socio-economic and varied agricultural situations in the country. However, they can be broadly classified into the followings-

a) **Tethering:** This is common in the sub-humid and humid zones, where probably because of intensive cropping; it is a convenient means of rearing sheep from the stand point of control, minimum labour input and utilization of feed *in situ*.

b) **Extensive production:** This involves low carrying capacity in situations where land is marginal and is plentiful. The system is used by nomadic people, usually in very low rainfall areas or during winter months when crop resides are available.

c) **Intensive production:** The sheep are fed in confinement with limited access to land. It involves high labour and cash inputs. Cultivated grasses and agro-industrial byproducts are fed *in situ*. This system also has the advantage of allowing control over the animals.

d) **Semi-intensive production:** This system is practiced to some degree in most of the situations but the nature and extent of integration depend on the type of crops grown and their suitability to sheep. The advantages of this system are increased fertility of land via the return of dung and urine, control of waste herbage growth, reduced fertilizer usage, easier crop management, increased crop yields, and greater economic returns.

Breeds of Sheep

Total number of sheep breeds present in the world is more than 1200. India has 44 breeds of sheep. Amongst them 5 can be classified into medium or fine wool, 14 into coarse carpet quality wool and rest into hairy meat type breeds. They vary from the non-woolly breeds of sheep in the southern peninsular region mainly kept for mutton and manure to reasonably fair apparel wool breeds of Northern temperate region. These breeds have been discussed in detail in chapter 3 of this book.

Housing of Sheep

Sheep do not require elaborate type of housing but the provision of the minimum housing requirement will definitely increase productivity. It is therefore necessary to provide cheap houses with thatched roofs made of locally available materials and fences nearby. Only a very simple cheap shelter to protect the sheep from rains, snowfall and extremes of weather is all that is required. Advance pregnant ewes, ewes at the time of lambing and young stock need the special care. Different types of housing and various space requirements have been discussed in detail in chapter 5 of this book.

Feeding and Watering of Sheep

Feed may be defined as any dietary substances that nourish the animal body for maintenance, reproduction and production. Two main types of feeds are roughage and concentrate. Roughage includes fodder crop, crop residue, hay, silage, straw and stover etc. Concentrate is the mixture of different ingredients like maize, ground nut cake, wheat bran, rice polish, fish meal, meat meal, mineral mixtures, salt etc. The straw is made more palatable using delignification by soaking for 30 minutes in lye (3.3 percent sodium hydroxide) and then washing thoroughly. Other way of increasing the feeding value of straw is to impregnate it with a mixture of urea molasses and trace minerals (1.5-2 percent feed grade urea, 10 percent molasses and 0.5 percent mineral mixture). Minerals are required at the rate of 2-3 percent of ration.

Depending upon the agro-climatic condition and breeds of sheep, the water requirement ranges from 3-5 litres per day. Tropical sheep breeds have different tolerances to both water stress and quality of water. For example the Marwari, Chokla and Pugal breeds are able to consume saline water. It is not clear whether this tolerance is genetic or due to conditioning. Sheep on exceptionally lush pasture can and will voluntarily go for extended periods without drinking water.

Feeding of Different Categories of Sheep

1. **Feeding during resting period:** Sheep can be maintained only on grazing because the requirements of the sheep are low during resting period (no reproduction and production). During this period, ewes are kept on a low level of feeding. The flock can be used as scavengers to clear up poor and over grown pastures, banks stubble's etc. The practical way to find out whether the non producing animals are receiving sufficient nutrients from their grazing is to weigh them periodically. If the animals are maintaining their body weight or gaining a little to the extent of about 0.02 to 0.03 kg/day they can be stated to be on maintenance rations. If they are continually losing bodyweight, then their grazing will have to be supplemented with additional concentrate mixture.

2. **Flushing :** Nutrition of ewes should be stepped up to promote their body weight about two weeks before the onset of the breeding season. Gain in body weight prior to mating likely to produce more lambs of better weight. About 5-10 percent (up to 30 percent) of lambing can be expected to increased by flushing. The breeding flock should be treated for internal and external parasite before starting the flushing period.

Benefits of Flushing Ewes are as follows

- Synchronisation of oestrus i.e. all the ewes will come in heat nearly at the same time.
- Ewes will come in heat promptly.
- The flushing act as body reserve.

Followings are some of the Flushing Rations for Ewes

- A good mixed pasture of legumes and grasses.
- A grass pasture + 150 g wheat bran/ head /day.
- Grass pastures + 250 g grains + 450 g oil cake.
- Legume hay + 100 g of wheat bran + 150 to 200 g of grain.
- Green fodder at the rate of 10 % of body weight + 100 g oil cake / head/ day.

Reduced fertility may be seen because of over fattening. The over fat ewes may be gradually brought down to the desirable lean condition by reducing the ration and exercising.

3. **Breeding:** Flushing ration can be continued during breeding season.
4. **Gestation**
   a. **Feeding during early and mid pregnancy:** For a healthy strong lamb crop and increase in productive life of a ewe, good feeding during gestation is the key stone.

Advantages of Proper Feeding of the Ewes are

- There will be increase in the number of birth of strong and healthy live lambs.
- There will be prolonged productive life of a ewe.
- Wool productivity will be improved.
- Milk yield by the ewes will increase and will result in healthier weanlings.
- Decreases the probability of ewe's and their lamb's death as a result of exhaustion and weakness.

During early and mid pregnancy, the extra nutritional requirements are not much and whole of their needs can be met entirely by grazing. Some ration recommended for exotic ewes are-

- Grazing on a good pasture.
- Sorghum silage (1 to 2 kg) + Legume hay (½ to 1 kg / head / day).
- *Ad. libitum* supply of plenty of maize + sorghum fodder + 50 g oil cakes / head / day.
- Grazing on stubble's and harvest field supplemented with 100 g oil cakes / head / day.

b. **Feeding during late pregnancy:** The foetus grows rapidly in the uterus during the last one month of pregnancy. The ewe cannot consume excess bulky roughage during the advanced stages of gestation as the space in the abdomen is reduced due to the growth of the foetus. Hence, they should be provided with extra concentrate. Pregnancy disease or ketosis in ewes may occur due to lack of enough energy in the feed. Therefore, molasses or grains may be fed at the rate of 225 g / head / day + green fodder - 1 kg + *ad. lib.* mature pasture grass to provide enough energy to the ewe.

5. **Feeding at lambing time:** The grain allowance should be materially reduced as lambing time approaches or immediately after lambing, but good quality dry roughage to be fed on free choice. After parturition the ration of the ewe may be gradually increased so that she receives the full ration in six to seven days time. In general, bulky and laxative feedstuffs may be included in the ration during the first few days. A mixture of wheat bran and barley or oats at 1:1 proportions is excellent soon after lambing; the ewe must be given just enough of slightly warm water.
6. **Feeding lactating ewe:** The lactation curve of sheep is quite typical and different from the lactation curve of cattle. The peak milk yield comes during first after lambing and then decline after wards. For rapid growth of lambs the milk production should be maintained at adequate level, and ewe must be supplemented with extra ration for the obvious reason. If they are provided good pasture, the requirements are more or less met. When good pastures are not available all care must be made to provide extra protein and minerals particularly of copper and cobalt other than common salt. About 16-18 % crude protein should be there in the supplemental feeding.

The shepherd at this stage must be cautious to increase the supplemented feed gradually up to 250 to 500 g either as gram *chunni* or a mixture of cereal grains and wheat bran in the proportion of 2:1 along with legume hay of 1-2 kg.

Care and Management of Different Categories of Sheep

Care and Management of Pregnant Ewes

- During pregnancy the ewe should gain 4-8 kg of body weight.
- Advanced pregnant ewes should be isolated from flock and kept in separate pen.
- Frequent handling of pregnant animals should be avoided.
- Ewes should not be allowed for fighting with flock mates.
- Inadequate feeding may result in pregnancy toxaemia, abortion and premature birth of fragile young ones.
- Extra ration over and above maintenance should be fed during the last trimester of pregnancy, keeping in view of milk production of ewes and quality lamb production.
- Pregnant ewes neither be crowded together nor be allowed to jump over hedges.
- The ewe approaching parturition should be placed in well ventilated lambing *pacca* floored pen/crates containing *pacca* manger and water trough for 4 to 6 days prior to expected date of lambing.
- They should not be allowed to pass through the narrow gate.
- They should be provided shades of asbestos sheet or tree during mid day especially in summer.
- The parturient animals must be saved from cold and chilly weather.
- To reduce stress to transition animal bedding material should be provided.
- During the last part of pregnancy the animals should be carefully viewed. It is observed that early maturing breeds have slightly shorter gestation length therefore management should be carried out accordingly.

External indications of last stage of pregnancy

- Ewes carrying a single lamb are heavier than at mating time but leaner when handled.
- Ewes carrying twins are the heaviest and leanest.
- Depression in croup region because relaxation of sacrosciatic ligaments.
- There will be enlargement of vulva.
- Vulva will become flabby, red and waxy.
- Tightening of teat and enlargement of udder.

Care and Management of Lambing Ewe

External indications of lambing

- There will be vertical movement of the tail.
- The ewe will try to lag behind or become separated from the flock.
- The ewe will scratch the ground with the help of fore legs.
- She will constantly change her position by standing and sitting alternately.

Care of ewe after lambing

- Placenta should be disposed of immediately.
- The hind quarter including udder and should be cleaned with fresh & clean water.
- The ewe should be provided with free access of water all the time.

Care and Management of Lambs

- Lambing pen must be equipped with portable manager and water trough.
- Usually mother will lick her lamb dry. However, under certain situations she doesn't do this make the lamb dry by using clean, soft, dry towels.
- Lambs will be on feet within10-15 minutes and reach up to udder within 20-25 minutes after birth. If lambs fail to do so,

provide assistance to reach to the udder for colostrum consumption.

- Teats of mother are tight and turgid immediately after parturition hence lambs face difficulty to take colostrum. Remove few strips of milk make and the teats loose.
- Keep the lambs along with the mothers in lambing pen for at least three days.
- Birth weight of the lamb should be recorded before colostrum feeding.
- Lambs should be protected from getting chill by wrapping in curtains made up of jute / blanket.
- In case of orphan lamb, foster mother's milk is rubbed on the rump of lamb and nose and allows smelling the mother.
- Whole skin of the dead lamb can be put over the body of orphan lamb for easy acceptance.

Feeding Management of Lambs

The lamb up to 2 weeks of age entirely depends on mother's milk. Colostrum must be given to the lamb as it is a rich source of fat, protein, vitamin and antibodies to protect it from several infectious diseases in the initial stages of life. For feeding of lamb, milk of ewes and cow should be mixed to supply all the required nutrients by the lamb. The lambs are allowed to suckle in the morning before the ewes are let out for grazing and again in the evening when the ewes are brought back to the pen. Under intensive production system of rearing, lambs are allowed to suck the milk thrice in day during 1st month, twice in 2nd month and once during 3rd (last month of weaning) month.

At one week of age, a handful crushed and compounded concentrate feed should be given to lambs for nibbling and required quota of concentrate should be given after 15th day onward. Gradual increase in concentrate mixture should be done. After about 1.5 months of age, lambs should be let loose nearby night pen for grazing or they should be provided with *ad lib* dry fodder and limited green fodder. The pastures available are often of poor quality or limited grazing area is available. Under such circumstances, the rations of these lambs may be supplemented with grains and oil cakes in addition to their mother's milk and pastures.

The grains which are to be fed to lambs should be cracked before feeding up to six weeks of age. After that period grains can be fed as such except in the case of hard grains, which may be cracked, crimped or rolled.

Complete pelleted ration consisting of roughage and concentrate has been found to be advantageous. Lambs had been found to consume more and grow faster. At the start of pellet feeding, 65 to 70 % roughages has to be incorporated in feed which should be gradually decreased to 50 % by ten to twelve weeks of age. Composition of conventional and non conventional concentrate feeds has been shown in Table 8.1.

Table 8.1 : Composition of conventional and non conventional concentrate feeds.

Ingredients	Conventional		Non - conventional	
	Pre weaning	Post weaning	I	II
Maize	38	15	-	-
GN cake	25	22	23	-
RP	24	15	12	-
Mango seed kernel	-	-	15	20
Babul pods chunni	-	-	17	19
P.J. pods	-	-	20	23
Corn steep liquor	-	-	-	15
Cassia tora seeds	-	-	-	10
Molasses	10	10	10	10
Mineral mixture	3	3	3	3
Total	100	100	100	100

Orphan lamb and chilled lamb are the two important conditions which require proper care and management for the survival of the lambs.

Orphan /Disowned /Bomb lamb

The orphan lamb is the result of disowned by the ewe or death of the ewe. Similarly, if the ewe has twins or triplets or may not have enough milk which results into disowning of lambs. These lambs can reared either on goat milk or by foster mothers. Liquid paraffin and antibiotics should be added to the hot goat milk, and fed at regular intervals. However, best effort would be to have a foster mother. To encourage owning of such lambs either kerosene oil should be applied on the nostril of the mother or ewe milk should be sprinkled on the orphan lamb; by smell of milk ewe would accept the lamb.

Chilled Lamb

Sometimes, the lamb becomes chilled when the lambing takes place during extreme cold weather. The care should be taken during the lambing time to protect the ewe and lambs from inclement weather. Fresh bedding should be provided to avoid chilling of lambs. In case lamb is chilled, the entire body should be placed in luke warm water except the head for few minutes. Then the body should be rubbed and wrapped in dry cloth or gunny bag and put it in a dry pen. The warm milk should be given at frequent intervals.

Care and Management of Breeding Sheep

Mating Seasons or Breeding Seasons

In India, there are three main breeding seasons. In summer season (March and April) about 60 to 80 per cent ewes come in heat. Another season of mating is autumn (June, July and August). This is the most important season of mating as about 15 to 20 per cent ewes come in heat during this season. The third season of breeding is the winter season (October to December) in which very few ewes come in heat. In the western region, ewes are mostly bred in autumn season. The lambs from this mating are born during the period of early spring (November to January). In general the fertility is high during autumn in the plain whereas in hilly areas high fertility is observed in summer.

The rams are allowed to run with ewe flock only for 40 to 50 days during the breeding season. This results in uniform lambing. Because of uniformity in lambing proper care, protection and feeding required during this critical period of life can be provided easily. Labour and other expenditure to be incurred for this purpose are restricted only to a short and limited period of about 40-45 days.

Heat periods: The duration of the heat period in sheep ranges from 1 to 3 days. The average for indigenous sheep is 21 to 39 hours. Generally ewes are bred near the end of the heat period. The oestrus occurs every 15 to 19 days (Average-17 days) during the season of breeding. However, females of tropical breeds cycle throughout the year unlike the temperate breeds which are affected by hours of day light and breed with declining day length.

Heat detection and mating: The ewes can be detected in heat by (i) use of appronized ram (ii) teaser/vasectomised ram (iii) estrus symptoms. The Sings of estrus are **(i)** following of ram by ewes, (ii) flaking of tail and (iii) scantly vaginal discharge. The ewe remains in heat for long period of time (36-39 hours) and ovulation take place at the end of /before the cessation of oestrus, the mating is done twice at the interval of 24 hours.

Gestation: *The* gestation period varies from 142 to 152 days with an average of 147 days. The medium-wool breeds and meat-type breeds ordinarily have a shorter gestation period than do the fine-wool breeds. High temperatures and high nutrition levels may shorten the gestation period two or three days. Ewes bred to white-faced, wool-breed rams may have a slightly longer gestation period than those bred to black-faced, meat-type rams.

Care and Management of Lactating Ewes

It is usually preferable to feed lightly on the day of parturition, but allow plenty of clean and cool drinking water. In addition to this the animals may be provided with bran and leafy hay. The concentrate level may be increased gradually. For lactating animals in addition to grazing, supplement with concentrate mixture @ 250- 350 g/day/animal for 2 months and adequate amount of leguminous green fodder is required.

Essential Farm Operations on Sheep Farm

Wool growth in different parts of the body especially around eye and penis may prevent sheep for easy mating. Problem of overgrown hooves and improper marking of rams may also be there. Ram as well as Ewe requires special management during breeding season. Following are some of the management points should be adopted for the preparation of ewe and ram for breeding.

1. **Tagging:** The cutting of wool from around the dock, vulva and inner portion of thighs (breech region) is called tagging. Ewes are tagged prior to breeding and lambing. It is carried out to prevent blowfly strike in breech region, to facilitate mating and for general cleanliness at lambing. In Australia, this practice is known as crutching.
2. **Eyeing:** In some breeds of sheep, wool grows around the eyes hence thorny grass seeds and other materials get accumulated in this wool. This result into wool blindness and as a result of this sheep can't see while grazing. So, the wool should be clipped from around the eyes. This process is known as "Wigging" or "Eyeing".
3. **Ringing:** Shearing of rams prior to breeding season will make the sheep more active but if this is not practiced, ram should at least be clipped from neck and from belly in the region of penis. This process is known as "ringing". It will help in

preventing the chances of infection at the time of mating and facilitates easy contact with ewes while mating.

4. **Care of feet:** Trimming of hooves is not necessary when sheep are raised under range conditions. These sheep have to walk sufficiently distance to keep their hooves smooth and clean by wearing, but when sheep are either stall fed or do not have sufficient exercise, the wearing of feet does not take place sufficiently. Due to this, the hooves over grow and facilitate accumulation of dung and dirty soil. Some time decaying of hoof takes place. A ram will have difficulty in mating if he will have over grown hooves. For the reasons mentioned above, trimming of hooves prior to breeding season is necessary. It can be done with the help of knife or pruning shears.

5. **Marking of ram:** When we are not using a teaser for detection of heat and rams are directly used for mating, it is advisable to mark ram. Marking of a ram will be helpful in recording of breeding date, for knowing the sexual activity of ram and from knowing which ewes are bred. Breast marking of a ram is done with different oil colours which may be changed after 17 to 18 days. Special marking harnesses are also available, which can be fitted to the rams in breeding season. Breast paint will be stamped on the rump of the mated ewe. It is useful to identify the ewe bred.

Management of Ram

The ram lamb attains sexual maturity at 5-10 months of age and can normally be put to light service at 12-18 months. After 2 or more years of age, they can be put to full breeding use. A mature ram should be able to serve 30-60 ewes, although in tropics one ram is kept for 10-20 ewes.

Mating Systems

Broadly the mating system can be classified in to two type i.e. Natural and Artificial insemination methods of breeding. Under the natural breeding, pen and flock mating are commonly used by farmers while hand mating is exclusively used for experimental purposes. Adoption of artificial insemination method in sheep is not easy as in case of cattle and buffalo due to certain problems connected with dilution and preservation of ram semen.

A. Natural Breeding

1. **Hand mating:** In this system teaser ram detects the ewes in heats, which are separated out for breeding with desired ram. The females are allowed to mate on by one. In this system a ram may not be allowed to mate more than three ewes a day.
2. **Pen mating:** In this mating system the ewes are divided in different batches varying from 20-25 ewes. Males are turned into the flock only during the night time and separated during day time. Which gives enough rest to the male and they can be fed properly.
3. **Flock mating:** This system of mating is practiced normally by the farmers. In this system males are allowed to run along with the females throughout the day and night. This system has its own demerits. The ram may lose its most of its body reserve in chasing the ewes and they may lose their body condition. Some time the ram may develop attraction for particular ewe in heat and serve it a number of times while other remains unattended resulting in empty ewes and low fertility rate. Moreover the ram some time exhaust itself overnight by serving more than a dozen times and the last served ewes not receive optimum number of spermatozoa and remain non pregnant.

B. Artificial Insemination

Artificial Insemination offer the best means of distributing germplasm from nucleus breeding flock to many small flocks within each ecosystem. Fresh as well as frozen semen is used. The vaginal speculum method of insemination is used for ewes. Generally artificial insemination leads to lower reproductive rate than natural service and frozen semen gives even much low pregnancy rate that is around 40 per cent. Cervical insemination is generally followed for better conception rate.

Embryo transfer requires surgery in sheep, therefore it may not become practical in production, but it offer great promise in enhancing genetic progress.

Methods of pregnancy diagnosis: Pregnancy diagnosis is important for rebreeding of empty ewes and for better management in case of pregnant ewes. Pregnancy diagnosis in sheep can be done by following methods.

- **External symptoms:** Cessation of heat, enlargement of belly from three months onward, animal becomes more docile and affectionate to owner etc.
- **Chemical method:** Urine sample of 5 ml is taken in a beaker and to that 5 ml of 1% Barium chloride is added and mixed properly. If the turbidity is noted that means the ewe is non-pregnant. Clear solution indicates pregnancy.
- **Other methods:** Rectal probe, laparoscopy, vaginal biopsy, ultrasonic fetal pulse detection, radiography, ballottement of abdominal region and progesterone hormone analysis are some other methods of pregnancy diagnosis in sheep.

Reproductive Performance: The reproductive performance or breeding efficiency is the number of young born and raised to a marketable age i.e. 3-5 months, corresponding roughly with weaning. The breeding capacity is expressed as a percentage of the number of breeding ewes in the flock during the breeding season or seasons for 1 year. Lambing more than once in a year, multiple births and less age at sexual maturity increases the breeding efficiency. The ewes that have been flushed and gain in weight just before breeding will usually shed more ova, assuming they were not already overweight, and thus deliver more lambs. Practices which are expected to enhance reproductive performance include disease control, good nutrition, mating at proper age and maintaining the correct number of vigorous fertile rams within the flock. Careful selection for increased reproductive efficiency using life time or repeated records can also improve performance. Crossing with a breed that has exceptional fertility is another method of using quality germplasm to improve breeding capacity.

Health Care of Sheep

Keeping the animals in healthy condition is indication of proper management. Disease, the opposite of animal health, may also be defined as any abnormalities in the functions and tissue of the animal body. Diseases are classified mainly into two i.e. hereditary (genetic) or acquired (contracted after conception or birth). According to etiology the diseases are classified into viral (virus), bacterial (bacteria), fungal (fungi), protozoan (protozoa) and parasitic diseases. Parasites are classified into ectoparasite (flies, mosquitoes, lice, fleas, ticks and mites) and endoparasites (flat worm: trematodes, tape worm: cestodes and round worm: nematodes). Some of the diseases of sheep and their control measures have been presented in Table 8.2.

Table 8.2 : Common diseases of sheep and their control measures.

Disease	Causative agent and clinical signs	Control and prevention
Anthrax or Charbon or Spleenic fever	*Bacillus anthracis.* Tar coloured dark bloody discharge from natural opening of body and sudden death.	Vaccination, carcass suspected of anthrax should never be opened. Burning or burial of carcasses. Common disinfectant in use is 5% lye solution.
Black quarter or Black leg (Less common in sheep)	*Clostridium chauvoei.* High fever, characteristics swelling under skin (thigh region) that crackles to touch. Soon after death a dark, frothy discharge flow from natural body openings.	Vaccination is good for 12-18 month. Burning or burial of carcasses.
Brucellosis/ Malta or Undulant fever	*Brucella melitensis.* Abortion and infection of uterus; orchitis in males. Lameness.	Good hygienic condition and vaccination, regular blood test and isolation of affected animal.
Johnes disease or Para tuberculosis	*Mycobacterium paratuberculosis.* Intermittent diarrhoea and constipation, foul diarrhoea being more prevalent, animals lose body weight.	Segregation of healthy animals, strict quarantine for new animals
Haemorrhagic septicaemia	*Pasteurella multocida.* Fever, depression, rapid breathing, sneezing, coughing; death sometime after 24 hours.	Segregation, avoidance of infected pasture, feed and water resources, pre-monsoon vaccination.
Tuberculosis	*Mycobacterium tuberculosis.* Usually lungs and lymph gland affected, loss of weight, swelling of joints, chronic cough and laboured breathing.	Periodical tuberculin test, BCG Vaccination.
Foot rot	*Fusiformis necrophorus.* Lameness and pus formation between hoof and underlying tissue.	Hoof trimming, and application of antiseptic. Cupric sulphate solution
Mastitis	*Steptococcus agalaccia.* Udder tender, swollen, abnormal milk secretion	Antibiotic infusion; sanitary measures. Plastering the teats with mud or dung.

Contd.

Tetanus	*Clostridium tetani*. Rigidity of muscle especially of head, "lockjaw". Respiratory muscle affected and fever sets in.	Injection of anti-toxin, muscle relaxant and penicillin. Management of hygiene especially in connection with castration procedures.
Blue tongue/Catarrhal fever	*16 antigenic strain(Insect borne)*. Fever, depression, lack of appetite, inflammation and ulceration of mucous membranes. Tongue swollen and bluish.	Vaccination,Control of biting insect.
Foot-and- mouth diseases/Aphthous fever/Epizootic stomatitis	*7 serotypes*. Eruption of vesicles in mucous membranes of mouth and on skin of digits.	Vaccination, slaughter. Native animals less affected.
Sheep pox	*Filtrable virus*. Disease is very virulent- lesion on cheeks, lips, nostrils, around the eye, on the vulva, udder and scrotum, respiratory and GI tract too in very severe cases.	Vaccination with appropriate vaccine. Careful nursing is helpful.
Rinderpest	*Filtrable virus*. High fever, blood stained shooting diarrhoea, high mortality.	Vaccination
Ring worm	*Trichophyton verrucosum*. Lesion on the skin as circular, raised crusts and local thickening of skin. Irritation on affected part.	Cleaning and disinfection of sheds, use of fungicides.
Coccidia	*Eimeria arloingi*. Loss of appetite and diarrhoea in lambs; spore in faecal examination.	Sanitary measures, segregation of affected animals.
Babesiosis	*Babesia bigemina*. High temperature, rapid breathing, red to dark brown urine, yellowish visible mucus membrane.	Control of ticks by dipping during tick season.
Trypanosomiasis.	*Trypanosoma evansi*. Fatal disease, pale mucus membrane, swelling often seen on legs.	Eradication of tse tse fly (Glossina spp.); breeding of trypano-tolerant animal
Theileriasis	*Thelleria hirci*. Haemoglobinurea- red to dark brown coloured urine	Control of ticks by dipping during tick season.

Contd.

Roundworm	*Hoemonchus contortus, Oesophagostomes.* Unthriftiness, loss of weight diarrhoea, anaemia, pale mucous membrane of eye and mouth, potbelly and swelling under the jaw.	Phenothiazine, copper sulphate and nicotine sulphate. Pasture rotation, proper stocking rate
Tapeworm	*Cysticercus ovis, Taenia saginata, Coenurus cerebralis.* Lambs most adversely affected, Larval stage called bladder worm, *Cysticercus ovis* affect mutton giving it a spotted or measly appearance.	Control difficult due to various species involved, pasture rotation and sanitation
Flukes	*Faciola hepatica.* Sheep are highly susceptible, death is rapid on heavy infestation, diarrhoea, anaemia, watery swelling on the dependent part of the body	Control of intermediate host snail; treatment with thiabendazole, tetrachloromethane and tetrachloroethane is recommended.
Mite	Mange miteRestlessness, scratching and rubbing, disturbed hair or wool and skin inflammation or lesions	Lime-sulphur/nicotine sulphate dip is a standard treatment, lindane and BHC is also effective.
Ticks	Blue tick, red tick, brown tick, black pitted tick Visual observation, common tick born diseases are babesiosis, theileriasis	Insecticide dip or spray (BHC, DDT, toxaphene, lindane)
Flies	Midges, mosquitoes, horse fly, stable fly, common house fly, tse tse fly etcVisual observation of the parasites and abnormal behaviour of the host animals, common diseases transmitted by fly are Anthrax, Trypanosomiasis	Periodic spraying in and around animal house with suitable insecticides, hygiene, proper manure disposal

Marketing and Transport of Sheep

Marketing of live animals and wool have been discussed in chapter 1 while methods of transport of sheep have been described in chapter 2 of this book.

Sheep Products

Meat is the principal products from sheep farming in the tropical countries. However, secondary products like wool, hair, milk and manure are also obtained from them. In certain areas they also act as pack animal and cash reserve.

1. **Milk:** Sheep contributed around 7.76 and 3.17 million tonnes of milk in world and Asia, respectively. In India, average daily milk production from sheep is very less (usually less than 1 kg/day) hence milk produced from them is generally used as source of feed for lambs and virtually little milk remains available for sale/consumption.
2. **Meat:** Along with goat meat (chevon), mutton is the chief source of meat supply in India. The share of Indian sheep is about 5.71% (0.2 million tonnes) of Asia (3.5 million tonnes) and 2.66% of the world (7.5 million tonnes) mutton production. Contribution of sheep in world and Indian meat pool has been discussed in chapter 1 of this book. In general, mutton has more fat and less protein than the chevon and average dressing per cent varies between 40-50%.
3. **Wool production:** Wool production at the end of 11th Plan (2011-12) is anticipated to be 44.4 million kilograms. It has decreased from 43.9 million kilograms at the beginning of 11th Plan (2007-08) to 43.0 million kilograms in 2010-11.
4. **Skin**: Skins are valuable for leather industry. After removal of wool, the skins are prepared by washing and soaking in salt pickle solution. Sheep skin is used for manufacture of leather coats, jackets, shoes, gloves, slippers etc. Lamb skin fetch better price than the others. Sheep produces 15542 thousand of skin/ year.
5. **Manure:** The sweeping of sheep sheds is placed in a pit for decomposing and it is applied later to the farm. The nutrients present in the urine are wasted in this method. The other effective method is sheep-penning (night folding) where sheep

are kept whole night in the fields and urine and faecal matter added to soil are incorporated to a shallow depth by cultivator. Faeces of sheep contains high amount of nitrogen, due to this reason it is called as golden hoofed animal. Sheep manure contains 2.8, 0.42 and 2.0% of Nitrogen, Phosphorus and Potash, respectively.

Wool

Fleece obtained from sheep is known as greasy wool or raw wool. It is an animal fibre which grows from follicles in the skin and has a greasy coating. Followings are the components of wool fibre

1. **Epidermis or Cuticle:** The cuticle cells comprise about 10% of the mass of the wool fibre and overlap each other with the exposed edges pointing towards the tips of the fibres. The cuticle cells of merino wool fibre are basically rectangular and 20x30x0.5μm in size. The cuticle, in textile use, is responsible for the surface properties of the fibres and also the barrier for diffusion of chemical reagents into the cortex of the fibres.
2. **Cortex:** The cortex of wool fibre forms about 90% of fibre volume. It has major influence on the mechanical properties of the fibre. There are two types of cortex cells i.e. ortho and para. The location of these two types of cortical cell within the cortex has a profound effect on the crimp of the wool fibre. In general, fibre for fine merino wool, the ortho-cortical cells usually account for over 50% of the cross-section. On staining with methylene blue, ortho-cortex absorbs more dye than para-cortex.
3. **Medulla:** In fine wool medulla is absent but in medium and coarse wool it is third layer known as medulla and is undesirable. Coarse wool fibres usually has diameter greater than 35μ and have a third type of cell, those of medulla. This is central core of cells arranged either continuously or intermittently along the fibre axis.
4. **Wool grease or fat content of fleece:** Wool grease is secreted by fatty glands around the fibre follicle of the skin. Wool grease or fat is part of yolk which is soluble in fat solvent and different from of suint.
5. Factors affecting grease in wool
    1. Age: Increases up to 6 years and then declines.

2. Sex: Rams seems to have more grease than ewes.
3. Body part: Wool from shoulder is more greasey followed by back, belly, side and legs.
4. Care and housing: Well fed, cared and housed have relatively more grease.

6. **Chemical composition of wool:** Wool contains chiefly keratin which is also a primary constituent of hair, feathers, nails, horns and hooves. Keratin in the mixture of nitrogen and sulphur compounds and amino acid chemically it contains- carbon - 50%, oxygen-22-25%, nitrogen-16 -17%, hydrogen:-7% and sulphur-3 - 4 %.

7. **Wool yolk:** It is used for all extraneous material adhered to wool as impurities and includes both grease and suint. Wool contains:
   1. Wool fibre: 15 to 70 %.
   2. Grease and suint: water soluble fraction of yolk: 12 to 47%.
   3. Moisture: 5 to 24%.
   4. Dirt: 3 to 24%.
   5. Vegetable matter.

8. **Types of fibre:** Wool is the natural protective covering of sheep. However, whatever grows on the body of sheep is not wool. In addition to wool, some other fibres also grow on the body of sheep viz. Hairy fibres, kempy fibres and heterotype fibres etc. in different proportion according to breed of sheep.
   1. **Kempy fibre:** Kempy fibres are coarse fibres of the outer coat. They have a long gradually tapering distal end or tip and slightly bulbous basal end or root. They are present in birth coat of lamb.
   2. **Heterotype:** It is a hairy type of fibre contains interrupted medulla. This type of fibres are present more in number in Marwari sheep.
   3. **True wool:** It is non-medullated, small in fibre diameter. The cuticular scale pattern of the fine, true wool fibre is of the irregular mosaic type throughout its whole length. Two scales are sufficient to encircle the fibre. True wool fibres also have a greater overlap of scales (serrations) than either kemps or medullated heterotype.

Wool Qualities and Factors Affecting It

Following are some of the factors which affect the wool quality

1. **Staple length:** Numbers of wool fibres which naturally form themselves into cluster or lock, the length of staple varies in different breeds of sheep. According to length, Merino wool is used either in worsted trade or woollen trade. The staple length is not uniform throughout the fleece. Staple length depends upon breeds, crimp formation, no of second cuts and method of shearing.
2. **Spinning quality:** Spinning qualities refers to the diameter of the fibre. Spinning quality includes elasticity, softness, pliability and felting power. Spinning quality is denoted as S′. The spinning quality is inversely proportioned to fibre diameter of the wool fibre.e.g. 64 S′ means 64 hangs can be spun from 1 pound clean wool yield. More the spinning quality, more hangs can be spun from 1 pound of clean wool yield.
3. **Soundness:** It is necessary tensile strength to undergo the combing process economically. Unsound wool is frequently referred to as tender wool which will break relatively easily when tension is applied. Unsoundness is caused by a sudden check or change in the feed supply to the growing fiber. Tender wool is due to copper and sulphur deficiency. Which will leads to shorter staple length and finally affects the price of fleece.
4. **Elasticity:** Elasticity refers to the property of the wool fibre which regains its former length after being stretched considerably. The crimpy formation of the wool fibre is largely responsible for its elasticity. The finer the fibre, the greater is its power of elasticity. Normally a wool fibre can be stretched up to 30 percent of its length and come back. Fine wool fibre is more elastic than course/carpet wool.
5. **Softness:** It is the characteristic which imparts great value to woollen clothing. Merino wools are softer than the wool from crossbred or down breeds. The softness in the wool is indirectly indication of crimp formation in the wool. Merino wool is softer than Indian wool.
6. **Condition:** Condition refers to the amount of wool grease or yolk plus other foreign matter such as dust and vegetable. The yolk is a mixture of sweat and wool. Grease or fat wool has no

value for manufacturing purposes until cleaned of all foreign matter. The heaviest conditioned wool is found amongst the finest grades of Merino sheep. Rams wool possesses a sticky type of yolk which has a strong odour and is harder to remove from the wool in the scouring process for this reason rams wool should not be bailed with other fleece wool.

7. **Felting property:** Felting refers to the adhesive qualities of wool and is due to serration present in the wool. The felting property of wool is directly related to the number of serration present per unit area. It is possessed to the greatest degree by superfine merino wools and because of its action wool can be spun into very fine fabrics.

8. **Density:** The term density is mainly used in referring to the growth of the fibres per unit area on the skin; a dense fleece is much more valuable than one which is slack or lacking in density. Density as a rule means added weight of fleece.

9. **Power of absorption:** Wool possesses the power of absorbing moisture to a greater extent (up to 18% of its weight). The capacity of absorbing moisture is one of the reasons why woollen underwear is healthier to wear than either cotton or silk because wool garments absorb perspiration readily without becoming damper clammy.

10. **Trueness:** Trueness refers to the wool fibre which is true in diameter throughout its length. Changes of feed often affect the trueness of the individual fibre. Trueness is usually judged by the appearance of the crimp - an irregular uneven crimp in a fibre indicates a lack of trueness.

Factors Affecting Various Quality Parameters of Wool

Some genetic and non genetic factors affect the quality of wool. Out of these some of the important factors are as follows-

1. **Sex of the animal:** Wool shorn from a ram is coarser as compared to the wool shorn from a ewe.

2. **Breed:** Breed of animals affect so many qualities of wool viz. staple length, grease content, softness, density, fibre diameter etc. Merino produces finer, shorter, stapled wool as compared to Lincoln breed. Wool from Merino contains higher amount of grease and is having more and uniform crimp as compared to the coarse wool.

3. **Nutrition:** Level of nutrition affects the quality of wool. Whenever poor quality feed deficient in sulphur and copper is given to the sheep may leads to tender wool and stringy wool, respectively. Fibres break easily and affinity to dye is also least.
4. **Health:** Sickness of the sheep may also cause weaker pointer in wool (Tender wool).
5. **Region of sheep body:** The quality of wool varies from region to region of body. The wool of breech region is coarser than that of the shoulder region. Modulated fibres are more in numbers in the breech, the belly and the head regions.
6. **Season:** Season of shearing also affect the colour of wool, staple length and the fineness.
7. **Shearing practices:** Method of shearing affects the staple length of wool. When wool is shorn by a machine the staple length of the wool will be uniform than hand shearing.

Shearing of Sheep

Shearing is the process of removing wool from sheep quickly, completely, easily and with minimum discomfort to sheep and the operator. Sheep in India are generally shorn immediately after the end of winter season when warm weather commences and when sufficient grazing is available in the fields. In places where two shearing's are obtained, the second one is carried out at the end of rainy season.

Methods of shearing- by the use of

1. Hand shearing (for the flock of 20 or less)
2. Hand machine (for the flock of less than 150)
3. Power machine shearing (for the flock of more than 150)

Procedure

1. All the sheep to be shorn are collected in a holding pen.
2. Sheep is placed on its rump, back to the shearer.
3. The sheep head is held back with the left arm while keeping forelegs in left hand.
4. The shaft to wool shears is taken in right hand and the fleece is opened up first on belly and then on brisket.

5. The wool is sheared down of right arm pit and down to right flank. The wool from udder or scrotum is sheared down.
6. The left hand kept on left stifle and the wool is sheared of left leg.
7. Keeping the hand on left stifle, the sheep is turned right side and the wool is removed on outside of left rear leg from foot, towards thigh.
8. The sheep is turned to lie on right side in such position that back of sheep remains towards shaft of machine and the fleece opened up on the back along the back bone.
9. After making the sheep stand and raising its head with left hand the wool of head and neck is sheared.
10. Standing on left side of sheep the wool from right side of sheep is removed.
11. Then with convenient position the wool from right foot and towards the tail is sheared off to obtain complete fleece of sheep.
12. The number of sheep is recorded before the sheep is released.
13. The entire fleece is collected and tied it with a cotton rope.
14. The fleece is weighed and recorded.

Precautions

1. Pulling of the wool should never be tried for shearing purpose as it may cause cuts on skin.
2. The sheep is hold firmly to prevent kicking and struggling to prevent injuries.
3. The skin is kept tight in front of shears.
4. Always the sheep should be turned from left to right to shear off wool from right to left.
5. Extra care is taken to prevent cuts on sheath, withers, teats, flank, ears, etc.
6. Mixing of dung, sand and other foreign matter should be prevented with fleeces.
7. Shearing should be performed at such times when climate is neither very hot nor very cold.

Chemical method of Shearing: Department of Agriculture Station, Bestville (USA) has developed the chemical shearing. Cyclophosphamide (CPA) @ 24mg per kg live weight is used for chemical shearing. After 12 days of giving the injection, wool starts falling and completes the shearing process within 3 days. The wool comes off in patches. There are certain disadvantages of chemical shearing. Those are (i) The sheep has to be confined for the days the wool is shedding (ii) the drug used for chemical shearing is adversely affecting the growth and development of foetus and (iii) drug leaves residual effect on meat and removes wool too close to the skin exposing the animal more seriously to environmental stresses.

Handling of the Clip

In woollen industry immediately after receiving the raw wool, processing is started which includes sorting, scouring, carbonization etc. Following are some of the important steps in handling raw wool.

1. **Skirting and rolling:** The first operation in the handling of the clip of fleece is skirting and rolling. The fleece after it is shorn is picked up and thrown on to the wool table. The belly wool is kept separate. The object of skirting is to remove any inferior wool and to leave the remaining portion of the fleece as even and as uniform in length and quality as possible. The inferior wool is usually found around the edges of fleece. Along with inferior wool, some urine and dung stained wool is also removed. Over skirting should be avoided. After skirting and grading wool is rolled and tied. Rolling of a fleece is done in such a manner that the most attractive wool (shoulder wool) will remain on the top side of the bundle. Each fleece should be treated individually during skirting. The skirting depends on
    - Type of breed shorn.
    - Quality of wool.
    - The number of sheep of the one class being shorn.
2. **Sorting:** Wool is subdivided in various groups according to spinning quality. Sorting is done in the spinning mill or in the warehouse prior to processing.
3. **Scouring:** Wools with high dirt, suint, wool grease and sand content are passed through duster and then scouring of wool is done. If wool is not having high dust content, it is scoured

out to remove dust and dirt present in the wool. Unfortunately, these impurities are embedded with grease and can be not removed without removing wool grease. Scouring is commonly carried out by passing the wool through a succession of long tanks containing suitable detergents in lukewarm water. This is gently agitated to loosen the dust. Finally, wool is washed with plain water and squeezed between the rollers and dried by spreading on the open floor.

4. **Carding:** If the amount of burr present is not great and the wool is of sufficient length then it may be removed by a special attachment to the carding machine. The spikes on the roller allow the wool to go between them and burr is left on the surface of the spikes. The card serves to homogenize the wool fibers and to distribute them in a smooth strand of card silver.

5. **Carbonization:** The seeds (burrs) from clover and tree foil are a nuisance in the wool so such very burry wool has to go through a process called carbonization. This process involves treatment of wool with dilute acids (H_2SO_4). Carbonizing converts vegetable matter into hydrocellulose which may be taken out afterwards.

Wool Grading

There are two systems of wool grading i.e. American system or blood system and British system or numerical system/spinning count system. American system or blood system is based on fineness of merino wool. Fleeces of the same fibre diameter shorn from full blooded Merino are called "Fine". The other grades are as follows: Half blood, 3/8 blood, ¼ blood. These are described on relative fineness of the wool obtained from sheep containing fractional amounts of Merino blood. Lower quarter blood, commons and braid are the wool of descending order in quality from quarter (1/4) blood wool.

British system or numerical system/spinning count system is based on finest count to which it can be spun. This is used in most countries of the world. This system is based on number of yarn or hanks that can be made from 1 pound of scoured or combed wool. If fineness of fibre is more the length of yarn is greater. A hank is equal to 512m in length. A grade of 60's would mean that 60 hanks could be made from 1 pound and it would be superior/finer to wool of 50's. Similarly, a grade of 50's would be finer than a 40's wool. In case of 40's wool, the number of hanks is 40 and so on. Two systems can be correlated as follows:

Blood system/American system		English/Numerical system
Fine	:	80's, 70's, 60's
Half blood	:	60's and 58's
Three eighth blood	:	56's,
Quarter blood	:	50's and 48's
Low quarter blood	:	46's
Commons	:	44's
Braid	:	40's and 36's

Wool grading in India: Indian wool is not graded extensively because wool produced is only coarse type (carpet type) type and also wool produced by an average shepherd is very less in quantity so it is very difficult to make several lines.

Wool classing: Classifying of wool according to length, quality and condition into different lines so that the wool may be placed on the market in reasonably even lines is called wool classing. This will help the broker or merchants to judge its utility efficiently. As a result the shepherd will get the highest possible price of his wool.

Advantages of Wool Classing

1. To display the clip to the buyers in as attractive manner as possible.
2. To grade the wool so that the wool clip will meet the requirements of the trade.
3. To place the wool clips on the market so that the best prices will result.

Manufacturing System

After all the processing of wool, finally it has to be manufactured. Following are the two methods of wool manufacturing-

1. **Woollen system:** Wool suitable for woollen clothing's is scoured, carbonized then oiling of wool is done. After these preliminary operations carding is done to prepare card silver. From the card silver directly yarn is prepared by the process of spinning. Here, combing of wool is not done. The wools used in the woollen process are shorter and are usually coarser than those used in worsted processing and are used for making blankets, sweaters and tweeds. Woollen yarns are larger in diameter and coarse to touch.

2. **Worsted manufacturing:** Wool suitable for worsted system of manufacturing is passed through preliminary operation like scouring, carbonizing, oiling etc. and is also passed through carding. Then card silver is combed on a special machine that arranges the fibres parallel to each other. During the process of combing the short fibres will come out which is called noil. Men's suit is the typical example of worsted product. Wool tips are spun into yarn, woven and dyed to make the finished fabric.

Table 8.3.1 : Difference between wool and hair.

Sl. No.	Parameters	Wool	Hair
1	Fibre softness	More	Less
2	Thickness/Diameter	Fine/Less	Coarse/More
3	Growth	Continuous, if not sheared	Reaches a maximum and then shed
4	Elasticity	More	Less
5	Medula	Almost absent	Present and more pronounced
6	Heat retaining property	Retains more warmth	Relatively less
7	Moisture	Retain more moisture (12-17%)	Retains less moisture(7-12%)
8	Cuticle	Irregular	Regular and smooth
9	Dyeing property	Permanent	Temporary
10	Inflammability	Slower and gives pungent smell	Faster and gives less pungent smell

Table 8.3.2 : Difference between medullated & non medullated fibre.

Sr. No	Non medullated fibre	Medullated fibre
1	Immerges out from secondary follicle	Immerges out from Primary follicle
2	Absence of medulla	Presence of medulla
3	Smaller in diameter	Larger in diameter
4	More no of crimp per unit area	Less no of crimp per unit area
5	More no of serration per unit area	Less no of serration per unit area
6	More felting property	Less felting property
7	More elastic in nature	Less elastic in nature

Preparation of Animals for Show

For preparation of animals for show the following points should be followed-

1. The selection of animal should be made on the basis of some guidelines i.e. Animal should be true to the breed characteristics and free form defects.
2. Feed extra quantity of concentrate mixture depending upon the condition of the animal. Linseed meal is quite good feed for improving the condition of coat as it adds a gloss to the hair.
3. The horns of the animals should be rasped to make them look symmetrical and smooth.
4. Hooves should be carefully trimmed, shaped and make them smooth.
5. Show animal should be accustomed to use a leading rope so that the animal is led easily and stand in a position that will display its good points in the judging. For this training is needed.
6. The animal should be thoroughly washed with mild soap. The excessive dust should be scrubbed out. After washing, all the soap should be removed by using plenty of water. The animal should be dried in the sun (in dry, warm weather). In wet weather it should be cleaned with cloth.
7. Sometimes people use colour, ghoongroo, flowers, chains, clothing etc. to make them more attractive.
8. The training to stand in the show should be given well in advance.

Judging Sheep

Judging and grading of sheep

For selection of animals judging and grading is the first pre requisite. Judging of sheep is done on the basis of body characteristics.

Parameters and points

1. **General:** 40 points: (a) Weight, (b) Conformation, (c) Condition and (d) Quality of animal (10 points each).

2. **Carcass:** 60 points: (a) carcass (10 points), (b) shoulders (7 points), (c) ribs (6 points), (d) Loin (10 points), (e) Legs (15 points) and (f) Neck (2 points).

Some Common Terms Related to Sheep Production

Apparel wool: Wool suitable for manufacture of apparel fabrics.

Bale: A highly compressed package of either grease or scoured wool.

Bleating: Sound produced by sheep and goat.

Blending: Mixing together of various grades and / or lengths of wool in either the raw or semi-processed state to obtain a specific kind of yarn.

Break: A distinct weak place along the lock in fleeces due to restricted diameter in wool fibres, which is usually due to illness and fever or severe stress.

Bright wool: Grease wool that is nearly white, showing very little yellow colour due to excessive yolk.

Britch or breech wool: Wool from the hindquarters of sheep, usually the coarsest wool in the fleece.

Clip: Removal of hair or wool – the entire amount obtained at one time from a single sheep or whole flock.

Crimp frequency: Number of crimps per inch/cm, largely determined by genetic factor.

Crone: An old broken mouthed ewe retained in breeding flock beyond normal time due to excellent breeding performance.

Crutching: Wool removed from the area around the dock and/or udder of sheep.

Degras: Commercial name of crude wool grease.

Delaine wool: Long stapled fine wool.

Density: The number of fibres grown on a given area of the skin of sheep : the more fibres, the greater the density.

Dingy: Wool lacking brightness possibly due to excessive yolk, presence of parasites or unfavourable environment.

Dinmonts : Ram lamb after first shearing.

Fellomongering : The process of removal of wool fibre from sheep skin through use of chemical application (sodium sulphide) or administration (10-14 mg of thallium) for artificial moult in sheep.

Gimmer/theaves : Ewes after first shearing till the first lamb are weaned.

Grease wool: Wool as it is removed from live sheep, prior to scouring.

Hogg: A lamb after weaning but before first shearing (weaning to shearing)

Hoggest/hoggs/tegs: Wether lamb ready for sale

In-lamb: Ewe pregnant

Kemp: Thick slightly curved hair usually mixed with wool on a fleece.

Lanolin: Purified wool grease used extensively as an additive in cosmetics and skin medications.

Pelt: The skin of sheep, including the wool.

Raddle: Dye marked brisket of ram used for heat detection.

Rejects: Fleeces that are rejected for normal processing due to severe faults such as cotted, tender, black, kempy, or extensively stained fleeces.

Rig: Male sheep not castrated properly.

Seggy: A ram castrated after first service.

Shrink: The non wool portion of grease wool, including grease, vegetable matter, dirt, suint, paints, insects, etc.

Shrinkage: Weight loss in the raw wool when it is scoured, expressed as percentage of original weight.

Skip: A worthless lamb.

Stag: A male sheep castrated after it has developed full secondary sexual characteristics.

Stained wool: Wool permanently discoloured from urine, manure, some anthelmintics, etc.

Suint: The dried up perspiration that is found in the natural fleece. It is soluble in water.

Tags: Dungy locks of wool, described as "dags" in international wool marketing.

Tender wool: Wool with a weakness producing break in staple.

Virgin wool: Wool that has been used in making fabrics for the first time.

Yearling mutton: Meat of sheep of either sex between one to two years of age.

❑❑❑

Chapter - 9

Swine Production

Rana Ranjeet Singh

Origin and Domestication of Pigs

Pigs are nonruminant, even-toed ungulates and belong to suborder Suina which includes hippopotamuses, peccaries and pigs. All true pigs are members of the family Suidae which includes five living genera namely *Sus, Babirousa* (babyrusa), *Hylochoerus* (giant forest hog), *Phacochoerus* (warthog), *and Potamochoerus* (bushpig and red river hog) that can be called as pigs. European breeds are believed to be descendants of the European wild pig *Sus scrofa* and Asiatic or oriental breeds originated from the Eastern Asiatic banded pig, *Sus vittatus* a successor of *Sus crystatus.*

Phylogenetic Classification of the Pig

Kingdom	:	Animalia
Phylum	:	Chordata
Class	:	Mammalia
Order	:	Artiodactyla (even-toed ungulates)
Suborder	:	Suina (hippopotamuses, peccaries, pigs)
Family	:	Suidae
Genus	:	*Sus*

Species : *barbatus* (bearded pig of Philippines, Malaya, Island of Sumatra and Borneo, which is fairly large (<150 kg) animal with a red or black coat).

: *celebensi* (warty pig of Sulawesi other eastern islands of Indonesia which is medium-size (<70 kg) animal with a red/ brown coat).

: *verrucosus* (Javanese warty hog of Java, Sulawesi, Philippines and has large body size (<120 kg) with yellow, red or black hair).

: *salvanius* (pygmy pig of Southeast Nepal, Assam. They weigh less than 10 kg.

: *crystatus* (Indian crested pig or Asiatic wild pig of India).

: *vittatus* (banded pig of Malay Archipelago).

: *scrofa* (wild boar of Europe, Asia); domestic pig.

Sources : Grooves (1981), Payne & Wilson (1999) and McGlone & Pond (2003).

Archaeological evidences point out that domestication of pigs was in progress about 9000 years ago in Southwest Asia. Later on domestic pigs gradually diffused to ancient Syria, Sudan, Egypt (6000-7000 years ago), Greece and Europe. Meanwhile, domestication of pig was in progress in eastern Asia and their history of domestication in China goes back to about 7000 years ago (McGlone & Pond, 2003). Though, there is another theory which suggests that pigs were separately domesticated at places located in many different regions.

Characteristics of Pigs

Pigs are monogastric, omnivorous animal and their body remains well insulated with layer of fat except in baby pig. They use snout to unearth grubs, earthworms and other materials from the soil, known as rooting behaviour. There are four toes on each foot, only two of each is functional. Pigs have coarse and sparse bristly hair coats with a small number of sweat glands as a result they are vulnerable to heat stress. They prefer swimming or rolling in mud and wallowing as behavioural means of thermoregulation. Domestic pigs vary significantly in their size, body shape, colour, ear carriage, prolificacy and other traits. Pigs are considered as most intelligent animal among the domesticated farm animals. They can smell, see, hear very well and has good learning potential. The chromosome number in swine is polymorphic. It is generally agreed that the domestic pig has 38 (2N) chromosomes whereas the European wild pig has 36 (2N) and the hybrid pig has 37 (2N) chromosomes. Pigs are primarily raised for meat (pork) production and secondarily for the production of lard, pig skin, bristle and manure etc.

Pig Population and Distribution

Pigs are found throughout the world but they are more in number in grain producing areas or in the regions where large quantities of by-products or offals are available. In addition to that climate (lesser number of pigs are found in the arid areas of the world), cultural & religious beliefs of the indigenous people (few pigs in countries with a predominantly Muslim population) affects the size of the pig population of any given region. Highest numbers of pigs are found in Asia followed by Europe and Americas. Pig population of the world, leading countries in pig population and distribution of pig numbers among different states of India has been discussed in the chapter 1 of this book.

Advantages of Swine Farming

1. Pigs are prolific breeder with short generation interval. A sow can be bred as early as 8-9 months of age and can farrow twice in a year. They produce 6-12 piglets in each farrowing which may weigh around 80 kg at about 6 months of age.
2. Swine are very efficient converter of feed nutrient into pork. They produce more live weight gain from a given weight of feed (1:3/3.5) than any other class of meat producing animals except broilers (1:2).
3. Hogs can utilize wide variety of feed stuffs *viz.* grains, forages, swills (kitchen wastes from hotels, hospitals, schools, colleges and army barracks) and agricultural by-products and convert them into valuable nutritious meat.
4. Dressing percentage is higher in pigs (65-80%) as compared to cattle (50-60 %), sheep and goat (40-45 %) because skin is also used for eating and due to smaller bone size percentage of edible meat in pigs is higher.
5. Besides pork, a range of other useful products are obtained-blood may be used in sausages preparation as well as in blood meal, pig skin for leather, hair for brushes, hooves are used to manufacture glue and bones are crushed into bone meal.
6. Hogs are well adapted for self feeding and hence labour requirement is less.
7. Pigs can be kept in a relatively smaller space.
8. Initial investment is small and returns come quickly.

9. Housing for pigs is generally cheap in the tropics so that little capital is required to set up a pig unit.
10. Pork provides essential nutrients which is required for good health. Energy value of pork is higher than beef or lamb. It is rich in vitamins like thiamine, niacin and riboflavin.
11. Pig manure contains on average 0.70, 0.68 and 0.70% of nitrogen, phosphorus and potassium, respectively. It can be used as a fertilizer for the soil or fish ponds, production of methane gas and culture of *Chlorella* and *Azolla* that can be used as an animal feed.
12. Pig farming provides an opportunity to integrate with other enterprises like crop, poultry cum fish culture etc.
13. Pig stores fat rapidly for which there is an increasing demand from poultry feed industry, soap industry, paints and other chemical industries etc.
14. Pigs are adapted to both intensified and diversified farming.
15. Sows have good mothering ability which is essential for survival of baby pigs and to rear them to a good weight before weaning.
16. Pork can be more effectively preserved with salt than other types of meat.

Disadvantages of Swine Farming

1. Requires maximum concentrate and minimum roughage in their diet.
2. When raised mainly on grain, production cost becomes higher.
3. Pig is in direct competition with man for cereals.
4. Sensitive to unfavourable ration and careless management.
5. Highly susceptible to many diseases and parasites.
6. Sow needs more attention at the time of farrowing to avoid crippled birth.
7. Hard on pastures due to rooting and close grazing habits.
8. Not suitable to sparse vegetation.
9. Keeping pigs in or near urban areas may pose a health problem owing to the flies which they attract.
10. Have religious taboo (in Muslims).

Scope of Swine Farming in India

Pigs are widely distributed across different agro-ecological regions of the country. However, more than one fourth of the total pig population of India is concentrated in North-Eastern Region. Pig rearing has been traditional occupation of weaker section of rural society particularly the tribal people. They lack means to undertake intensive pig farming with good foundation stock, proper housing, feeding and management. Production performances of these animals are usually poor.

Pigs are multipurpose (meat, manure, bristle etc) animals with high fecundity, high feed conversion efficiency, early maturing, short generation interval which can be raised under a variety of production systems ranging from simple backyard pigs to large scale integrated pig industries with sophisticated bio-safety measures. Further, it can be raised on relatively small space.

As a result of various research and development efforts, this sector has gained slow but steady momentum during the recent past and is changing from backyard activity to that of a commercial endeavour but still consolidated efforts are required to fully exploit this sector.

Pig Production Systems

There are different systems of pig production that are in use throughout the world *viz* free range (scavenging system), semi-intensive and intensive system.

1. **Free range (scavenging) system**: In this system pigs are free to move around the houses and surroundings. They get their food by scavenging, picking up food offals here & there. Frequently the food they collect is supplemented with kitchen refuse or agricultural waste products. Few or no arrangements exist to provide shelter to them. Due to the higher resistance to diseases and ability to make use of low quality feeds generally local breeds are utilized. Pig act as a savings account' or insurance policy' as they are generally sold when extra cash is needed (e.g. for buying various agricultural inputs, at times of illness or marriage etc) for the family. This system requires a minimum amount of inputs and financial risks involved are also small as little money and little time being spent for looking after the pigs. Performance of animal can be improved by constructing simple deep litter pens/ conventional pen with concrete floor/timber pen with thatch roof, providing

supplementary feeds once or twice daily, use of improved sires for upgrading purposes and better health management practices.

2. **Semi-intensive system:** There are a lot of variations of the semi-intensive system in vogue and this system can be practiced in areas where incidences of internal parasitic infestations are low. In this system animals are confined to a smaller area than the previous system. Generally, breeding pigs are maintained outside on pasture and fattening pigs are raised intensively in buildings. In open area fencing is done preferably with netting and is provided with adequate shade and water supply. This system of pig keeping opens up possibilities for improved feeding and disease control, which in turn can result in faster growing and healthier pigs and/or in larger litters. Further, this system requires only low financial inputs but it is more labour intensive and requires technical knowledge. Improvements to semi-intensive systems can be made by giving attention on feeding and health care practices and by following selective breeding.

3. **Intensive system:** In this system of pig rearing, pigs are kept in complete confinement on a concrete or other form of flooring (slats made up of wood, concrete, steel and/or aluminium) which can be cleaned daily. Separate arrangements are made for fatteners, boars, sows, and sows with their litters. Bedding may be provided on concrete floors and provision of feeding trough and water trough is made inside the house. In this system larger numbers of pigs may be kept and managed well, internal parasites can be adequately controlled and labour costs are reduced to a minimum. Sufficient amount of technical knowledge is required to make the right decisions at the right time and investment is also high. In this system of pig keeping, kitchen refuse and agricultural waste products are not sufficient to attain a higher rate of growth in order to market the animal at a lower age, so certain feeds will have to be bought. The animals no longer function as a savings account but are grown to be marketed.

Breeds of Pigs

Characteristic features of Indigenous or *desi* pigs are- small body size, pot bellied, usually black in colour with short ear and elongated face. The average birth weight, litter size at birth and at weaning is low. Tail reaches nearly to the hocks which has a tuft of hair. Females have 6-12 teats. Adult pigs weigh up to 150 kg. Indigenous and exotic breeds of pigs have been discussed in detail in the chapter 3 of this book.

Housing of Pigs

A proper housing system should meet all the essential requirements (adequate ventilation, space for free movement, exercise, sunlight, cleanliness and protection from inclement environmental conditions etc) of pigs so that they remain healthy and can be raised economically. In addition to that it should increase feed & labour efficiency and should also provide better working condition to the farmer. Due to poorly developed thermoregulatory mechanism pigs are more prone to extremes of heat and cold hence, maintaining proper temperature inside the house is of utmost importance. Further, to restrict the pigs from rooting, housing structure must be robust and gaps should be avoided.

There are two main system of housing-open or loose housing and indoor or closed housing system. In the most places a combination of two is in vogue. Choice of a particular type of housing system depends on number of animals, type of pigs to be produced, availability of land and climatic conditions. Housing requirement for pigs vary with its category. So at organized pig farms different types of sties are constructed for different categories of animals such as boar sty, weaning pig sty, farrowing sty, fattening sty, sick animal sty, dry sow & gilt sty. Besides these, various ancillary facilities like wallowing tank, loading ramp, holding area, manure pit and feed store etc are also constructed at pig farm. Individual sty should be provided to boars and pregnant animals. For most parts of India usually a double row system of covered area with a common passage in between them and a linked open area at the back side of each of the covered room is ideal design. Further, housing of pigs has been discussed in detail in chapter 5 of this book.

Feeds and Feeding of Pigs

Pig is an omnivore and has simple digestive system with a functional caecum. Pig likes to graze or chew forage but their fibre utilisation ability is limited. Hence, like ruminants they cannot be maintained entirely on roughages. Further, an increase in dietary fiber depresses the digestibility of gross energy. Due to their characteristic mouth and teeth they can eat food on the surface of the soil or dig it out from the soil. Pigs compete with human beings for its food but they can also thrive on by-product feeds, grasses, silage, animal and vegetable based concentrates etc. Feeding alone accounts for around 70-75 per cent of total cost of pork production. Therefore, pig ration should be well balanced, palatable, free from anti-nutritional factors, suitable for the category of pigs for which it is formulated and must be economical. Some of the commonly used feedstuffs in pig production are

A. **Carbohydrate rich feedstuffs**
   - **Cereals**: maize, wheat, rice, barley, oat, sorghum, millet etc.
   - **Root crops**: cassava and cassava peelings, potato, sweet potato carrot, beet etc.
   - **By-products**: rice / wheat shorts or middlings, rice/wheat bran, molasses, bread, biscuits etc.
   - **Others**: cane molasses, citrus molasses, sugar, bananas etc.

B. **Protein rich feedstuffs**
   - **Animal products**: whole-milk powder, skim-milk powder, fish meal, casein etc.
   - **Plant products**: peas, lupins, beans etc.
   - **By-products**: milk and milk by-products, meat meal, meat and bone meal, blood meal, feather meal, fish offal meal, cottonseed meal, soybean meal, coconut meal, safflower meal, sesame seed meal, sunflower meal, groundnut meal, linseed meal, rapeseed meal, brewer's yeast etc.
   - **Single cell protein**: dried yeast, algae, bacteria etc.

C. **Fat and oils**: tallow, sheep fat, pig lard, fish oil, coconut oil, cotton seed, sun flower, rape seed, soya bean, groundnut etc.

D. Others
   - **Plant products**: papaya, tomato etc.
   - **By-products**: brewer's grains, fruit pomace, garbage, wheat bran, rice bran, kitchen, hotel waste etc.
   - **Roughages:** alfalfa meal, alfalfa hay, silage and pasture etc.

Methods of Feeding

Feeding method affects the productivity of the animal. Choice of feeding method depends on intended production type, rationing system and type of feed available etc. Different methods of feeding pigs are

1. **Complete diets:** In this method all feed ingredients are mixed together prior to the feeding generally in the dry form and can be fed as dry, pellets, or wet form.
2. ***Ad libitum* feeding**: Diets which are based largely on bulky feeds can often be fed *ad libitum* because the pigs' gut capacity will limit their energy intake. Advantages of this feeding method are - it saves labour and allows the pigs to fulfil their full growth potential. However, this method results in wastage of food, lower food conversion efficiency and finished carcass tend to be relatively fat. The young animals are fed *ad libitum* up to 45 kg of body weight.
3. **Restricted feeding or limited feeding:** In this method, 75 to 80 percent of full feed is provided to the animal. This method brings improvement in feed conversion efficiency and carcass quality (more leanness) but reduces growth of animal and consequently increases time to market. Further, it requires extra time, labour, and equipments. Restricted feeding should never be used with pigs weighing less than 45 kg.
4. **Split-sex feeding:** In this method growing-finishing barrows and gilts are fed in separate groups as per their nutritional needs.
5. **Phase feeding** : In order to meet the rapidly changing nutritional needs of pigs during the first weeks after early weaning (3 weeks of age) this method of feeding is followed. Phase I lasts for 7 to 10 days post weaning and during this phase piglets are provided with pelleted diet containing 20-22 per cent crude protein. Phase II followed for 1 to 2 weeks after Phase I. Diet containing 18 to 22 per cent crude protein should be fed either in pelleted or meal form. Phase III is followed 3 to 5 weeks post weaning and during this phase diet should contain 18-20 per cent crude protein (Gillespie and Flanders, 2010).

Nutrient Requirement of Pigs

Factors like breed, sex, age, health status of pigs, environmental temperature, energy concentration of the diet, availability and absorption of dietary nutrients and level of feeding etc affects nutrient requirement of pigs. The nutrient requirements are expressed as per cent of the diet or amount in per kg of ration. Feed energy is utilized by pigs for various body functions like- maintenance of normal body processes, fetal load, growth, milk synthesis, reproduction, thermoregulation etc. Feed protein is the essential source of amino acids and amino acids are required for building protein and to replenish lost protein in the body. As mentioned earlier that pig has limited ability to utilise fibre so for growing-fattening pigs 6-8 percent fibre may be used while the sow ration may contain 10-12 percent fibre. The other nutrients consist of 10 essential amino acids, 17 vitamins, about 13 or more minerals, essential fatty acids and certain unidentified factors. Generally, feed requirements of weaner pigs (56-120 days), grower pigs (121-180 days), adults and sows are about 0.25-0.75, 0.75-1.5, 1.5-2.5 and 2.5-3.5 kg/day, respectively.

Water: Water requirements of pigs vary with season, dry mater intake, feed ingredient, health status, stress level and stage of growth. Pigs should have plenty of fresh, clean water available at all times. More water is consumed in summer than in winter. In general, young pigs need more water per kilogram of body weight than older animals. Increase in the fibre, salt and protein content of diet increases water requirement. Water requirements of different categories of pig with free access of water have been given in Table 9.1.1.The relationship between feed intake and water consumption by weaners was described by Brooks *et al.* (1984) using the following equation.

Water intake (L/day) = 0.149+ (3.053× Daily dry feed intake in kg).

Table 9.1.1 : Water intake allowances for pigs.

Categories	Water intake daily (kg)*
1. Lactating sows	25–40
2. Pregnant sows	10–20
3. Growing pigs 20–160 kg	Dry feed intake ×5
4. Young pigs	Dry feed intake ×6

*total dissolved salt in water should be <3000 parts per million, of which the content of calcium should be <1000ppm, sulphates <1000ppm, nitrates <100ppm and nitrites <10ppm.

Source : Kyriazakis & Whittemore (2006).

Care and Management of Different Categories of Pigs

General management practices like control and handling, age determination, identification methods and castration etc has been already discussed in chapter 2 of this book.

Care and Management of Piglets up to Weaning

- Soon after the birth all mucus from snout must be cleaned to facilitate breathing.
- The navel cord should be tied off and cut with disinfected pair of scissors at about 1.0 to 1.5 inches from the body and dipped in tincture of iodine.
- It is essential that all piglets must get colostrum shortly after birth. If the number of piglets born are more than the number of teats a sow posses then surplus piglets may be raised by hand or on a foster mother.
- Weak piglets should be assisted in getting colostrums and should be placed on front teats and healthier piglets to the rear teats.
- Piglets are likely to be affected by chilling (as they lack hairs, low body fat and glycogen storage, larger surface area) during winter season particularly in northern part of the country. Hence, some source of supplemental heat should be provided to the piglets during first few days after birth to prevent losses due to chilling. Generally heat lamps are fixed in the creep area at about 45 cm from the floor and after 4-5 days height of the heat lamp is raised to have lower temperature under the lamp.
- Piglets can be ear notched or ear-tattooed or ear tagged for identification.
- Trampling by the sow should be prevented during first two weeks.
- During first few weeks of their life newborn piglets depend on mother's milk. Colostrum feeding is essential for the first three days as it provide immunity against various diseases.
- Baby pigs will start to nibble on a feed within a week after they are born. When piglets are weaned earlier, pre-starter ration is supposed to replace milk. The pre-starter ration should be palatable and fortified with proteins, vitamins, minerals, antibiotics.
- **Clipping of needle teeth:** Piglets are born with eight (four incisors and four canines) very sharp needle or wolf teeth located on the sides of the upper and lower jaws. These teeth

may cause injury to the sow's udder, other piglets and attendant. Therefore, it should be clipped soon after the birth. They should be clipped at the gum line in piglets less than two days old while in more than two days old piglet one-third to one-half of the tooth should be clipped. Care must be taken to avoid injury to the gum.

- **Tail docking:** It is cutting off the pig's tail and is intended to prevent tail biting when pigs are raised in total confinement. It is usually performed within 1-3 days after birth. Instruments like side-cutting pliers, special cutting instruments, or a chicken debeaker may be used. The usual location is about 12-18 mm from the base of tail. Tail stub should be treated with some antiseptic.
- **Piglet anaemia:** Piglets are born with a small reserve of iron in their body (50 mg in liver) and their mother's milk (poor source of iron) usually does not provide sufficient iron as per the requirements of the piglets (5-10 mg/daily). Therefore, piglet anaemia occurs due to iron deficiency. Pigs raised on concrete floor are more prone to this condition. Anaemic piglets are pale in the regions of the ears and belly, are lethargic, breathe rapidly and frequently show diarrhoea. Anaemia results in slow weight gain and increase susceptibility of pigs to disease and infections. Piglet anaemia can be prevented/ cured by
    - Allowing access to clean area of soil or providing fresh, clean earth in the piglet's pen daily.
    - Using soil soaked with a solution made from 500g ferrosulphate, 75g copper sulphate in 3 litres of water.
    - The daily administration of 4 ml of a 1.8 percent ferrosulphate solution.
    - Swabbing/painting of the sow's udder with ferrous sulphate solution and sugar daily from birth until the piglet start eating creep feed.
    - Safest and easiest method of preventing piglet anaemia is by giving intramuscular Iron-dextran injection @ 100 to 150 mg per pig in the neck or ham area. Care must be taken since overdoses may cause shock in the piglets. First injection may be given 3 days after the birth and if required second injection (somewhat smaller dose) may be given at about 3 weeks later.

- **Raising of orphan pigs:** Conditions like death of a sow after farrowing or larger number of piglets born in a farrowing than number of teats of a sow or conditions like mastitis/ agalactia etc results into orphan piglet. Such piglets may be raised by hand or on a foster sow. In any case piglets must get colostrum as their first feed. For this purpose colostrum from other sow may be used, if it is not available cow's colostrum can be used and if no colostrum is available injectable gamma-globulins can be used. While raising the orphan piglets on a foster sow there should not be more than 3 days' age difference between the foster mother's own piglets and the orphan piglets. Before transferring the orphan piglets, it is advisable to smear them with placental fluid of foster sow so that sow accepts them thinking as their own offspring. In artificial rearing, milk substitutes can be reconstituted with cow milk or water and fed to the piglets either through shallow trough, or bottle four times daily for first 10 days and later thrice daily. Fresh drinking water should always be available. They should be housed individually for first two weeks otherwise they may start chewing of tail or ears due to suckling instinct.
- **Creep/starter feeding:** The practice of self feeding concentrate to baby pigs in a separate area away from the dam is known as creep feeding. As nutritional requirements of young piglet are different from the sow hence, it is advantageous and economical to follow this practice. It may be used either as a follow up feed (after 3 weeks) to a pre-starter feed or as a creep feed from first week up to weaning (56th day). At the end of this period young pigs weigh around 14-18 kg. To get maximum gain through weaning young pigs are provided with creep ration along with dam's milk. Creep feed should contain 19-20% crude protein (CP) and a most of which should be of animal origin. If piglets to be weaned earlier than 8 weeks then ration with an even higher CP content of 20-22% should be fed. A creep ration should be of high quality, complete mixed feed, highly palatable and high in mineral and vitamin content. The creep ration should have low fibre content. To keep the ration fresh, feed (pelleted or crumbled) is provided frequently in small amounts and placing it in a shallow pan or sprinkling feed on the floor may encourage the pigs to start eating. Water should always be available.

- **Weaning:** Normally piglets are weaned from dam at about 8 weeks of age when the milk yield declines in sows. Now-a-days, there is a trend toward earlier weaning of pigs, it means weaning at 3 weeks of age or less. Pigs should weigh at least 5.4 kg at 3 weeks, 4.1 kg at 2 weeks and 2.4 kg at 1 week of age at weaning (Gillespie and Flanders, 2010). Advantages of early weaning are- it increases the probability to get more litters from a sow per year, piglets will grow faster, sow will not lose too much weight, feed costs should be less as the process of double conversion of food (first by the sow into milk and then from milk into live weight by the piglet) will be removed. Disadvantages of early weaning are that it requires better management and well fortified costly feeds.

Larger pigs in the litter may be weaned first and smaller pigs may be weaned a few days later. While practising weaning sow should be taken away from piglets not piglets from the sow. Separation of sow should be a gradual process. Initially, sow should be separated from piglets for few hours each day, then for a day and finally for all time to minimise the stress of weaning. Generally in India, the average litter size is of 9 pigs per litter and average number weaned per litter is 7. The piglets should be gradually shifted from creep feed to grower ration. At the time of weaning they should be dewormed, vaccinated against swine fever and H. S. and weight of individual pigs are recorded.

Care and Management of Lactating Sow/Gilt

Nutrient requirements of lactating sows are much higher than pregnant sows. Hence, total quantity of the sow's ration should be steadily increased after farrowing based upon the number of piglets she is nursing, composition of diet, condition of sow, environmental temperature and farrowing facilities. The natural lactation length in sow is about 12 weeks and peak milk yield is attained at around 3 weeks after farrowing. For maximum milk production she should be full fed. During this period a sow will usually eat about 4 to 7 kg of feed per day. A ration for lactating females should contain 16% CP.

Care and Management of Growing-Finishing Pigs

Rearing of pigs from weaning to market is less demanding than the rearing of suckling pigs. During this period, pigs should be grouped in uniform size lots by weight and are generally segregated by sex as males gain weight faster than gilts. The size of the group shouldn't be larger than 30 to 50 animals and weight range should not be more than 20 percent above or below the average weight of the group. Objective of the grower-finisher pig management should be-good health, curtail losses, development of good meat carcass quality, good food conversion efficiency, low feed cost per unit of live weight gain, low total cost per unit of live weight gain and good return on investment and capital.

Growing-finishing pigs can be raised either in confinement or on pasture. Confinement system of rearing is capital intensive, requires balanced diet and proper disease and parasite control. However, generally rate of body weight gain remains a little higher under confinement conditions. Raising pigs on pasture can reduce the need for protein supplement in the ration. Putting a nose ring in the snout prevents the pigs from rooting in the pasture. Some provision of shade must be there so that pigs can take rest during hotter part of the day. Growing pigs should be dewormed at about 15-16 weeks of age. All-in/all-out method of raising finishing pigs is practiced at many farms. In this system, pigs are moved as a group (about the same age and size) through each stage of production to market. Facilities are cleaned and disinfected between groups of pigs to reduce the chances of transmission of diseases.

Growth rate depends upon inherited abilities, feeding and managerial practices. Under good feeding and managerial conditions, pigs will usually attain about 19, 24, 37, 54, 72, 90 kg of live weight at around 10, 12, 15, 20, 24 and 28 weeks of age, respectively. Pigs show most rapid growth when full fed on complete mixed diet. Faster growth rate shortens the time to reach slaughter weight thus income on capital investments can be obtained quickly. Further, it reduces overhead and labour cost. In the tropics pigs are mainly raised for pork and only few are raised for bacon or for lard purposes. If leanness is the objective pigs can be fed 75-80% of the *ad lib*. Additionally, restricted feeding helps to achieve superior feed conversion efficiency by reducing feed wastage associated with *ad lib* feeding. The starter ration for weaned pigs contains 18% CP and may be fed until the pigs are 14-23 kg of body weight. Pigs weighing around 23-55 kg are provided grower ration containing around 16% CP while a fattening ration containing 14% CP is provided to the pigs weighing around 45-90 kg of body weight. Compositions of ration for different categories of pig have been shown in Table 9.1.2.

Table 9.1.2 : Composition of ration for different categories of pig.

Ingredients	Per cent according to the category				
	Creep	Weaner	Grower	Market	Breeding boar /pregnant sow
1. Maize	60	55	50	45	50
2. Wheat bran	14	17	20	25	18
3. Groundnut cake	14	16	18	20	20
4. Molasses	5	5	5	3	5
5. Fish meal	5	5	5	5	5
6. Salt	1	1	1.5	1.5	0.5
7. Mineral mixture	1	1	0.5	0.5	1.5

Feed additives as per the recommendations.
Source: ICAR (2002).

Care and Management of Breeding Pigs

It is better to separate boar and sow immediately after weaning and could be reared either indoor or outdoor.

Boar: In order to reduce the incidences of fighting among them boars are housed separately. Boars are considered to be sexually matured at about 15 months of age though a well grown boar can be used for the first time when they are 7-8 months old (junior boar). Maximum numbers of services which can be recommended for a junior boar per day, per week and per month are 2, 8 and 25 while for mature boars it is 3, 12 and 40, respectively. The boar ejaculates between 100-500 ml of semen containing 30×10^9 to 150×10^9 sperm over a period varying between 1 to 15 minutes (average 5 min). The ratio of boars to females may be 1:50. To avoid low conception rate, boars are rotated on a group of females. Outdoor exercise throughout the year is very much essential to keep the animal in thin, thrifty condition and sexually active. Boars should not be over fed as it results in overweight, sluggishness and reduced libido. The boars should be conditioned by increasing his ration at least six to eight weeks prior to the breeding season. Breeding boars should be given meals containing 14-16% CP and may be fed @ 2-2.5 kg/head/day. Boars should be fed twice a day preferably after service. Number of females which can be served by a junior or mature boar in a breeding season have been shown in Table 9.2.1.

Table 9.2.1 : Number of services per boar as per the breeding season.

Length of breeding season	Hand mating		Pasture mating	
	Junior boar	Mature boar	Junior boar	Mature boar
1. 3 weeks	20	30	10	15
2. 6 weeks	30	40	20	30

Source : Sharda (2005)

Gilt/sow: Pig is recognized as non-seasonal polyestrous species. Gilts usually reach puberty at about 6-8 months of age and are usually mated after one or two estrous cycles because first oestrus is accompanied with higher proportion of immature oocytes and mating at first oestrus results in lower conception rate. Reproductive traits of female pig have been shown in Table 9.2.2. Signs of oestrus are characterized by changes in both the behaviour and anatomy of the gilt/sows.

Methods of heat detection: Heat detection in sows/gilts can be done by following methods

- In the period 2-6 days prior to oestrus reddening and swelling of the vulva is seen, usually reaching a maximum at or before the start of the estrous period. These changes are more pronounced in gilts and are difficult to observe in sows.
- Mucous discharge from the vulva.
- Decreased vaginal pH.
- Proliferation of vaginal leucocytes.
- Reduced appetite.
- Increased restlessness.
- Characteristic vocalization.
- Oestrus gilt/sows frequently mount other sows or allow themselves to be mounted.
- Standing heat reflex or lordosis (mating stance of females comprising immobilisation, arching of the back and in some breeds the cocking of the ears).

Among all the above mentioned signs of oestrus, lordosis is the most reliable one and is exhibited in the presence of a boar (boars releases olfactory and auditory stimuli, olfactory stimuli in the form of aerosol containing the salivary pheromones-5α androstenone and 3α androstenol) but can also be elicited by applying downward pressure on the back of the gilt/sow hence also known as riding or back pressure test. However, in absence of boar only about 50-60% of females exhibit this reflex.

Table 9.2.2 : Reproductive traits of pig.

Parameters	Time/weight
1. Age to breed gilts	6-8 months of age.
2. Weight to breed gilts	Approximately 100-115 kg.
3. Period of estrous cycle	21 day intervals (19-24 days).
4. Length of heat period	2-3 days.
5. Ovulation time	38-42 hours after onset of oestrus (2nd day).
6. Best time to bred	12-14 hour prior to ovulation to 2-6 hour post ovulation.
7. Number of services per sow	2 services (first service on the first day on which heat is observed and second 12-24 hours later).
8. Occurrence of heat after weaning	2 - 10 days.
9. Gilt or sows are to be farrowed	12 - 14 months.
10. Breeding after farrowing	Second heat after weaning of piglet.

Gilts tend to have a shorter heat period than sows. The number of ova released at a single oestrus depends on age, breed, parity and nutrition. Ovulation rate increases from first post-pubertal heat to third heat and sows usually ovulate between 15 and 20 eggs within a time span of many hours. Due to the release of large number of ova over a period of many hours maximum fertility and fecundity can be obtained by mating sow twice during the oestrous period.

Flushing: About 10 days prior to mating sows and gilts are given a good grower ration which is known as flushing. It influences the regularity of heat, the number of over shed, fertilisation and subsequent implantation of the zygote particularly in females which are in poor condition. After mating sows and gilts should be fed low rations immediately after mating because of inverse relationship between feed intake and concentration of circulating steroids in the blood.

Methods of pregnancy diagnosis: Pregnancy diagnosis in sows/ gilts can be done by following methods

- Pregnancy (<18-24 days post-mating) diagnosis can be best accomplished by the use of boar to recognize females that returned to oestrus.

- Plasma prostaglandin concentrations can be estimated on days 13-15 post mating.
- By recognizing changes in the depth of the epithelial cell layer of the vagina of sows at about 18-25 days post-mating.
- Rectal palpation can be practiced to determination of both the size and condition of the pulse in the middle uterine artery after 30 day post-mating.
- Ultrasound techniques like Doppler technique (detection of foetal heart rate/foetal pulse), amplitude-depth technique or A-mode (foetal fluid in the uterus) and real time or B mode scanner (the identification of cross-sectional images of foetuses) can be used to detect pregnancy accurately at about 30-60, 30-90 and 20 days post-mating, respectively.
- Plasma progesterone concentration remains normally above 5ng/ml in pregnant sows on 17-24 day post-mating whereas negligible in non-pregnant sows.
- Oestrone sulphate concentration remains higher in the blood of pregnant sows between 23-30 days post-mating whereas only basal levels are seen in non-pregnant sows.

Care and Management of Pregnant Sow/Gilt

The gestation period of a sow varies from 112 to 120 days with an average of 114 days (3 months, 3 weeks and 3 days). It is advisable to house pregnant sows and gilts in group in a separate pen and should not be mixed with new animals to avoid fighting which may result in abortion. In order to avoid them becoming over fatty they should be given optimum exercise. Wallowing in ditches, driving through narrow gates and overcrowding in pens should be avoided. During gestation period it is important to provide sow/gilt a well balanced ration. Various workers have suggested that maintaining them on limited feeding (energy) programme will restrict them from becoming too fat. Owing to their ensuing growth, gilts have higher nutrient requirements than the sows. Ration for gilts should have 16% CP whereas for sows should have 14% CP during the final months of gestation. One week prior to their transfer to the farrowing pen they should be dewormed with a suitable chemical agent. Before transferring the sow into the farrowing pen they should be thoroughly scrubbed with soap and a mild antiseptic particularly to the side that remain in contact with the floor (sides and under surface of the body) to remove dust, microbes and parasite eggs etc. Before placing the sows in farrowing pen walls and floors of farrowing pen should be scrubbed with phenyl water. In addition to that a standard disinfectant should be pumped into crevices and angles. For *Kutcha* floor 2 to 3 inches of top layer should be replaced with fresh earth. After drying off the shed bedding material should be provided. Cut straw (rye or wheat), saw dust or wood shavings are good litter material as they absorb moisture, lie down close to the floor and posses little interference to the piglets to reach out the mother's teat.

Care and Management of Sow/Gilt at Farrowing

To provide more quite and comfortable environment sows should be segregated and transferred into farrowing pen 2-3 days and gilts 4-5 days prior to the expected date of farrowing. Few days before farrowing clean litter material should be provided to pregnant females so that she can build a nest. Farrowing pen should be clean, dry, well ventilated and no loose water trough should be present. To protect the piglets from crushing by their mother a guard rail or farrowing rail made up of iron pipe (5 cm in diameter) or hardwood timber (8 cm in diameter) should be provided in the farrowing pen. The rail should be raised 8-10 inches from the floor and should be 8-12 inches from the wall. Instead of guard rails farrowing crate can be used alternatively. It is a device which confines a sow at farrowing and the piglets have space of their own on either side to run around freely. It provides enough space so that sow lie down and get up comfortably but restrict to turn around. Farrowing crate can be set permanently or temporarily.

One to two days before farrowing the diet should be reduced and more bulky and laxative feed should be given to avoid constipation. On the day of farrowing no feed should be offered for about 10-12 hours but plenty of fresh, clean water should be provided. From the next day bulky diet is gradually replaced with normal diet.

Usually around 36 hours before parturition the vulva swells and the teats become firm. On the last day of pregnancy sow becomes more restless and vulva become greatly enlarged and udder appears to full and turgid, by slightest touch milk can come out. Generally, farrowing occurs within around 6 hours after sow shows a period of Intensive activity and nest building occurs. Intensive activity includes increase in frequency of standing up and lying down activity and rooting & pawing at the pen floor. In the last 90 minutes before parturition the sow stays lying on one side. If well fed and managed properly mature sows of good breed usually farrows 8-14 piglets. However, gilts normally farrow fewer piglets than the sow. To have a uniform breeding schedule many farms practices induced farrowing where a group of sows are induced to farrow within a short period of time by using some drugs. This practice also helps in even-up the litter size using cross-fostering pigs and efficient utilisation of labour.

Normally sows don't require any help during farrowing. However, a familiar pigman may be present there without unnecessarily interfering the normal process of birth in any way. Farrowing is completed in about 2 to 5 hours and the interval between each piglet may be few to 60 minutes and afterbirth is expelled either approximately within 4 hours after the last piglet is delivered or in part when all piglets from a uterine horn comes out.

Pig Health and Diseases

Incidences of disease outbreaks and subsequent loss of productivity on a farm can be avoided by increasing the natural immunity of the animals, reducing the factors which increase the susceptibility of animals to a disease by maintaining good standards of feeding, housing, care etc. Prevention is better than cure hence a well thought-out good herd health programme considering sanitation, isolation, disease free breeding stock and proper care and management of different categories of animals should be followed. Depending upon the condition of the farm vaccination and deworming programs should be followed.

A normal healthy pig has good bloom, moist snout, warm ears, pink colour mucous membrane of eyes and vulva and curl in tail and is always alert. Signs of ill health are- pig stand or lie separate from the others, off fed, sunken eyes, cold ears, tail hangs, hairy appearance, signs of diarrhoea, lameness, respiratory distress and change in vital parameters of the animal (temperature, pulse rate, and respiration rate). These early signs of ill health are generally noticed by experienced attendants and pig keepers. Some of the diseases of pigs, their control measures and vaccination schedule have been presented in Table 9.3.1 & 9.3.2.

Table 9.3.1 : Common diseases of pigs.

Diseases	Causative agent and clinical signs	Prevention &control
Bacterial diseases		
Leptospirosis	*Leptospira pomona* Acute, sub acute infection- high fever of 40°C for 3 to 5 days, anorexia, diarrhoea and icterus with haemoglobinuria, abortion, still birth and heavy mortality may occur.	Good hygiene, serological testing to remove reactors and vaccination in endemic area.
Brucellosis	*Brucella suis, B. abortus and B. melitensis.* In female –infertility, abortion at any time during gestation or stillborn piglets, metritis. Male- epididymitis and orchitis.	Disinfection of shed, slaughter and removal of serological reactors.
Swine erysipelas	*Erysipelothrix rhusiopathiae.* Hyper acute form (common in adult)- high temperature, reddish purple patchy skin and sudden death.Acute form (young & adult animals)- high temperature, reddish purple patchy skin and pathogonomic diamond skin lesions appear within 24-48 hr of the onset of clinical symptoms.	Disinfection of shed, antibiotic, antiserum treatment and vaccination.
Swine dysentery/ bloody scours/ black scours	*Serpulina hyodysenteriae.* Bloody diarrhoea. After sometimes, the faeces become watery and contain thick mucus. Slow growth rates and poor feed efficiency.	Treatment with antibiotics, maintaining isolated herds, controlling rodents, reducing stress, avoiding overcrowding in the facilities, maintaining strict sanitation, and clean, dry environment.
Clostridial diarrhoea	*Clostridium perfringens.* Listlessness and yellow, bloodstained, watery diarrhoea. Death usually occurs within a day and a half after the symptoms are first seen.	Sanitization of farrowing crates or pens and washing of sows before putting them in the crates or pens. Vaccination of sows and gilts.

Contd.

Atrophic Rhinitis	*Bordetella bronchiseptica & Pasteurella multocida.* Affects the nasal passages and the tonsils. The more severe form generally affects the nasal passages, tonsils, and lungs.	Good sanitation and management practices must be followed. Vaccination can be done. Sulfonamides or oxytetracycline may be included in the sow diet during the last month of gestation to reduce the shedding (discharge) of bacteria.
Viral diseases		
Foot-and-mouth disease	Several types and sub types of aphthovirus. Fever, formation of vesicles on the coronary band and on mucosa of mouth (lips, gums, palate and tongue).	Slaughter, disinfection and strict control on the movement of pigs. Vaccination should be undertaken to prevent the disease.
Swine influenza	Swine influenza virus. Sudden out breaks involving all or large part of herd showing symptoms, cough, dyspnoea, prostration, fever, anorexia, conjunctival discharges.	No specific treatment, Good housing and sanitation are important. To control secondary infections antibiotics may be given.
Swine fever or Hog cholera or Pig typhoid	Hog cholera virus or *Togoviride pesti virus suis.* Apathy, anorexia, blotchy reddening of skin, diarrhoea, high fever, drop in leukocyte count, ataxia, convulsion and death.	No effective treatment, antibiotics may be to control secondary infections. Vaccination is done to control the disease.
Swine pox	Vaccinia and swine-pox-virus. Red circular eruptions on entire body.	No specific treatment. Better sanitation may prevent occurrence of disease, after recovery immunity develops.
Aujeszky 's disease (pseudorabies)	Aujeszky's disease virus. Symptoms vary with the age of the animal. Sneezing, coughing, slight fever, muscle spasm, involvement of nervous system is turning point and once it develops animals die within couple of days.	Antiserum is given intra-peritoneally, disinfection and vaccination.
Porcine Reproductive and Respiratory Syndrome (PRRS)	PRRS virus. Late term foetal death, abortion, weak pigs, and severe respiratory disease in young pigs.	Isolation, sanitation and vaccination by Modified-live virus vaccines and killed virus vaccines.

Contd.

External parasites		
Sarcoptic Mange	*Sarcoptus scabei var suis*. Initially small red pappules and general erythrema around the eyes, around snout, on the concave surface of the auricles of the external ears etc, irritation and thickening of skin of affected body parts (hyperkeratosis).	Spraying of pig gently with a snap sack or pressure washer using various mange dressings & proper sanitation may be followed.
Lice	*Haematopinus suis*. Irritation leading to rubbing of affected body parts against available object.	Pigs should be dusted with an insecticide, such as DDT powder or sprayed with a 0.5% DDT solution as and when required.
Internal parasites		
Ascariasis	*Ascaris suis*. Irregular appetite, thirst, vomiting, mild colicky pain, obstruction of masses of ascarids, peritonitis, coughing, pneumonia and swelling of abdomen.	Good sanitation and systemic use of medicines.
Cocciodiosis pens	Diarrhoea, dehydration, faeces may vary in colour as well as consistency from yellow to grey green or bloody depending on the severity.	Good hygiene, disinfection of farrowing houses and should be as dry as possible.
Kidney worm	*Stephanurus dentatus*. Adversely affect growth.	Anthelmintics, sanitation and rearing on concrete or in confinement can be used to control kidney worm.
Miscellaneous diseases		
Mastitis-Metritis-Agalactia	Cause is unknown. Lack of milk just before farrowing. The udder is hot and hard. The sow may have fever, depressed and go off feed.	Reducing stress during pregnancy may help in prevention. Provision of plenty of water and proper ration. Milk letdown may be stimulated by washing the sow's udder with warm water and a mild disinfectant.

Contd.

Mycoplasmal pneumonia	*Mycoplasma hyopneumoniae.* Feed intake is reduced, poor growth rate and coughing.	Sanitation, isolation vaccines are the preventive measures.
Hypoglycemia	Lack of sugar in the diet. Shivering, weakness, unsteady gait, dullness, and loss of appetite. The hair coat becomes rough. Diarrhoea may develop in some pigs. Death occurs within a day and a half after the symptoms appear.	Treatment with dextrose solution or dark syrup either injected or given through oral route.
Parakeratosis	Lack of zinc in the diet. Rough, scaly skin and poor growth rate. The condition has the appearance of mange.	Control and treatment consists of supplying adequate zinc in the ration.
Porcine Stress Syndrome	Genetic factors. The sudden death of heavily muscled hogs is referred to as porcine stress syndrome. Constant movement, tail twitching, and trembling ears. The skin tends to develop splotches when the hog is excited. The splotches are red on white pigs and purple on black pigs.	Avoid stress when handling or moving hogs. Do not crowd hogs in pens, and provide plenty of feeder and watering space. Do not handle or move hogs in extremely hot weather. Provide plenty of bedding in cold weather.

Table 9.3.2 : Vaccination schedule for pig

Disease	Age and booster doses	Dose & Route
Swine erysipelas	Around 3-4 weeks, repeat 3-6 weeks later and later every 6-9 months	1ml s/c
Swine fever	Around 2 months of age, booster dose after 4 weeks followed by regular vaccination at 6 months interval.	1 ml s/c
Foot and mouth disease	Around 2 months of age, repeat after 4 weeks and followed by regular vaccination at 6 months interval.	1 ml (polyvalent tissue) /2 ml (Inactivated tissue culture)

Marketing and Transport of Pigs

Professional pig producers plan their production operations in such a manner that they are able to sell their hogs as per the need of the market. Farmer should analyze the requirement of the market and decide on its own marketing strategy to get highest price for their produce. In India pig/pork marketing mainly takes place through the informal sector in the form of locally raised fresh pork mostly in the north eastern India where pork consumption is more common. There are three main products: weaner piglets, slaughter pigs and fresh pork in the pig sub-sector's output market. Usually marketing of pig/pork takes place as describe below.

Piglets are produced by households (with one or more sows) that keep breeding sows and piglets are marketed in one among the many options available *i.e.* i) they may directly sell these piglets to a household that rears piglets for slaughter. This type of business generally takes place within a village or with a nearby village from a reputed breeder. ii) They may sale their piglets to the weekly local market on the days fixed by the local authorities. iii) They may sell piglets directly to the traders visiting the villages for piglets. The purchased piglets are transported by suitable transport means to the traders' homes where they were kept until the next weekly market. A trader may deal directly with breeder, rearers and among themselves for local or district/state/national markets. The price of piglets varies depending on breed, age, sex, growth, source and season of purchase. iv) Moreover, producers directly or through middleman get order from meat production unit and gather required number of pigs from various pig producers in the area and transport them to the meat production unit in hired transport vehicle where they are paid on the basis of live weight taken before slaughter after resting pigs for 24 hours.

Slaughter pigs and fresh pork can be marketed in many ways-i) Rearing households may slaughter their pig and sell the pork in the village. This system is advantageous when markets are far away from the village, poor roads to the market, high transportation cost and during festival season. ii) Producers sell all slaughter pigs to the local retailers visiting the villages and procured pigs are transported to the retailer's stocking yard and where they are offered with feed and water. The pigs are slaughtered near the market place. iii) Pork wholesalers also play a role in the slaughter pig supply chain by procuring meat from another market and selling it through local retailers.

In developed countries hogs may be sold through: (1) direct marketing, (2) terminal markets, or (3) auction markets. Sometimes hog producers form a group and sell their animals through one of the three above mentioned methods. Prices are generally fixed on the basis of carcass measurements and premium prices are paid for hogs with more lean muscle and less fat.

Transport of pigs by various means has been already discussed in chapter two of this book.

Pork Production

Pork is the term given to the meat derived from the pigs. It is the most widely consumed meat in the world. It comprises around 37 per cent of world's total meat production and around 50 per cent of the total meat produced in Asia and 48 per cent in Europe. Pork production in different regions and leading countries in pork production has been presented in Table 9.4.1 & 9.4.2.

Table 9.4.1 : Pork production in the different regions of the world.

Region	Pig meat (million tonnes)		
	2000	2005	2010
World + (Total)	90.00(38.45)	99.25(38.01)	109.25(37.33)
1. Africa + (Total)	0.76(6.73)	0.94(6.84)	1.23(7.12)
2. Americas + (Total)	15.36(20.71)	17.00(20.32)	18.62(20.65)
3. Asia + (Total)	48.00(52.53)	55.60(52.52)	61.95(50.29)
4. Europe + (Total)	25.37(49.09)	25.17(48.48)	26.96(47.89)
5. Oceania + (Total)	0.48(9.07)	0.52(9.00)	0.47(8.36)

Figures in parenthesis indicates per cent of world's total meat production.

Source : FAOSTAT (2010).

China is the world's leading pork producing county followed by USA, Germany, Spain, and Brazil. India produced only 0.43 and 0.30 per cent of world pig meat in the year 2005 and 2010, respectively.

Table 9.4.2 : Leading pork producing countries of the world.

Countries	Pig meat (million tonnes)		
	2000	2005	2010
1. China	40.75(45.28)	46.62(46.97)	51.71(47.34)
2. USA	8.59(9.55)	9.38(9.45)	10.18(9.32)
3. Germany	3.98(4.42)	4.49(4.53)	5.48(5.02)
4. Spain	2.90(3.23)	3.16(3.19)	3.36(3.08)
5. Brazil	2.60(2.89)	2.80(2.82)	3.07(2.82)
World + (Total)	90.00	99.25	109.25

Figures in parenthesis indicates per cent of world's total pork production.

Source: FAOSTAT (2010).

Pig meat may come from suckling pig at one hand and from animal weighing 180 kg on the other hand. National consumption patterns determine the weight at which pigs to be finished and consumed. Prime meat is generally obtained at around 30–60% of mature size, at below 30% of mature size pork may lack flavour and above 60% of mature size pork may lack tenderness. Market classification of swine is based on their use, sex and grade. Based on their utility some of the market classes of pigs are

- Roaster: pigs weighing under 5-10 kg body weight.
- Market/slaughter/butcher pigs: swine under 55 kg of body weight which are killed and sold as meat.
- Butcher sow: sows weighing above 200 kg and are no longer supposed to be used for breeding.
- Feeder pigs: are the animals which are sold to be fed to higher weights before slaughter.

As per their sex market pigs are classified as - barrows, gilts, sows, boars and stags. Most of the pigs are harvested as barrows and gilts at about 100 kg of body weight.

The USDA has established grade standards for barrows and gilts based on carcass quality and the yield of the four lean cuts. The four lean cuts are: ham, loin, picnic shoulder and boston shoulder. The five official USDA grades for barrows and gilts are: U.S. No. 1 (highest percentage of lean cuts at over 60), U.S. No. 2, U.S. No. 3, U.S. No. 4, and U.S. Utility. As the grade number increases, the percentage of lean cuts decreases. Other factors like muscling and back fat thickness determines the grade.

Pork like other meat is a very good source of protein, essential amino acids. Protein in pork is of high biological value. It is also a good source of energy (Table 9.4.3). Pork is an excellent source of calcium, phosphorus, iron and some of the trace minerals. Thiamine content in pork is many times higher than beef and lamb. Liver is a good source of Vitamin A, C and some of the vitamins B-complex. Pork is a very good source of polyunsaturated fatty acids and it contains all essential amino acids in higher amount than beef and lamb. The major objection to boar meat is due to the boar odour or boar taint. Skatole and androsterone have been identified as the main cause of this problem.

Table 9.4.3 : Water, protein, fat, ash (in percent) and calorie content of some raw and processed livestock products.

Product	Water(%)	Protein (%)	Fat (%)	Ash (%)	Calories /100g
1. Fresh					
a. Beef (lean)	75.0	22.3	1.8	1.2	116
b. Beef carcass	54.7	16.5	28.0	0.8	323
c. Pork (lean)	75.1	22.8	1.2	1.0	112
d. Pork carcass	41.1	11.2	47.0	0.6	472
e. Chicken	75.0	22.8	0.9	1.2	105
2. Processed					
a. Beef, lean, fried	58.4	30.4	9.2	–	213
b. Pork, lean, fried	59.0	27.0	13.0	–	233
c. Lamb, lean, fried	60.9	28.5	9.5	–	207
d. Milk (pasteurized)	87.6	3.2	3.5	–	63
e. Egg (boiled)	74.6	12.1	11.2	–	158

Source : Gunter & Peter (2007).

Preparation of Pig for Show

Pig show provides an opportunity to the farmer to exhibit their best animals, brings fame to the farmer and popularises best breed as it act as an advertising media. It provides a platform where comparison among superior animals within and between breeds can be made. It further make possible to exchange experiences among the breeders, chance to learn about better feeding and management practices. It provides an opportunity to sell the limited number of animals. It helps in building a better competitive spirit among the participants.

The process of getting swine ready for the show is known as fitting. It enhances the features of the animal and includes selection of animal, feeding, exercising, training, grooming, washing, and clipping.

1. **Selection of show pig:** While selecting the animal- age, pedigree, weight, sex, conformation and soundness of the animal should be considered. Hogs are generally selected at 2 to 3 months of age, weighing around 50–60 pounds so that they should attain a show weight (about 235-270 pounds) in approximately 100 days thereafter. If possible those lines are selected which has earlier history of winning the show in the past. Desirable physical characteristics are- long-body with sound feet, well developed hams and joint should be well angled.

2. **Feeding:** Initially they are fed through self feeder and later on to control weight gain they are hand fed.

3. **Exercising:** Regular walking exercise will help in maintaining the animal in good condition and make the animal more attractive in the show ring.

4. **Taming/gentling:** Animal is trained to move in desired direction. They are trained to turn to the right by gently tapping on the left jowl and vice versa. Train them to move forward by tapping lightly in front of shoulder and to stop by holding the bat in front of his nose.

5. **Trimming of toes:** Three to four weeks prior to show the toe of the pig is brought to proper shape with the help of hoof trimmer and hoof rasp is used to square-up the sole.

6. **Washing and rinsing:** This is the first step in the preparation of a pig for show and is also necessary to wash the animal before clipping. To accustom the animal to washing, it is generally started about 3 weeks prior to the show and frequency is increased to once a week before show and done again the day before the show. First the pig is wet with water to remove of the large piece of dirt and debris then soap is applied and rubbed & washed. Head of pig is washed in the last. Rag is used to wash the snout and inner side of ear instead of brush.

7. **Clipping:** It should be done at home within a week prior to the show. Clipping make a youthful, fresher appearance and enhances muscle tone acuity. It is begin from back against the grain of the hair then one side of the pig clipped at a time from the ham to the side to the shoulder up to the face.

8. **Oiling:** It makes skin soft, smooth and give shine to the hair coat. It is applied after the pig is bathed the first time and a further light application is just before the show.

9. **Showmanship:** Following guidelines should be followed during show

   - Exhibitor should be well dressed and groomed.
   - Keep the animal well groomed and watered prior to their class show begins.
   - It is better that exhibitors should watch ring layout, other classes and activities of the judge before their class show begin.
   - Enter the ring without delay when class is called.
   - The pig should be kept between the exhibitor and the judge.
   - Exhibitor should always look calm and confident, must keep one eye on the judge and another on the pig.
   - The exhibitor should foresee movement of his/her pig and the judge and should take action accordingly.
   - The pig should be kept moving at a steady pace around the ring.
   - Never touch the show pig with the hands. Show stick should be used to tap on the point of the shoulder in the region of the jowl to encourage movement.
   - The exhibitor should place himself/herself on the side of the pig away from the judge near the flank and ham/loin juncture.
   - Be prepared to follow the instructions of judge as well as to answer the queries of judge.

Judging Swine

Judging is an art and expertise can be achieved by repeatedly practicing it with patient over period of time. For appropriate judging it is crucial to have meticulous understanding of ideal physical features of swine of different breeds, age group and reproductive characteristics etc. The parameters of judging swine varies as per the purpose for which they are reared *i.e.* market hogs, breeding hogs.

Procedure

- Usually numbers are marked on the body of animals to identify them for judging and are observed from a distance of about 15 feet when they move about the ring.
- From side view balance, length & depth of side, ham development, gait and physical characteristics are observed.
- From rear view depth of ham, depth of loin and body and position of hind legs while from front view meatiness, trimness of head and shoulder, spring of rib and position of forelegs can be observed.
- These traits are observed and compared with the ideal hog and the other hogs in the class. During judging points are noted down with furnishing the reasons for particular points given.
- Traditionally only visual assessment of number of traits of the hogs were used to rank the animals. Now-a-days performance data (breed, sex, feeding, production environment, marketing plan etc) of hogs in combination with visual assessments are used to rank the animals.
- For market barrows parameters like conformation (general appearance, hind quarters, fore quarters & middle), finish (ham, rump, back, loin, jowl and belly) and quality (bones) of the animal while for breeding swine general appearance, conformation (hind quarters, fore quarters & middle) and breeding qualities (underpinning, mammary system & breed characters) are evaluated on a 100 point scale for judging them.

Chapter - 10

Laboratory Animals

Amit Kumar, P Kumar, S Kumar and P P Dubey

The group of animals constantly used in laboratories for general research in all discipline of science is called as laboratory animals. Commonly, it includes rats, mice, rabbits and guinea pigs. However, under certain circumstances, additional animal species like hamsters, non-human primates, amphibians, fowl, sheep and pigs can also be included. This term incorporates traditional laboratory animals, farm animals, wildlife and aquatic animals. Nontraditional animals, inclusive of invertebrate species are also included where they are relevant to the unit's mission. Generally, these animals come under vertebrate but broadly laboratory animals may be classified in two major groups *i.e.* laboratory mammals and non-mammalian laboratory animals. The laboratory mammals include mouse, rat, guinea pig, rabbit, cat, dog, ferret, gerbil, bats, hamster, mink and monkey (Rhesus). Beside these mammalians the large domestic animals/livestock are also used as laboratory animals (horse, cow, sheep, goat *etc*) for fulfilling the demands of various laboratories. Poultry is also used as non-mammalian laboratory animals and Specific Pathogen Free (SPF) chicks are at utmost demand for production and testing of biologicals. The other non-mammalian laboratory animals includes reptiles (tortoise, lizzards and snakes), insects (cockroach, drosophila, house fly, house cricket, locusts, tsetse fly, stablefly *etc*) pisces (fishes and snails) and amphibians (toad and frog). Laboratory animals have been used extensively for basic research in drug and bacteria screening, toxicology, antibiotic sensitivity testing, animal models of diseases and production/testing of biologicals. Laboratory mammals are also valuable for diagnosis of infectious diseases in man and domestic animals. In this chapter the most commonly used laboratory animals *i.e.* mice, rats, guinea pigs and rabbits will be discussed.

Laboratory Mice

The most popular model organism in mammalian is laboratory mice, which were derived from wild mice belonging to the *Mus musculus* species that included the generation of 'fancy' mice in both Asia and Europe. The laboratory mouse owes small body size and high fecundity. The wild or house mouse belongs to the sub-family-Murinae, family-Muridae, sub-order -Myomorpha of the order-Rodentia. All existing laboratory strains of *Mus musculus* have been developed by selective breeding of wild type. Mice have 40 chromosomes in their diploid genome. The strains of laboratory mice are genetically classified into outbred and inbred category. The Swiss albino is most popular strain of laboratory mouse whereas the Balb c and C57BL/6J are the most frequently used inbred strain of laboratory mice. Laboratory mice are unique in that there are a large number of inbred strains. During recent past, laboratory mice are widely used for research and testing the potency of biologicals. The first transgenic animal was produced in mouse model. Genetically modified mice are used extensively in research as models of human diseases. The most common type is the knockout mouse, where the activity of a single (or in some cases multiple) genes are removed. They have been used to study obesity, heart disease, diabetes, arthritis, substance abuse, anxiety, ageing and Parkinson`s disease. Transgenic mice generated to carry cloned oncogenes and knockout mice lacking tumor suppressing genes have provided good models for human cancer. Beside this the laboratory mice are being used for education purpose in various pharmacies, medical and veterinary colleges.

Size and body parts: The body weight and size depends on the strain of mice. Generally the inbred strains posses lower body weight than the outbred strains. Also indred strains of mice do have smaller fecundity over outbred strains. The pelage of wild mouse is agouti with a lighter color on the ventral surface. The eyes are black and skin is pigmented. However, frequently available strains of laboratory mice are albinos having similar body size of wild mouse. Although other body color are also available in laboratory mice. Adult laboratory mice weigh up to 30 grams have blood volume of 2 ml with resting heart rate of 500-700 beat per minute. The birth weight of laboratory mouse is approximately 1-2 gram with average weaning weight of 9-12 gram at the age of 3 weeks. The weighing must be done on fine electronic weighing balance for improving the accuracy of weight. The head of laboratory mice are small (in relation to body) and triangular in shape with with sharp muzzle. Ears are large (in relation to body). Mouse has thin tails where length of tail is longer than body. Legs of mice are smaller relative to body size.

Laboratory Rats

There are different kinds of rats. Often the literatures about wood rats and kangaroo rats, black rats and Norway rats are available. In articles about scientific research, reports about Wistar, Fisher and Sprague Dawley rats are available. Among pet rat owners, Dumbo, Rex, Hairless, Berkshire, Platinum and Tailless rats are known. In a nutshell, wild rats belong to many different species, while almost all domestic pet rats and laboratory rats belong to a single *species*, the Norway rat (*Rattus norvegicus*). A *strain*, according to Webster's, is a group of individuals which share a presumed common ancestry and have clear-cut physiological but not usually morphological distinctions. An *inbred strain* is defined as the product of over 20 generations of brother-sister matings, which results in individuals that are 98% identical to each other.

A laboratory rat is the member of species *Rattus norvegicus* which is bred and kept for scientific research. *Rattus norvegicus* belongs to order-Rodentia, family-Muridae and subfamily -Murinae. The rats do have 42 diploid numbers of chromosomes. Laboratory rats have served as an important animal model for research in psychology, medicine and other fields. Wistar and Sprague Dawley rats are commonly available outbred strain of rats. Among these the albino Wistar laboratory rat with its red eyes and white fur is an iconic model organism for scientific research in a variety of fields. Wistar rats are an outbred strain of albino rats belonging to the species *Rattus norvegicus*. This strain was developed at the Wistar Institute in 1906 for use in biological and medical research and is reported to be first rat strain developed to serve as a model organism at a time when laboratories primarily used *Mus musculus*, or the common house mouse. The Sprague Dawley rat and Long-Evans rat strains were developed from Wistar rats. Wistar rats are more active than other strains like Sprague Dawley rats. The *Sprague Dawley rat* is an outbred multipurpose breed of albino rat used extensively in medical research. Its main advantage is its calmness and ease of handling because of its docile nature. This breed of rat was first produced by the Sprague Dawley farms/company in Madison, Wisconsin. *Bio-breeding Diabetes Prone rat* is an inbred rat strain that spontaneously develops autoimmune Type 1 diabetes. Long Evans rats are white with a black hood, or occasionally white with a brown hood. They are utilized as a multipurpose model organism, frequently in behavioral and obesity research. Zucker rats were bred to be a genetic model for research on obesity and hypertension. They are named after Lois M. Zucker and Theodore F. Zucker, pioneer researchers in the study of the genetics of obesity. Hairless rats, Royal college of surgeons (RCS) rats and Shaking rat Kawasaki are another strains of rats used in bio-medical research.

Size and body parts: The Wistar rat is currently one of the most popular rat strains used for laboratory research. It is characterized by its wide head, long ears and having a tail length that is always lesser than its body length. The muzzle of Wistar rats are large and blunt with wide muzzle. The tails of Wistar rats are thick and legs (especially hind legs) are comparatively longer relative to the body. The ears are small relative to the head. The birth weight is of 4-5 gram and adult body weights vary from 200-300 gram in female and it may go up to 500 gram in males. The weaning weights of rats are 40-50 gram at the age of three weeks.

Identification of Mouse and Rat: Ear clipping or punching, as described by Short & Woodnott (1969) is a convenient way of marking individual rats. Tail tattooing with more advance equipment is another approach of identification. For short duration marking the water proof marker pen may be used on tail.

Handling of Mouse and Rat: Always wear rubber gloves before handling the mice and rats. Mice and small rats can be picked up by tail. Larger rats can be handled similarly like mice but one should hold the root of tail otherwise the skin of tail may be slipped. Larger rats especially pregnant ones should not be picked up using tail. For handling pregnant and larger rats the weight should be taken by a hand, either underneath the body, or grasping the rats over shoulder and chest.

Figure 10.1 : Rat with young pups.

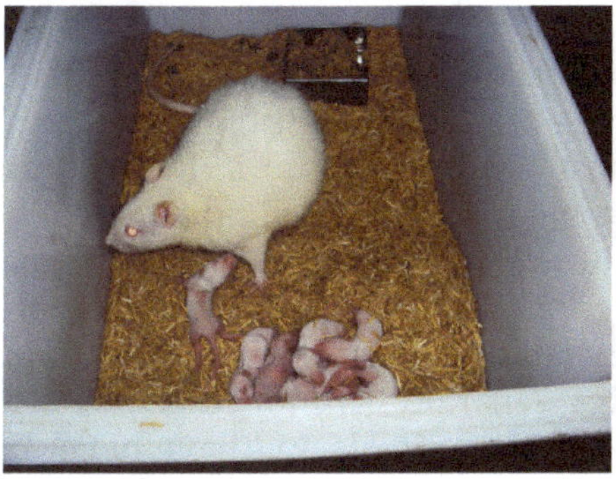

Figure 10.2 : Guinea pigs in Housing tray.

Guinea Pigs

Guinea pigs (*Cavia porcellus*) were known by a variety of common names: "Sea Pig," "Barbary Rabbit," or "Little Sea Pig". The laboratory guinea pig was derived from domesticated stocks of wild guinea pigs (*Cavia cutleri*) which originated in Peru, South America. In South America the guinea pig is raised as a source of meat. Before their introduction into biomedical research, guinea pigs were bred for fancy and kept as pets for about 300 years. Many of the laboratory strains used today originated with the breeding work of Dunkin and Hartley in 1926. The Dunkin Hartley is major outbred stocks of guinea pig known with scientific name *Cavia porcellus*. The *Pirbright-Hartley* is an outbred stock used in England, which is the same type of animal. It belongs to order-Rodentia, sub-order-Hystricomorpha and family-Caviidae. So guinea pigs are *hystricomorph* (hedgehog-like) rodents. Guinea pigs do have 64 diploid number of chromosome. Till date 20 different inbred strains have been produced and only two strains (strain 2 and 13) are existing now. Mostly inbred strains of guinea pigs are being used for studying oncogenic leukemia viruses. The guinea pig has been widely employed in biomedical research since 1780. The guinea pig has been used in antibody production, tumor genesis, nutrition, genetics, radiation research and dental studies including antibacterial action of saliva and production. The guinea pig is used extensively in development of sera. Lavoisier used the cavy for measuring heat production. Much of the work relating to the discovery of Vitamin C, the diagnosis of tuberculosis and the research on anaphylaxis has been performed in this species.

Size and body parts: Female guinea pigs are referred to as sows and the males are called boars. The average life span of guinea pig is 4-5 years but exceptionally they may survive up to 8 years. The adult male weight varies from 400-500 grams whereas in adult female weight varies from 800-900 grams. The birth weight varies from 80-100 grams. The bodies of guinea pigs are stocky and compact and their legs are short. Unlike most of the other rodents they have no tail. Four coat types have been recognized in the guinea pig. The "English" guinea pig has short, smooth hair and is the most commonly used in the laboratory testing. The "Abyssinian" guinea pig has a coarse, short hair coat with the hair forming whorls or rosettes. The "Peruvian" guinea pig has a coarse, long coat, while the "Angora" or "silky" guinea pig has a fine, medium-length coat of hair. Guinea pigs may be of single color or up to three spots of colors. Guinea pigs generally live together peacefully. Guinea pigs are most docile laboratory animals and have never been noticed biting tendency. However, they are sensitive and are easily excited by loud sudden noises, fast movements and changes in their environment. Guinea pigs are active during daylight hours. Guinea pigs tend to freeze when they hear unfamiliar sounds and may remain immobile from several seconds to 20 minutes. They also exhibit a *Preyer or pinna reflex* that is why they will cock their ears in response to the sound. Conversely, they tend to run and scatter when they see sudden movements. This scattering behavior can involve "stampeding", jumping or rapid circling of the cage. Introduction of a new male into a stable group can lead to fighting and should be avoided whenever possible. Dominant animals will frequently *"barber"* or chew off the hair of subordinate animals which results in a patch of hair loss with the underlying skin. Barbering is also associated with boredom or the stress of overcrowding. Vocalization among group members is important in their social interactions. These sounds consist mainly of squealing, whistling and grunting sounds. Guinea pigs whistle in anticipation of feeding.

Handling and Identification: Guinea pigs are restrained by grasping the animal around the thoracic area from above with the thumb and forefinger right behind the front legs for lifting. Do not hold a guinea pig too firmly around the thorax or abdomen as breathing can be impaired or the lungs or liver may be damaged. Cage cards may be used to identify singly-housed animals. Permanent identification of an individual can be accomplished by the use of ear notching, ear tagging, tattooing or microchips. Dyes and markers can be used for temporary identification.

Laboratory Rabbits

Rabbits are member of family Leporidae of the order-Lagomorpha, found all around the world. Rabbit have 44 chromosomes in their diploid genome. The male is called as buck and female as doe; a young rabbit termed as *kitten* or *kit*. The presence of rabbit with human being remained from the time immemorial. Romans were the first to domesticate wild rabbits. Wild rabbits (*Oryctolagus cuniculus*) domesticated in Roman Empire as game animals. Such rabbits were increasingly populated in and around southern Europe, North Africa and Spain. Rabbit keeping in commercial base started in the last century. The scarcity of food following first and Second World War, rabbit rearing as meat animal became popular. During those days rabbit's flesh was denoted as 'Underground Mutton' and served as a source of food for rural families. The most important use of New Zeeland White rabbit is in production of biological/vaccine, drug testing and use as research material.

Breeds of rabbit: Three types of breeds are available for rabbitry unit. Heavy weight breeds (4-6 kg) includes White giant, Grey giant, Flemish giant; the medium weights breeds (3-4 kg) are New Zealand white, New Zealand red, Californian whereas the low weight rabbit breeds are Soviet chinchilla and Dutch. The most popular broiler rabbit breeds are New Zealand White, Soviet chinchilla, Grey Giant and Dutch and their crosses.

Size and body parts: The average life span of rabbits is about 4-5 years. The body weight of adult laboratory rabbits varies between 2 to 2.5 kg. The birth weight of kits varies between 40-60 grams. The rabbit is known for its long ears that provide it excellent hearing capability. Rabbits also use their ears to let out heat because they cannot sweat. The large eyes are positioned on the sides of the head, giving the rabbit a visual range of 190 degrees. The rabbit's nose and mouth are close together forming a small muzzle. Rabbits also have open-rooted incisors that grow continuously. The scent glands are positioned under the chin. In female rabbits skin around the neck is loose and is referred to as the dewlap. In rabbit small chest area is located in the front of the body running from below the neck to just above the forelegs. The forelegs are small limbs in comparison to the powerful hind legs. The hocks are located at the heel of each foot. The stomach is located along the underneath of the rabbit's body. The body is slimmer around the chest area and thicker around the hind quarters where the spine curves. This part of the back is called the saddle. A rabbit enjoy food with its head lowered to the ground. Rabbits mostly sleep during the day. They often sleep with their eyes open so it is hard to tell when they are sleeping. Rabbits thump their hind feet to signal danger. Some rabbits will thump one foot on the ground while others will thump both hind feet at the same time. Rabbits sit up on their hind feet to get a better view of the surroundings. Usually periscoping occurs when a rabbit hears something is trying to figure out about the sound. Rabbits will tiptoe around when they are not confident of their surroundings. It smells different, so thumper tiptoes up to it and takes a sniff. Rabbits can make a purring noise by grinding their teeth. A quiet tooth purr means rabbit is happy. A louder crunchy type of teeth grinding indicate rabbit is in pain. Single male rabbits will often try to mount soft toys, cushions and even people's legs. This is a sign that your rabbit is ready to be neutered. Male/female bonding is usually the easiest to achieve and in this case, being fixed is a must.

Handling and Identification: Rabbits are especially susceptible to the effects of stress and should always be approached in a calm and confident manner. The handler should restrain the rabbit firmly by the scruff with the other hand ready to support the animal's hindquarters. The rabbit should be held by its head tucked under the handler's arm and with the back and hindquarters supported by the handler's forearms. Rabbits may also be restrained for injection by wrapping the animal in a drape or towel. If the rabbit is securely wrapped, it will not struggle. Intravenous injections into the rabbit's marginal ear vein can be readily made if the rabbit is restrained with a drape.

Temporary marking can be done using water proof marker pen for a limited time span. However, for everlasting identification, especially for rabbit breeder it is desirable to follow tattooing in rabbits. A legible tattoo in the domestic rabbit's right/left ear is very important for permanent identification of the animal. The skin of the ear is penetrated with the needles of the tattoo digit. The Animal Identification & Marking Systems (AIMS™) tattoo identification system is a humane and permanent method for identifying rabbits with tattoo markings. This allows the proper placement of tattoo markings within the ear to avoid trauma to important ear veins, arteries and nerves.

Transportation of Laboratory Animals

The transport of animals from one place to another can be avoided and animal must be undertaken with care. The main considerations for transport of animals are the mode of transport, the cages, floor space requirement in cages, food and water during transport, protection from transit infections, injuries and stress. The mode of transport of animals depends on the distance, climatic conditions and the species of animals. Animals can be transported by road, rail or air taking into consideration above factors. In a single cage maximum 25 mice, 25 rats, 12 guinea pigs and 2 rabbits must be placed with almost triple recommended floor space than housing.

Record Keeping and Handling of Laboratory Animals

Proper animal handling and housing are critical components to successful research. The researchers require proper handling of mice during blood collection from tail vein/orbital sinus/cardiac puncture/ saphenous vein, intraperitoneal injection, subcutaneous injection, oral feeding and sexing. A record-keeping system for inbred lines and various experimental colonies must easily enable the investigator to trace the collateral relatives, ancestors and progeny of any animal. Ledgers are usually book keeping ledgers or spiral-bound notebooks. A ledger may be used for each strain or mutant stock or in small colonies, for all animals of several stocks. The ledger must include the following information about the sequence number assigned to the them, sex, phenotype, dam and sire along with their strain or genotype, generation and subline, number born in litter, date of birth, litter weight at birth and subsequent ages.

A pedigree card is designed to fulfill the needs of the researcher and the type of colony under study. One side of the card may be used for all information about breeding performance of the female and the reverse side of the card may be used for auxiliary data. In some systems, two cards are used-a mating card and a litter card. Cage tags are mainly of two types: those providing only location on the shelf and those providing information such as the identification numbers of the animals in the pen, or number of mice in the pen, and any other information desired along with location.

Housing of Laboratory Animals

Many factors should be considered in planning for adequate and appropriate physical and social environment. Housing and environmental requirements for commonly used laboratory animals are summarized in Table 10.1. The species, strain, and breed of the animal and individual characteristics such as sex, age, size, behavior, experiences and health are important factors in decision making about housing. The design and construction of housing are also important for housing of laboratory animals. It is important that they are housed in an isolated building located as far away from human habitations as possible and not exposed to dust, smoke, noise, wild rodents, insects and birds. The animal rooms should occupy about 50-60% of the total constructed area and the remaining area should be utilized for services such as stores (8-10%), washing (8-10%), office and staff (8-10%), machine rooms (4-5%) quarantine and corridors (12-15%). The health and safety of personnel engaged in laboratory animal house may be affected if proper housing is not practiced. Male mice can have musk like smell. They sent their territory and if you have more than one male the smell can be offensive to some people. The temperature and humidity of mice breeding colony is very important. The mice stop breeding during too hot or too cold weather. If the temperature is above 40°C the mice may die. The ideal temperature for maintaining mice colony is between 22 and 24°C. This range of temperature can be maintained using air conditioner or a controlled electric fan heater. An animal room must be kept clean, quiet, and uncluttered. Research procedures which are permitted within the room include injections, blood collection, examinations and other noninvasive techniques. Surgery, euthanasia and necropsy are not permitted in the animal room. Light levels should be adequate for the animal to perform normal behaviors and for the worker to perform their duties. Diffuse lighting in the range of 130-325 lux is normal. A diurnal 12 hour light cycle for most species and 14 hour light /10 hours dark for rats and mice are standard. Room ventilation must be adequate to provide oxygen and remove heat, chemical, biological wastes. If the ventilation is not proper the high ammonia concentration causes respiratory irritation to both animals and attendants, predisposing them to infection by lowering their resistance. The standard rate is 10-15 air changes per hour. Fresh air is desirable, although recycled air may be provided if it has been conditioned to remove excess heat, carbon dioxide, moisture, odors, or biological or chemical hazards. Room ventilation should normally be adjusted to maintain room pressure negative to the corridor to protect adjacent personnel areas or positive to the corridor if the room is serving as an isolation room. Ventilation ducts and filters should be cleaned at least monthly. The *microenvironment* (cage) of an animal is the physical environment immediately surrounding it. It includes temperature, humidity, and gaseous and particulate composition of the air. The room, barn, outdoor habitat constitutes the *macro-environment (animal room).* Temperature in rooms should be maintained in a range suitable for the species of animal and the animals should be protected from abrupt changes. A range of 68°F to 74°F is recommended standard for mammals. A 4°C change in temperature can cause 10 fold alterations in biological responses. The temperature also affects fertility and lactation. The relative humidity should be controllable within the range of 30% to

70% throughout the year. Noise in animal rooms should be minimized whenever possible. Some species may experience reproductive problems when exposed to excessive noise so noise from mechanical equipment in adjacent areas should be avoided. Room wall and floor surfaces should be constructed of materials that are easily sanitized. Bedding material should be clean, dry, dust-free, absorbent, non-aromatic, non-toxic and preferably soft. The bedding material must be fumigated with potassium permanganate and formaldehyde. The hydrogen peroxide may be good for sanitation of animal cages. Bedding must be changed when it is visibly wet. Cages may be sanitized in a commercial cage washer with a soap wash and a high temperature (180^{0}F) rinse. This service can be contracted with research animal resources (RAR). Cages may be hand washed with detergent, rinsed in water, then dipped in a sanitizing agent (1/2 oz bleach per gallon of water) and allowed to dry. The floor space and recommended environmental requirements for different species of laboratory animals have been given in Table 10.1.

Table 10.1 : Housing and environmental requirements for commonly used laboratory animals.

Requirements	Mouse	Rat	Guinea pig	Rabbit
1. Avg. adult weight	25-30g	200- 300g	400-500g	2.0-2.5kg
2. Type of housing	Cage	Cage	Cage/Pan	Cage
3. Floor area per animal (sq.cm)	65-100	100-150	300-600	3700-4600
4. Cage height minimum (cm)	12	14	18	36
5. Room temp (°C)	22-24	22-24	22-24	22-24
6. Relative humidity (%)	45-60	50-60	45-60	45-60
7. Suitable bedding material	Paddy husk saw dust	Paddy husk saw dusk	Paddy husk Saw dust	-
8. Nesting material	Paper cutting	Paper cutting	Wood shavings dry grass	Cotton papercutting /dry grass
9. Ventilation air changes (per hour)	10-12	10-12	10-12	10-12
10. Light intensity (LUX)	300-400	300-400	300-400	300-400
11. Photocycle (Light : Dark)	12:12	12:12	12:12	12:12

Source : Guidelines for care and use of animals in scientific research (2000).

- **Mice and Rat housing:** Materials for cages include galvanized and stainless steel, plastic, wood, glass and paper. The most common type of mouse cage for a breeding pair is a box approximately 12 x 6 x 6 in. deep. Cages are available commercially with many dimensional variations of this basic design. The cages should be seamless and designed for easy cleaning and there should be about 8 sq in. of floor space per adult mouse. They should be light in weight, yet sufficiently sturdy to withstand repeated washing and rough handling. Also caging materials should be resistant to the corrosive effects of mouse urine. The polypropylene/polycarbonate cages are now frequently being used as mice and rat cages. Fully automated Individually Ventilated Cage (IVC) is recent housing arrangement in modern laboratory animal houses.
- **Guinea pig housing:** Guinea pigs are housed in solid-bottom cages with adequate bedding materials (paddy husk, chopped paper etc.) such as shoebox cages or tubs, or on wire mesh cages. Bedding materials should be as dust-free as possible and wood shavings, corn cobs or shredded papers are acceptable. The use of wire mesh should be avoided since guinea pigs unaccustomed to wire mesh and may lacerate or break limbs by walking or running on the mesh when their feet get entangled. Tops are not required for the cages as guinea pigs seldom jump or climb however the cages should be at least 25 cm high. Rectangular cages are preferred over round ones to decrease the incidence of stampeding. High ambient temperatures without adequate airflow predispose the animals to sterility and heat stress as compact, stocky body of the guinea pig does not dissipate heat very well. Pregnancy toxaemia often results if temperature goes beyond the limits. The relative humidity of animal housing areas should be between 30% to 70%. A 12 hours light and dark cycle is commonly used in the guinea pig housing areas.
- **Rabbit housing:** For scientific and profitable rearing of broiler rabbit, housing plays an important role. Housing for rabbits depends on the location of rabbit farm, local climate and size of the enterprise. To protect rabbits from the climatic conditions such as in hot summer, rain and predators construction of shed is necessary. There are three types of housing systems individual cage, hutch and colony system. Mostly housing arrangement for laboratory rabbits are done in caging system.

- **Cage system**: The space requirement in cages for adult male rabbit is 4 $feet^2$, female dam 4 $feet^2$ and bunnies 1.5 $feet^2$. Adult rabbit cages should have 1.5 feet length, 1.5 feet breadth and 1.5 feet height. This cage is suitable for one adult rabbit or two growing rabbits. Accordingly the cage size can be increased keeping in view the space requirement of bunnies. To provide safer and calm environment during kindling the nest box is essential. Nest boxes can be made up of galvanized iron or wood. Nest box size should be sizable to keep inside the kindling cage. Feed and water troughs for rabbits are usually made up of galvanized iron. The feed trough should be designed in the shape of "J" and they are usually fit outside the cages.

Figure 10.3 : Individual rabbit in cage.

Figure 10.4 : Rabbit cage with kindling box.

Nutrition of Laboratory Animals

- **Nutrition of rat and mice:** Good nutrition starts with a diet that meets all of the minimum daily nutrient requirements. The mouse recommended daily allowances are very similar to that of the rats. The requirement of calcium, manganese, vitamin K, and vitamin B_{12} in rats are 130%, 130%, 490%, and 290% more than the requirement in human, respectively. Calcium deficiency results in stunted growth, osteoporosis, rear leg paralysis and internal bleeding. Manganese deficiency results in birth defects, reduced food intake, stunted growth, skeletal deformities and early death. Vitamin K deficiency results in major blood losses from minor injuries and spontaneous internal bleeding. Vitamin B_{12} deficiency results in stunted growth. Vitamin E deficiency causes sterility in male rats. The diet of rats, moisture content should not be more than

10 %. In the diet of mice protein and fat percentage should be 15 and 5 %, respectively. The diet of rat must contain 12 % protein and 3-4% fat. However, in pregnant rats and mice the protein content may be increased up to 18 %. The above mentioned requirement may be achieved by formulation of ration using 62 parts wheat crushed, 30 parts maize crushed, 7 parts wheat bran and 1 parts common salt. The mix ration may be cooked in milk at the rate of 5 ml milk per adult mice. In case of rats 10 ml milk per rat must be used for cooking the ration in semisolid food. The mice and rats must be offered 5 and 10 grams semi solid food per adult animals. The guinea pig requires 18 % protein and 1-2 % fat in ration. The laboratory rabbits must be supplied with 17 % protein and 2 % fat in ration. The energy requirements for mouse, rat, guinea pigs and rabbit are 3000, 3800, 3000 and 2500 Kcal, respectively. However, during pregnancy these animal requires 4-5 % more protein and 200-300 Kcal more energy in the feed formulation to meet the growth requirement of fetus. The fibre content in mouse, guinea pig and rabbits must be 5, 10, and 12 %, respectively. Pelleted feeds balanced for different species of animals are now available commercially. These are easy to procure and use without wastage. However one has to be careful on quality of the feed from batch to batch. It should be obligatory on manufacturer to mark each bag with the type of food, date of manufacture, the batch number, the ingredients used and chemical composition.

- **Nutrition of guinea pigs:** The guinea pig is unable to synthesize its own vitamin C, like humans and nonhuman primates. Hence, vitamin C must be supplied in guinea pigs ration from outside source. It is important to store food properly and use it within 90 days of the milling date. If available, fresh fruits and vegetables are rich in vitamin C may also be offered as treats otherwise ascorbic acid (50-75 mg) may be supplemented along with diet or drinking water. Guinea pigs are strongly *coprophagic*, eating feces directly from the anus or the cage floor. The feces contain essential vitamins produced by the fermentation of feed by microorganisms within the caecum. Guinea pigs consume an average of 6 grams of food/100 gram body weight daily. Rations for guinea pigs usually contain 20% protein due to the high requirement they have for certain amino acids. Fresh water should always be available and may be offered by bottles with sipper tubes or automatic waterers. Bowls of water placed on the cage floor quickly become contaminated with feces and bedding. Adult guinea pigs drink an average of 10 ml/100 gram body weight per day. The guinea pig ration can be formulated using 35 parts wheat bran, 25 parts maize crushed, 30 parts crushed gram and 10 parts fish meal. About 40 gram mixed ration may be offered to get daily protein and energy requirement. About 100 gram seasonal green fodder may be offered in daily fodder requirement of guinea pigs.

- **Nutrition of rabbits:** Rabbits can be offered all types of grains (sorghum, bajra, Maize, and other grains) and legumes. Green fodders such as Desmanthus, Lucerne, Berseem, Maize leaves, Agathi, kitchen wastes such as carrots and cabbage leaves and other vegetable wastes are also relished by rabbits. The energy and protein requirements of rabbits have been shown in Table 10.2. The required energy and protein can be obtained using 30 parts broken maize, 30 parts ground bajra, 25 parts wheat bran, 13 parts fish meal or gram or ground nut cake, 1.5 parts mineral mixture and 0.5 parts salt.

Table 10.2 : Energy, protein and fat requirements of rabbit.

Details of nutrients	For growth	For maintenance	For pregnancy	For lactation
1. Digestible energy (Kcal)	2500	2300	2500	2500
2. Protein (%)	18	16	17	19
3. Fibre (%)	10-13	13-14	10-13	10-13
4. Fat (%)	2	2	2	2

Adult male and female are fed with 100 grams concentrate whereas lactating doe are fed 250 grams concentrate daily. Young bunnies can be reared on 50-75 grams concentrate daily. Adult animals must be provided 250-300 grams of green fodder daily but lactating, pregnant rabbit and young bunnies must be offered 150 grams of green fodder daily. The teeth of rabbits are continuously growing. Therefore rearing rabbits with concentrate feed alone is impossible hence roughage must be always offered to rabbit for avoiding buck teeth in rabbit due to excessive growth of incisor. The time schedule for feeding should be maintained strictly. If there is late in feeding of rabbits, they become restless and which in turn reduces the body weight.

Breeding and Reproductive Management in Laboratory Animals

The details of breeding and reproductive parameters have been summarized in Table 10.3. while physiological norms & vital statistics of commonly used laboratory animals are in Table 10.4.

Breeding and management of Mice: The life span of mice also varies with strain but generally it ranges between 1.5-2 years. Mouse reproduction varies seasonally and the ovulatory cycle coordinates mating. The estrous cycle of female mice has a length of 4 to 6 days. The estrous cycle can become synchronized in female mice housed together continuously and can be suspended in the absence of exposure to male pheromones. Suspended cycles can be restarted by exposure to male pheromones. This phenomenon is known as Whitten effect was observed by Wesley K. Whitten (1968). The Bruce effect, or pregnancy block, refers to the tendency for female mice to terminate their pregnancies following exposure to the scent of an unfamiliar male. In mice, pregnancy can only be terminated prior to embryo implantation. The Bruce Effect was first noted in 1959 by Hilda M. Bruce. Long daily light periods in the daily cycle support robust reproduction therefore laboratory mice are maintained on a long (12 hours of light) light cycle. Female mice are only receptive to the mating when they have ovulated. The physical stimulation of the cervix in mating is required to make the uterus receptive to embryo implantation. The male produces a copulatory plug from secretions from the vesicular and coagulating glands that blocks the vulva for 8 to 10 hours after mating. Whether a female has mated can be determined by the plug, by inspection or with a blunt probe. The early embryo (pre-implantation embryo) is free in the oviduct and uterus of the female mouse for the first 4 days. Pre-implantation embryos can be manipulated in culture by injection, infection and by adding or removing cells. They can be cultured and transferred surgically into a recipient female (embryo transfer). If the uterus of the estrus female is receptive, then pups can be born. Recipient females are mated to vasectomized males prior to the embryo transfer surgery to ensure that the uterus is receptive. These pseudopregnant females are mated according to a schedule such that the age of the transferred embryos is matched. Gestation length in the mouse ranges between 18 to 21 days, depending on the strain. The litter size of laboratory mice varies between 6-10 pups. Most litters which are born can be successfully reared by their mothers. At birth mice are born hairless with their eyes closed. Pups which are not viable may be partially or wholly consumed by mice in the cage. When the mother is not happy, she may eat up entire litter. The cannibalism may be avoided by regular supply of 1 % salt in feed lot or ration. A key component of keeping mice happy is to keep them in stable social groupings. By three weeks of age, they have their adult hair, open eyes, teeth and can jump to the top of the cage for feed and drink. At this time they can be removed from their mothers and are typically housed together by sex. Sexing of laboratory mice is done on the basis of the distance between anus and genital opening. If the gap between anus and genital opening is shorter the mice will be female whereas male mice will have larger ano-genital gap.

Figure: 10.5 : Sexing of mice.

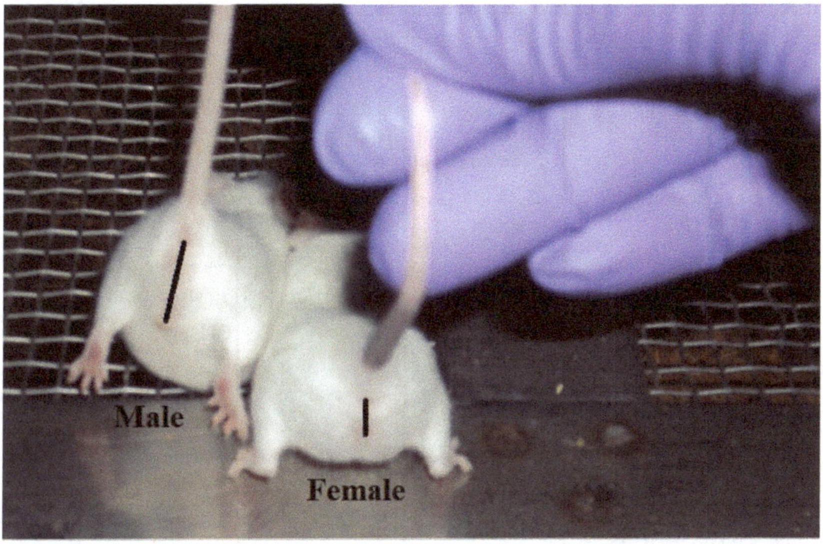

Sexual maturity can occur as early as 6 weeks in females and 8 weeks in males, depending on strain. The reproductive life span varies widely with strains but females typically have a reproductive life span of 6-8 months whereas males have a reproductive life span of about a year.

Breeding and management of laboratory Rats: Although, mice and rats are genetically varying in greater way but most of the breeding parameters are similar. The rats do have 42 diploid numbers of chromosomes whereas mice posses 40 diploid number of chromosomes. Pair or Harem mating in preferred in rat breeding. The pregnant rats are kept in a separate cage for safe kindling and nursing to young pups. The rats do have slightly higher pregnancy period of 20-21 days than mice. The litter size of laboratory rat is slightly higher than laboratory mice and varies between 8-12.The breeding ages for males are 65-300 days whereas for females it ranges between 65-200 days. The birth weight is of 5-6 grams and adult body weights vary from 250-300 in female and it may go up to 500 grams in males. Light is an important factor for rat breeding and minimum 12 hours/day light is mandatory for proper rat breeding. The Bruce and Whitten effect is not evident in rat breeding. The life span of rat is of 2.5-3.5 years. All other husbandry practices are similar to mice.

Table 10.3 : Breeding and reproductive parameters of commonly used laboratory animals.

Reproductive parameters	Mouse	Rat	Guinea pig	Rabbit (NZW)
1. Estrous Cycle (days)	4-5	4-5	16	NA
2. Duration of estrus	10h	13-15h	6-11h	-
3. Time of ovulation	2-3 h after estrus (spontaneous)	8-10 h (spontaneous)	10h (spontaneous)	Induced 10-11 h after mating
4. Average gestation period (days)	21	21	68	30
5. Litter size	6-10	8-12	2-5	4-6
6. Estrus after parturition	Postpartum	Postpartum	Postpartum	35 day
7. Reproductive life span (years)	1	1	3-4	2-3
8. Mating System (M:F ratio)	Pair/Trio/ Harem	Pair/Harem	Harem	Hand mating
9. Max. number of females per male	5	5	6	10
10. Mammary glands (T.A.P.) no. of pairs	3,1,1(Five)	3,1,2 (Six)	-,-,1 (One)	1,2,1 (Four)

A=Abdominal, P=Pelvic, T=Thoracic, h=hours, d=days.Source: Guidelines for care and use of animals in scientific research (2000).

Breeding and management of Guinea pigs: The unique characteristics of the guinea pig that differ from other experimental animals include the highly developed young at birth due to the long gestation period (65-70 days). Unlike the offspring of most other rodents, which are altricial at birth, newborn pups are well-developed with hair, teeth, claws and partial eyesight. They are immediately mobile and begin eating solid food quite earlier (2 to 3 days of age) though they continue to suckle and can be easily be weaned at 10 days of age. They are nervous, high-strung animals compared to the other laboratory rodents and are rather noisy. Unlike mice and rats, guinea pigs have little difference in the ano-genital distance and hence sexing may not be done using ano-genital gap. Male and female guinea pigs do not differ in external appearance apart from general size. The position of the anus is very close to the genitals in both the sexes. Female genitals are distinguished by a Y-shaped configuration formed from a vulvar flap; while the male genitals may look similar with the penis and anus forming A like shape, the penis will protrude if pressure is applied to the surrounding hair. The male's testes may also be visible externally from scrotal swelling. The two common methods for mating guinea pigs are "Monogamous Pairs" method and the "Polygamous Group" method. Mostly laboratory animal breeder follows the polygamous or Harem mating (one male and up to 10 females) system in guinea pigs. Mating may be detected by the presence of a vaginal plug. After a few hours the plug falls out and may be observed as a waxy mass on the cage floor. The presence of plugs is an efficient method of predicting pregnancy. Sows in advanced pregnancy may be removed from harem system and returned after the young ones are weaned. The male reproductive span is of 4 years whereas adult female reproductive span is of 1.5 to 4 years (4 litters). The sow is a non-seasonal, polyestrous breeder and estrous cycle is of 16 ½ days (range of 16 to 19 days) with an estrous duration of 6 to 11 hours. Ovulation is spontaneous and occurs approximately 10 hours after the beginning of estrus. A post-partum estrus occurs 2-15 hours following parturition and 60-80% of sows housed with males will become pregnant at that time if bred. The birth weight varies from 70-100 grams and litter size from 3-5 only. The abortions are most frequent if temperature of animal house goes over 30° C. Pregnancy toxaemia appears to be most common in hot climate. Sows are usually bred at 2-3 months of age, which generally corresponds to a weight of 350-450 grams. It is important that sows be bred prior to 7 months of age to prevent permanent fusion of the pelvic symphysis, which often results in dystocia. Boars are usually bred at 3-4 months of age or 600-700 grams body weight. Outbred stocks of guinea pigs have better reproductive performance than do the inbred strains. Inbred strains in general have decreased vigor, reproductive potential and fertility.

Breeding and management of Rabbits: Rabbits have a very good reproductive potential. Although, the breeding season for most of the rabbits is throughout the year but in tropical country like Indian the autumn stress is encountered during hot humid weather where male infertility is noticed due to azoospermia. The age of sexual maturity of doe is about 6-8 months and a buck gets sexually mature at about 7 months. Rabbits can be selected at the age of 5-8 months. There is no specific estrous cycle period in rabbits. The female rabbits allow the male rabbit for mating if female rabbit is in estrus. If a female rabbit is in heat, its vulva will be congested in red color. When a male rabbit is kept near to a female rabbit in heat shows depression of its back and raising of its back portion of the body. At the same time if the female rabbit is not in heat it will go to the corner of the cage and attack the male. Courtship and mating are very brief, lasting only 30 to 40 seconds. Courtship behavior involves licking, sniffing, and following the doe. Spraying urine is also a common sexual behavior. Female rabbits are reflex ovulators. Ovulation begins 10 hours after mating. The male and female ratio should be 1:10. Normal gestation period is about 30 days. Pregnancy detection can be done by expert hand after 14 days of mating by palpating the abdomen by finger. Pregnant rabbits should be fed with increased quantity of concentrate feed (100 gram to 150 gram excess) than the normal feeding. The pregnant rabbit should be transferred to the kindling cage on 25 days after mating. Five days before the expected date of mating nest box should be kept in the kindling cage. The average size of the litter varies but it usually ranges between 4-12 kits. Larger breeds have larger litters. A kit can be weaned at about 5 to 7 weeks of age and sexing must be done at the age of 5 months for putting male and female rabbits in separate pen or cage.

Figure: 10.6 : Sexing of rabbit.

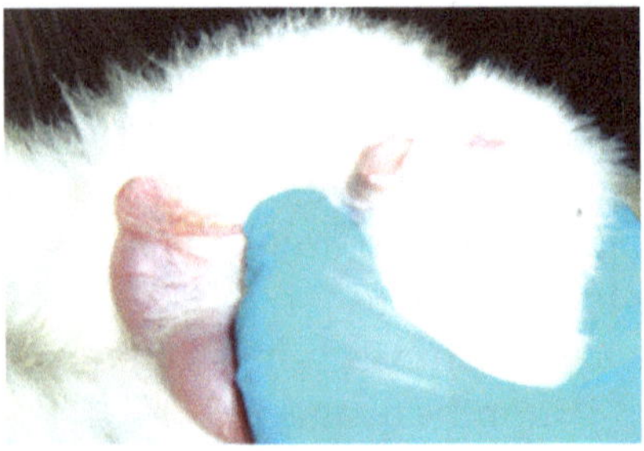

Male Rabbit

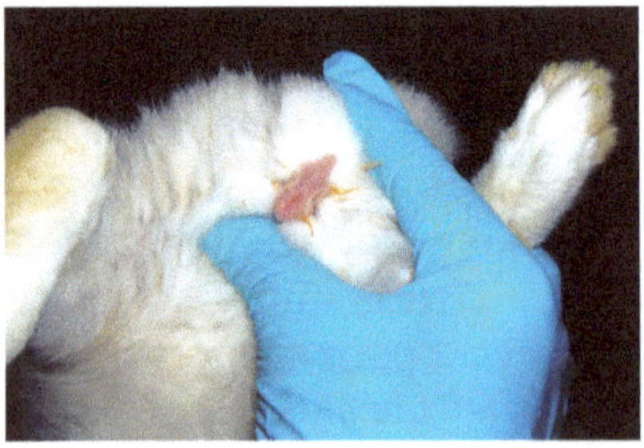

Female Rabbit

In case of death of dam, lack of maternal care and falling of young ones from the cage results in difficulty in the identification of the dams. Under these circumstances the step mother must be used for nursing the young ones. The females can be bred after one week of weaning. The female rabbit also may or may not lose clumps of hair during the gestation period. Before kindling, the female usually make a nest or burrow and line the nest with fur from the dewlap, flanks and belly. Kits are altricial (born blind), naked, and helpless. Eyes of baby rabbits open at about 10 to 11 days after birth and they start eating on their own at around 14 day of age.

Table 10.4 : Physiological norms of commonly used laboratory animals.

1. Physiological norms	Mouse	Rat	Guinea pig	Rabbit
2. Diploid Chromosome number (2N)	40	42	64	44
3. Weight at birth (grams)	1-2	4-5	80-100	40-60
4. Age at weaning (weeks)	3	3	3	8
5. Weight at weaning (grams)	9-12	40-50	250-300	800-900
6. Age at maturity (weeks)	6-8	10-12	16-20	24-32
7. Weight at maturity (grams)	18-22	150-200	250-400	1500-2000
8. Adult Weight (grams)	25-30	200-300	400-500	2000-2500
9. Average rectal temperature (^{0}C)	37.4	37.5	38.6	38.7
10. Respiratory rate (per minute)	90-180	80-150	60-110	35-56
11. Average pulse rate (per minute)	600	300	150	133
12. Life span (years)	1.5-2.0	2.5-3.0	4-5	4-5

Source : Guidelines for care and use of animals in scientific research (2000).

Care of Young Ones

The healthy weaned bunnies will have shiny hair & coat, very active, good appetite, eyes are usually bright and without any discharge and gradual body weight gain. The ill bunnies would be dull and depressed, lacking appetite, loosing hairs and may have mucus discharge from nose, eye and ears. The following care must be taken for young bunnies:

- The nest box should be removed after 5 weeks of kindling.
- Kits should be examined and dead ones should be removed daily.
- If the bedding becomes wet it should be replaced by a fresh and clean one.
- No sudden change in feed is advisable.
- Hydrogen peroxide @ 1ml per liter of water is added 20 minutes before offering water to the bunnies.

Anaesthesia: The scientists should ensure that the procedures which are considered painless are conducted under appropriate anaesthesia. It must also be ensured that at no stage the animal is conscious to perceive pain during the experiment. Neuromuscular blocking agents must not be used without adequate general anaesthesia. Before using actual anaesthetic agent the animal is prepared for anaesthesia by overnight fasting and using pre-anaesthetics, which block parasympathetic stimulation of cardio-pulmonary system and reduce salivary secretion. The choice of local or general anaesthesia should depend on the type of surgical procedure. Local anaesthetics (epidural and spinal) are used to block the nerve supply to a limited area and are used only for minor and rapid procedures. Along with nasal inhalation, general anaesthetics are also used in the form of intravenous or intra-muscular injections such as barbiturates. Appropriate dose of Ketamine hydrochloride (Intramuscular injection) and Pentobarbitone Sodium (Intra venous) is recommended in mice, rats, guinea pigs and rabbits. Prior to anaesthesia 0.02-0.05mg/kg Atropine for all species by s/c or i/m or i/v routes used to reduce salivary and bronchial secretions.

Euthanasia: Euthanasia means "easy death" and is resorted to in events where an animal is required to be sacrificed on termination of an experiment or otherwise for ethical reasons. The investigator must ensure that the animal is clinically dead before it is sent for disposal. As a humane method of euthanasia there must be an initial depressive action on the central nervous system for immediate insensitivity to pain.

The method should in all cases meet the following requirements:

(a) Death without causing anxiety, pain or distress with minimum time lag phase.

(b) Minimum physiological and psychological disturbances.

(c) Compatibility with the purpose of study and minimum emotional effect on the observer & operator.

(d) Location should be separate from animal rooms and free from environmental contaminations.

(e) Method should be reliable, reproducible and safe to the personnel involved.

A number of euthanasia methods have been recognized humane which could be physical, use of inhalant gases, injectable drugs and general anaesthetics in heavy dose. Cervical dislocation as physical method is recommended for euthanasia in mice, rats, guinea pigs and rabbits. Decapitation is approved method in mice, rats and rabbits but not in guinea pigs. Inhalation of carbon monoxide or carbon dioxide or carbon dioxide with chloroform is recommended in mice, rats, guinea pigs and rabbits. Alternatively over dose of Barbiturate may be used as anesthetic agent in mentioned laboratory animals.

Disposal of Animal Carcasses

All animal carcasses whether healthy, infectious or radioactive, must be packed in polythene bags before sending them for disposal. All healthy or infectious animals may be buried deep in the ground covered with lime and disinfectants or burnt in an incinerator. The staff handling such animals must be apprised of the hazards involved in their job and protective clothing, gloves and mask must be provided to them for their personal safety.

Specific Pathogen Free (SPF) Animal Production

SPF is a term used for laboratory animals that are guaranteed free of particular pathogens. It is always accompanied by a list of the absent pathogens. Use of SPF animals ensures that specified diseases do not interfere with an experiment. For example, absence of respiratory pathogens such as Influenza is desirable when investigating a drug's effect on lung function. SPF animals are special stock of animals that are kept in specific pathogen free facilities under rigorous monitoring system, which are subjected to sensitive and accurate diagnostic methods. SPF animals are neither disease free nor disease resistant. They may well carry other pathogens for which their health status has not been assessed. Buyers should be sure to obtain the list of pathogens for which a stock is specified to be free and ask for details about the date with detail of the most recent health screening, the health surveillance program used to monitor the stock, the diagnostic methods used and disease history of the SPF facility.

The animals are born through a caesarian section then special care taken so the newborn does not acquire infections. It can be performed using sterile isolation units with a positive pressure differential to prevent entering of outside air and pathogens. Everything that needs to be inserted into the isolator, such as food, water and equipment needs to be completely sterilized and disinfected and inserted through an airlock that can be disinfected before opening from the inside.

Gnotobiotic Animal Production

A **gnotobiotic animal** is an animal in which only certain known strains of bacteria and other microorganisms are present. Gnotobiotic animals are born in aseptic conditions, which may include removal from the mother by caesarean section. Such animals are normally reared in a sterile or microbially-controlled laboratory environment and they are only exposed to those microorganisms that the researchers wish to have present in the animal. These gnotobionts are used to study the symbiotic relationships between an animal and one or more of the microorganisms that may inhabit its body. This technique is important for microbiologists because it allows them to study only a select few symbiotic interactions at a time, whereas animals that develop under normal conditions may quickly acquire a microbiota that includes hundreds or thousands of unique organisms.

Disease of Laboratory Animals

Diseases of Mice: *Citrobacter rodentium,* Chronic Respiratory Disease, *Corynebacterium bovis, Helicobacter* Infections, Pasteurellosis, *Pseudomonas,* Septicemia, Salmonellosis, Staphylococcal Furunculosis and Tyzzer's Disease are common bacterial disease of mice. *Myocoptes musculinus* (A.), *Myobia musculi* (B.) and *Radfordia affinis* (C.) are common fur mites. Hymenolepid tapeworms and encysted tapeworms infestation are observed in mice. Mouse Hepatitis Virus (MHV), Parvoviruses (MMV, MPV), Epizootic diarrohea of infant mice (EDIM), Sendai virus, Mousepox (Ectromelia Virus), Mouse Norovirus (MNV) are common viral infection in mice.

Disease of Rat: Murine Respiratory Mycoplasmosis (MRM), CAR Bacillus, Streptococcal disease, Staphylococcal dermatitis, Pseudotuberculosis (Corynebacteriosis) and Tyzzer's diseases are common bacterial disease of rats. *Spironucleus muris* and *Giardia muris* are protozoa that colonize the upper small intestine. *Rodentolepis nana* (dwarf tapeworm) and *Hymenolepis diminuta* (the rat tapeworm) can infect rats. Rats serve as intermediate hosts for the cat tapeworm *Taenia taeniaformis*. The cysticercoid cyst (*Cysticercus fasciolaris)* embeds in the liver and may cause a palpably enlarged liver if cysts are numerous. *Syphacia muris* (Rat Pinworm) is common nematode infestation in rats. Coronaviruses (RCV and SDAV), Respiratory Viruses (Sendai and PVM), Parvovirus (RV, H-1, and RPV-1) and Rat Respiratory Virus (RRV) are viral infections generally noticed in rats.

Disease of Guinea pigs: *Bordetella* pneumonia, Streptococcal pneumonia, Hemorrhagic Typhlitis, Mastitis, Tyzzer's disease and Salmonellosis are common bacteria born disease in guinea pigs. Dermatophytosis and Pododermatitis (Bumblefoot) are fungal infections observed in guinea pigs. Guinea pig lice, *Gliricola porcelli* (A.) and *Gyropus ovalis (B.)* are chewing lice that commonly parasitize guinea pigs.

Chirodiscoides caviae is a fur mite. Fur mites may occasionally cause partial alopecia and pruritus in selected animals. *Trixacarus caviae* is a sarcoptic mange mite of the guinea pig. This parasite occasionally can cause an intense pruritus leading to self-mutilation, abnormal behavior, debility and death. The metazoan parasite commonly associated with guinea pigs is *Paraspidodera uncinata,* a roundworm which resides in the caecum. Respiratory viruses, Cytomegalovirus (Salivary Gland Virus) and Lymphocytic Choriomeningitis (LCM) are bacterial infection of guinea pigs. Dystocia most commonly occurs in primiparous dams older than six months of age due to the fusion of the pelvic symphysis and lack of pelvic inlet expansion needed for delivery of fetuses. Pregnancy toxaemia is a syndrome that occurs in late gestation or in the immediate postpartum period. Two causes have been identified; first pre-eclampsia or the circulatory form with uteroplacental ischemia and second fasting ketosis or the nutritional form with hypoglycemia and hyperlipidemia. Predisposing factors include obesity, large fetal loads, sudden dietary changes, lack of exercise and environmental stress/high temperature.

Diseases of Rabbit: The bacterial diseases in rabbits are Pasteurellosis, Enteritis, Tularemia Tyzzer's Disease, Wry neck disease and Mastitis. Skin infection is produced in rabbits by *Dermatopysis* fungus. There is loss of hair around the ear and nose of rabbits. Griseofulvin or benzyl benzoate cream can be applied on the affected areas. Ear canker is caused by ear mites *Psoroptes cuniculi.* Rabbits are commonly infested with *Ctenocephalides* sp., especially *C. felis. Haemodipsus ventricosis* are blood sucking louse in rabbit. Besides above mentioned diseases, Malocclusion (Excessive growth of incisor) Buphthalmia (Congenital Glaucoma), Splay leg (paralysis of limbs) and Hydrocephalus (skeletal deformities) are a few genetically inherited diseases in rabbits.

CPCSEA Guidelines for Good Laboratory Practices (GLP)

The goal of these guidelines is to promote the humane care of animals used in biomedical and behavioral research and testing with the basic objective of providing specifications that will enhance animal well being, quality in the pursuit of advancement of biological knowledge that is relevant to humans and animals. It focuses on the various corners of laboratory animal production and maintenance. The concerned areas of monitoring are veterinary care, animal procurement, quarantine, stabilization & separation, surveillance, diagnosis, treatment & control of diseases, animal care & technical personnel, personal hygiene, animal experimentation involving hazardous agents, multiple surgical procedures on single animal, duration of experiments, physical restraint, physical plant, physical relationship of animal facilities to laboratories, functional areas, physical facilities, environment, animal husbandry, feeding, bedding, water, sanitation and cleanliness, assessing the effectiveness of sanitation, waste disposal, pest control, emergency, weekend and holiday care, record keeping, standard operating procedures (SOP_S), personnel and training, transport of laboratory animals, anesthesia and euthanasia, maintenance, laboratory animal ethics, transgenic animals, disposal and breeding and genetics of laboratory animals.

The Concept of the $3R^s$

Russell and Burch (1959) introduced the Principles of Humane Experimental Technique by implementation of the three "R's" *i.e.* Replacement, Reduction and Refinement. Recently during 2004 the CPCSEA has introduced the concept of 4th "R" in existing concept of Russel and Burch. The 4th "R" stands for rehabilation of animals upon which the experiments is completed. The average costs of rehabilitation and aftercare of different species of animals were there after worked out based on the actual average expenses incurred by various institutions. The sub-committee was of the view that the following minimum amounts may be necessary for maintenance of animals after experimentation (for feed and husbandry but not including infrastructure and other overheads). *However, this may be reviewed once in two years, if required. The rehabilitation of small animals is not necessary.* Minimum cost per day per animal (in Rs.) was Rs.15 for sheep/goats, Rs. 20 for dogs/cats and Rs. 25 for cattle/horse/monkey.

Accreditation of Laboratory Animals

Accreditation programs have been developed in several countries to insure that commercial breeders produce uniformly healthy, well-fed animals of known genetic background. However, to minimize the transmission of disease and parasites, careful consideration should also be given to the planning, construction and operation of animal facilities. The accreditation also promotes scientific validity of the research animals and reliable results that exclusively depends on superior animal care. Also, the accreditation demonstrates a willingness to go above and beyond the minimum required by the law and assures public that institution is committed to the responsible use and treatment of animals in science. The functions of accreditation are:

1. Provides a confidential, third-party peer review.
2. Promotes scientific validity.
3. Represents quality.
4. Demonstrates accountability.
5. Assures funding sources.
6. Helps recruit quality people.
7. Shows a real commitment to humane animal care.

Commercial Rabbit Farming

Rabbits are small mammals in the family Leporidae of the order Lagomorpha, found all around the world. Rabbit keeping in commercial base started in the last century. Subsequently, Food and Agricultural Organization (FAO) through different programs encouraged the farmers for rabbit farming. The rabbit farming gradually spread to other countries. In India, rabbit rearing has attracted farmers since the last ten to 12 years. During early sixties the rabbit farming introduced at Himachal Pradesh, Jammu and Kashmir through introduction of German angora breed. In the year 1977, the government of India took much interest in rabbit farming in different states and established a Central Research Station to encourage rabbit farming and to envisage research program to mitigate the problems in rabbit farming in the country. The ever growing demand for non vegetarian proteins in developing countries like India helps the rabbit industry to grow further and further. But more attention is required to be made to popularize and commercialize the rabbit meat. However, at places like Kerala, Karnataka, Andhra Pradesh, Uttar Pradesh North-eastern states and Himachal Pradesh rabbit meat is available. The rearing of rabbits is not only for meat purposes but also it is important for their skin, hairs and production of biological/vaccine, drug testing and used as research material.

Advantages of Rabbit Farming

1. Rabbitry requires less space and it can be done anywhere at farm, backyard, on terrace or even at home. The capital investment for this is very less for starting rabbitry unit.
2. As Rabbits are vegetarian one can feed almost anything vegetarian to the rabbits including surplus vegetables from

the market, tree leaves or cattle grass etc. Hence, it is very easy and affordable to feed a rabbit.

3. No skilled labour is required. Thus, growing or keeping imported rabbits (broiler breed) is authorized by Indian Law.
4. The pregnancy is just of 30 days and we keep the baby rabbits with dam for 4-7 weeks. A female can have 6-7 litter in a year which indicates for high reproductive potential.
5. Growth rate in selected line of broiler rabbits is very high. They attain 2 kg at the age of 3-4 months. However, it's feed conversation ratio (FCR) usually lower than the poultry and swine.
6. It requires less care for disease than swine and poultry. Even the vaccination is not mandatory for rabbit farming.
7. Litter size (Number of young ones born/ kindling) in rabbits is high (around 8-12) in selected lines of broiler rabbits.
8. When compared to the other meats rabbit meat contain high protein (21%) and less fat (8%). So this meat is suitable for all age groups from adults to children. Rabbit meat is almost" cholesterol free" and therefore, heart-patient can eat this meat instead of other meats. It is the highest nicotinic acid containing meat.
9. Rabbits are midway between ruminants and monogastric animals and can convert cellulose into meat. They reingest their faeces in the early hours of morning directly from their anus through lips.

Chapter - 11

Equines

Md. Manzarul Islam

Equines are herbivorous monogastric animals. Though, simple-stomached, they can digest considerable amount of fibrous feed in the caecum and are called as hindgut fermenters. Equine includes horse, donkey, zebra and mule which are used as work and pack animal. All three species of equine belongs to the family *Equidae*. The scientific name of Horse and Donkey is *Equus caballus* and *Equus asinus*, respectively. The horse was probably last farm animal domesticated by the man. Horses appear to have been domesticated first in central Asia or Persia, more than 3000 years ago.

Equine Population in India

As per 18th Livestock census (2007), the total population of horse & ponies, donkey and mule are 0.61, 0.44 and 0.14 million nos., respectively. During the inter censual period from 2003 to 2007, there is decrease in the population of all three species. The population of horse and ponies are highest in the states of Jammu & Kashmir followed by Uttar Pradesh while mules and donkeys are present in maximum numbers in the state Jammu & Kashmir and Rajasthan, respectively.

Use of Equines in General

- Prior to the domestication, horses were used as a source of food by the man of Old Stone Age. Still in countries like France,

Belgium, Switzerland and many other part of the world it is used as a meat animal.

- Horses, ponies, mules and donkeys are the most valuable pack animals for transportation of men and material in many parts of India.
- Horses are widely used for races, polo, sports, hunting and various other purposes.
- Horses are also used by the army, paramilitary forces and the police for mobility, patrolling and riot control duties.
- The donkey is an important animal in the economy of landless labourers, small and marginal farmers and other poorer sections of the society, both in rural and urban areas.
- Mules have their use both in military establishments and civilian occupations.
- In the medical field horses are used as a factory for the manufacture of antitoxins that are used for rendering animals and people immune to certain diseases such as tetanus.

Breeds of Horses

There are more than 300 breeds of horses in world while in India only six recognized breeds are available. According to National Bureau of Animal Genetic Resources (NBAGR), the indigenous breeds of horses/ponies include Marwari, Kathiawari, Manipuri, Spiti, Bhutia and Zanskari (Table 11.1). Among these, Marwari and Kathiawari are considered as 2 distinct breeds or types although they have several characteristics in common.

Table 11.1 : Breeds of horses with their home tract and purpose.

Breed	Home tract	Purpose
1. Kathiawari	Gujarat	Pace & Speed
2. Marwari	Rajasthan	Pace & Speed
3. Bhutia	Himalayan	Ranges Hill purpose
4. Manipuri	Manipur	Speed, Polo & Sports
5. Spiti	Himachal Pradesh	Pace
6. Zanskari	Jammu & Kashmir	Hill purpose

Breeds of Donkey

Recently NBAGR has recognized Spiti as a breed of donkey in India, whose home tract is in Himachal Pradesh.

Peculiarities of Ass or Donkey

The asses were first domesticated probably in Egypt or Mesopotamia around 3000-4000 BC. Large and much longer ear in ass are the most noticeable characteristics which makes it different from horses. Brush like switch is present at the end of the tail. Scanty hairs present are present on mane and tail. Three distinct types of Indian assess viz., Indian donkeys, Indian wild ass and Kiang has been reported by FAO. In India there is an organized exotic donkey breeding farm at National Research Centre on Equines, Hisar and Bikaner, which is able to provide donkey semen for mule production as well as for upgrading of indigenous donkeys. Presently, china has raised about 10 million high bred donkeys for carting. Donkey carts in China carry about 400 kg loads. Indian wild ass is found in Rann of Kutch (Gujarat) and Kiang in Ladakh, Kiang is largest and heaviest of the entire Asiatic wild Ass.

The weight of the asses varies from 80 to 480 kg (180 to 1,060 lb) while the height at the withers ranges from 7.3 hands (31 inches, 79 cm) to 15.3 hands (63 inches, 160 cm). Donkeys have the life expectancy of 12 to 15 years. Large ears of donkeys help to pick up more distant sounds and may also help to cool the blood. Donkeys can defend themselves by biting, striking with the front hooves or kicking with the hind legs. The breeding age in donkeys is about 12 months and the pregnancy period is also about 12 months.

Peculiarities of Mule

Mules are hybrid born out of male donkey (jack) and the female horse (mare). One of the difficulties in mule production is to locate a fertile jack that can serve the mare. The mules are usually produced for draught, pack and transport particularly in difficult terrains including hills. The mule combines some of the superior qualities of both the species. It has size, speed, strength and spirit of the horse. The sound produced by a mule is not exactly like a donkey or a horse. Instead, a mule makes a sound that is similar to a donkey's but also has the whinnying characteristics of a horse (often starts with a whinny, ends in a hee-haw). The coats of mules come in the same varieties as those of horses. The draught mules stand from 155-172 cm in height and weight is 450-650 kg. There are two distinct types of mules, viz., Mountain Artillery (MA) and General Service (GS) kept by Indian Army. Mountain artillery mules are used for carrying heavy load and artillery in mountain region and general services mules are used for general transportation. Common colors are bay, black, and grey. Mules are reported to be highly intelligent.

General Management Practices

General management practices like handling, restrain devices and techniques, casting, methods of identification, castration, ageing, grooming, estimation of body weights, methods of transport and vices etc and identification of body parts, various terms related with equine husbandry have been already dealt in chapter 2 of this book.

Feeding of Equines

Equines are non-ruminant herbivore animals. Unlike monogastric animals, horses are capable of digesting plant fibres which they are getting from grasses and hay. Microbial fermentation takes place in the caecum of horses.

The mare has to pass adequate passive protection to foal which is provided through colostrum feeding. Besides, colostrum has laxative substance that promotes bowel movement and elimination of first faeces. Immunoglobulin's do not pass through dam's placenta and are absorbed only through first part of intestine during first few hours of life. The mare's colostrum contains protein up to 19 percent immediately after parturition which falls to 3.8 percent after 12 hrs and to a fairly constant level of 2.2 percent, after 8 days. The foal absorbs gamaglobulin as intact undegraded molecules throughout first 12 hours of life which gets reduced to a great extent in next 24 hours.

Foals should be allowed access to creep feed so that they achieve the maximum for their growth and bone development. At about 7-15 days of age, the foal starts to nibble on the feed given to the mother. To meet the inadequacy of nutrients in mare's milk and avoid-ance of imbalance diet to the foal, horse owners prefer to feed well balanced creep feed to the foals. It is usually recommended that creep feeding be started after 1-2 weeks of foaling. The creep should be located where the mare goes periodically during the day for water or shade. By weaning time, the foal should be consuming at least 2-3 kg of creep feed per day. Creep feeds also help in condi-tioning to the change of dependence on mother's milk to a manmade diet.

Only race horses and horses kept for shows are fed on scientific ground in India and most of the other horses used for different purpose are fed with dry grasses and small amount of grains. Grams are popularly fed by the owners of mules and Tonga ponies. Donkeys are fed only on dry grasses and no concentrates are fed to them. Ponies are fed with 2 to 2.5 kg feed consisting of oat and wheat bran in equal parts. Among cereal grains oat is the most popular for horses as it is forms a loose mass in the stomach which is well suited to the equine digestive system and its is more palatable and digestible than other grains like barley, corn, sorghum and wheat. Horses require dry matter 1.2% of its body weight; while those are in work, require somewhat higher dry matter of about 2 % of their body weight. For example horse weighing 500 kg will require 10 kg of dry matter when they are work. Foals below six months of age may require somewhat higher dry matter than the working horses. Protein requirement are 20, 14, 12 and 10% of total feed for new born foal, weanling, yearling and mature horses, respectively.

For horses, both over-feeding and under-feeding are harmful. Therefore, the diet has to vary as per the requirement of the horse and its physiological status. The time of feeding the horses should be strictly adhered to. Also no abrupt changes in the diet should be permitted. The feed of the horses must contain roughages (4-6% of body weight); otherwise the heavy feed concentrates are likely to pack the stomach. However, feeds too high in fiber (should not exceed 30%) may also cause discomfort or digestive disturbances. A source of salt or complete mineral mixture should always be available for horses at all times which give them an opportunity to take extra minerals depending upon their requirements. A clear, wholesome and fresh water source should always be available to the horses. Horses require about 2-4 liters of water per kg of dry matter intake.

Limited size of digestive tract is there in the horses and feeding large quantities of grain or concentrate feeds at one time results in higher frequency of colic and grain founder. Therefore, high performance horses, which require large amount of energy from grain or concentrate sources, should be fed three to four times daily instead of two times. Daily total concentrate feed should be fed in three times i.e. morning, noon and night in equal proportions. Hay should be fed 25% of total each in morning and noon, while remaining 50% is fed at night time.

Housing

The place where horses are kept is called stable. The area requirement and comfort varies from thoroughbred to country bred horses and from donkey to mule similarly. The housing of animals should be cost effectiveness with appropriate space, ventilation, lighting, nature of floor and roof, drainage, availability of clean water etc. Dust and pollen free atmosphere should be chosen for housing of horses. Te roof of the stable should be 15-20 feet high. Manger should be fixed at chest height in the corner of the stable. The height and width of door should be 7-8 and 4 feet, respectively. The dimension of an adult horse stable is 14 x 14 feet.

Equine Reproduction

Estrous varies according to the climate and the season in mares as they are polyestrous in nature. The breeding season is mainly confined twice a year. The spring breeding season commences from February till the end of June and autumn breeding season is confined to September, October and November. Among all the domesticated species of livestock the horse is having lowest reproductive efficiency. Seasonal occurrence of estrous, erratic and expanded oestrus behaviors in the early spring and uncertainty of the time of ovulation are some of the factors contributing to the low reproductive efficiency in mare.

The filly attain puberty at the age of 12 to 15 months, she starts to come into oestrus periodically. Optimum age at first mating should be 3-4 years and 2.5-3 years in male and female, respectively. The weight should be approximately 300-350 kg at first mating. Normally 30-40 mares are allotted to a stallion and the stallion has to cover 2 to 2 ½ times per mare in order to get them in foal. The new stallions in their first year of service in the stud should not cover more than 15 mares. During second year, the number of mares as well as covering quota should be gradually increased which should be limited to 50 coverings during the whole year. It is during the 3rd year of the stud service that a stallion is fully utilized and the covering quota should be increase to 80-100 in a year. Similarly, when the stallion gets older than 15 years, the covering quota should be accordingly reduced. Normally one covering should be allowed in 48 hours as replacement of semen takes place in this period. The productive life of mare is 14-16 years with approximately 10 foalings.

The average length of oestrous cycle is 21 to 23 days. Oestrus lasts for about 4 to 6 days. The optimum time for breeding is 2-3 days after appearance of heat. The signs of oestrus in mare are raising of tail, frequent urination, spreading of hind leg apart, contraction and relaxation of lips of vulva with winking of the clitoris and allows stallion to bite and chew her neck and flank. Ovulation usually occurs during the third to fifth days of cycle. The conception rates starts declining with advancing age. It has been observed that maximum conception rate is seen in mares from 9 to 15 years of age which slowly declines after that.

Gestation period in mare is about 340±10 days. Twinning is rare in mares with incidence just as low as 2.5 percent. Generally twinning is discouraged in stud farm (due to low birth weight of foal). Foal heat is seen in case of horse after 9-11 days of foaling, which is not suitable for mating as there will be less chance of conception.

Methods of Breeding in Horses

Following are the methods of breeding in horses.

- **Hand mating:** The stallion is allowed to cover naturally. One person holds the mare and two people handle stallion while mating to reduce the chances of injury during covering.
- **Pasture mating:** One stallion is turned to pasture with a band of brood mares for the entire breeding season. Both male and female are more liable to get injury in this method but the conception rate is higher than any other method of breeding.
- **Coral mating:** In a small paddock or coral the mare is turned in and when she is ready to be bred, the stallion is turned in. After successful mating, both mare and stallion are turned out.
- **Artificial Insemination:** Horses can also be inseminated by artificial insemination technique. It is helpful in those mares which are problematic such as such as having long oestrus period and silent heat.

Pregnancy Diagnosis

After mating of mare, it is necessary to go for pregnancy diagnosis. If a mare shows no signs of oestrus for 23 days after breeding, it is 90% sure that she is pregnant. Ultra sound scanner can be used for pregnancy diagnosis as early as 14-15 days post mating. Rectal palpation can be used as a tool for pregnancy diagnosis after 30-35 days of mating. It is advisable to go for pregnancy diagnosis at 30-35 days and at 100-120 days.

Care and Management at Foaling and of New Born Foals

Foaling generally occurs in the night. Symptoms of parturition like restlessness, frequent sittings and standings, urination, waxing of teats are seen in mares. Blind end of membrane sac with forefeet of foal comes out during actual foaling. Foaling procedure takes 15 - 30 minutes of time. Placenta normally falls within one hour after foaling. Administration of hormone or manual interference may be applied if there is delay in fall of placenta beyond 3 hours. In case, expulsion of placenta is delayed beyond three hours, it is liable to become infected with organisms and may lead to uterine infection and consequent sterility.

New born should be cleaned with straw or gunny bags. Usually the foal may be allowed to suckle the mother (colostrums) as soon as it stands after birth. Colostrum of mare is rich in protein, minerals and vitamins and contains antibodies produced in the mare's serum against most of the common harmful bacteria. Colostrum stimulates the release of meconium (first faeces) by its laxative effect. Navel cord should be cut at l to 1.5" by sterilized scissor and apply antiseptic. The cartilage of foot must be removed by sharp scissors. Warmness is provided by giving the light in winter season. Weaning in foals should be done at the age of 4-6 months.

Health Care for Equines

The normal temperature, pulse and respiration rate of horses at resting condition is 100-101°F (measured for 3 minutes), 28-40 and 8-16 per minute, respectively. Any deviation from normal rage may indicate some disease condition. The commonly occurring diseases in horses are Tetanus, Strangles, Glanders, Influenza or pink eye, African horse sickness etc. Colic, lameness and Tying up syndrome or Monday morning sickness are laso some of the important health problems of horses. Prevention is always better than cure, so regular vaccination should be done against important diseases. Always good quality food should be given to horses along with clean and safe environments.

Vices of Horses

Horses generally acquire different types of vices due to lack of work and exercise. Following are some of the important vices of horses.

- **Wind Sucking and Crib biting:** Horses arches their neck with swallowing of air and simultaneously grunts. The vice becomes incurable as the animals after developing it do not forget. As a result of wind sucking indigestion and colic may develop.

- **Weaving**: Animal move to and fro continuously at its standing space due to captivity/ confinement. The animals developed such vices should be given more exercises and the feed is given frequently in 4-5 divided doses.
- **Kicking**: Due to lack of sufficient work and exercise, this vice is commonly seen in mares.
- **Biting**: This habit is more in stallion then mares and can be controlled by inserting thick wooden bite into the mouth.
- **Sleeping while standing**: Due to over exertion and wrong bedding material and flooring, horses develop this vice.
- **Coporophagia**: It is the habit of eating own droppings. Once animal tastes its droppings, it is very difficult to eliminate the habit. In such cases droppings should be removed immediately. Generally horses developed this habit as a result of malnutrition and deficiency of salt in the feed.

Different Body Colours of Horses

- **Albino**: Skin is un-pigmented and covered with white hairs.
- **Bay brown**: Predominant colour is brown with bay muzzle, black limbs, mane and tail.
- **Bay**: Varies from dull red to yellowish colour. The bay horse has a black mane, tail and the limbs.
- **Black brown**: The predominating colour is black with muzzle and sometimes flanks have brown or tan colour.
- **Black**: The black pigments melanin in general is seen throughout the coat of the body, limbs, mane and tail.
- **Brown**: It is mixture of black and chocolate coloured hairs on the body coat; limbs, mane and tail are black.
- **Chestnut**: Body and legs are covered by yellow coloured hair. The shades vary from bright chestnut to golden red, dark chestnut to liver chestnut which is darkest of all. The mane and tail are chestnut, sometimes lighter and sometimes darker than body colour but never black.
- **Grey**: Skin is black with admixture of black and white hairs.
- **Skewbald:** Coat consists of large irregular patches of white and any other colour except black.

- **Stud grey**: Black hairs are predominant in the neck, head, shoulder and rump.

Different Body Markings of Horses

- **Blaze:** A white marking covering almost the whole of the forehead between the eyes and extending down the front of the face beyond the width of the nasal bone and usually involving the muzzle.
- **Conjoined star and stripe:** The stripe in continuation of a star.
- **Interrupted stripe:** Star and stripe appears as two distinct spots.
- **Snip:** Any isolated white mark in between the nostrils. Its position should be specified such as into the right nostril or the left nostril.
- **Star:** A white mark on the forehead either large or small.
- **Stripe:** A thin or broad narrow white marking running down the face.
- **Wall eye:** It is due to the lack of pigment either partial or complete, it gives bluish white or pinkish white appearance to the eye.
- **White face:** White colour covers the whole of the forehead and front of the face, extending laterally to the mouth either unilaterally or bilaterally.
- **White Muzzle:** Both lips will be white and extends to the region of the nostrils.

Different Types of Gaits in Horses

- **Walk:** The walk is the most useful gait of horse and has been termed as 'mother of all gaits' because many slow four beat gaits are modification of walk. The walk is characterized by four separate and distinct beats as each foot is placed in a regular 1-2-3-4 cadence.
- **Canter:** The canter is a fast, three beat gait done under restraint.
- **Gallop or Run:** The gallop or run is the fastest gait of the horse. It is an extended canter that results in a four beat gait is known as gallop.

- **Pace:** The Pace is a lateral two-beat gait. In the pace, the two legs on the same side of the horse move forward together, unlike the trot, where the two legs diagonally opposite from each other move forward together. In both the pace and the trot, two feet are always off the ground. However, the pace is faster than the trot and is a popular gait for harness racing, when speed is preferred to action.
- **Trotting:** The trot is a two-beat gait in which the diagonal fore and hind legs move together.

Important Points Regarding Management of Horses

- Post partum oestrus in mares occur 15 days after foaling. However servicing at post partum heat may cause abortion or still births and which may lead to low conception rate.
- Horses generally remain in standing while sleeping. This is because of 'check' ligament present in their hip joint.
- Shoeing is very important practice for horses. By the time the horses are taken to work for the first time (2-3years) the shoeing must be done. Horses require re-shoeing after every 6-10 weeks.
- Donkey use the hours of darkness for feeding.
- Equines must not be watered for at least 20-30 minutes before and after work.
- Adult animals must get at least 40-50 litres of water per day.
- Depending on breed, management and environment, the modern domestic horse has a life expectancy of 25 to 30 years.
- Temporary dentition: I=3/3, C=0/0, PM=3/3, M=0/0
- Permanent dentition: I=3/3, C=1/1, PM=3 or 4/3 or 4, M=3/3

Doping

To increase the speed, stamina and courage by administration of any substance, drug or food to get more marks than inferior animals in race is called doping. It is of two types i.e., Intentional (administration of doping agents such as methylxanthine, vitamins, hormones etc) and Unintentional (accidentally and therapeutically medicines administered as a treatment). Doping test can be carried out by examination of urine or saliva with the help of different tests like spectrophotometry, fluorometry, Elisa Test, RIA technique etc.

Important Terminologies Related with Equine Husbandry

Aged horse: More than 8 years

Ass: It is smaller in size than horse, longer ears and gestation period is one month longer than that of horse, short and erect mane.

Bishoping: It is a malpractice of artificially altering the teeth of older horses in an attempt to make them to sell as young horses.

Blemishes: This is the term used for abnormalities like shoe boils and capped hocks that do not affect the service ability of the horse.

Blood horse: A pedigree horse

Blood spavin: On the inner side of the hock enlarged varicose vein appears.

Bolting: Habit of horses of eating too fast

Broken horse: Well trained horse

Brood mare: Female horse used for breeding.

Cob: Short legged stocky horse.

Colic: Severe ingestion which causes abdominal discomfort and pain.

Colt: Young male of horse under three years of age.

Covering: Act of mating

Dental star: It is marking on the incisor teeth of horses which is used in judging

Donkey: Common name for the ass.

Doping: For increase or decrease of horse in race some drugs are administered.

Equine: Species of horse

Filly: Young female of horse under three years of age.

Foal at foot: Female with its offspring.

Foal: New born young one of horse either male or female.

Foaling: Act of parturition

Gelding: Castrated male of horse.

Halt-bred Horse: Race horse with no pedigree.

Hinny: When a stallion is crossed with a jennet, a resulting offspring is called hinny. Hinny is smaller in size than a mule.

Jack: Male ass is called jack.

Jennet: Female ass is called Jennet.

Light horses: Light horses are mainly used for riding, driving and racing. They measure 14.2 - 17 hands at withers (1 hand is equal to 4 inches); weight varies from 408 - 635 kg.

Mare: Adult female horse.

Mule: When jack is crossed with a mare and the resulting offspring is called mule.

Pack: Group of Animals

Ponies: It is mainly used for riding and driving. Height at wither is 14.2 hands. Weight is 136 - 400 kg.

Pure bred Horse: Entered in stud book maintained by a breed society.

Run / gallop: Fast, four beat gait of horse

Shelly: The newly erupted incisor teeth either temporary or permanent.

Spayed: Castrated female of horse.

Stallion: Adult male of horse.

Stud book: The permanent book for breeding records.

Stud: Male horse kept for breeding.

Thoroughbred horse: Crosses of warm blooded Arabian horses and cold blooded European horses. Mainly used for racing and crossbreeding programmes.

Tush: The canine tooth of horse present in stallion or gelding but absent or rudimentary in mare.

Wall eye: Partial/complete discolorations.

Whinny: The horse's sound that denotes happiness

Yearlings: 1- 2 years of age.

Chapter - 12

Camel Production

Md. Manzarul Islam

Origin and Distribution of Camel

A camel is an even-toed ungulate within the genus *Camelus* of the family Camelidae bearing distinctive fatty deposits known as "humps" on its back. There are two species of camel, viz. *Camelus dromedarius* (single humped dromedary) having habitat in dry hot arid lands of Africa and Asia, and *Camelus bacterianus* (double humped or *bactrian*) habitat of cold arid lands of Central palaerctic (a zoogeographical region consisting of Europe, Africa north of the Sahara, and most of Asia north of the Himalayas) region. Family Camelidae also includes other species which are camel like mammals. Those are Lama (*Lama glama*), Alpaca (*Lama pacos*), Vicuna (*Vicugna vicugna*) and Guanaco (*Lama guanicoe*). These four species are termed as new-world Camelidae having habitat in South America. The Indian camel belongs to *Camelus dromedarius.* A very small numbers of double-humped camels are also available in Ladakh area of Jammu and Kashmir.

Camel is an important component of desert eco-system. It is capable of traveling long distance in hot, arid and sandy regions carrying men, material and providing energy to agricultural operations. They provide milk, meat, hair and used as work animals. It is believed that Camels were domesticated during the period from 3000 to 2500 BC. Average life expectancy of a camel is 40-50 years. The height at the shoulder and hump of a fully grown adult camel is 1.85 and 2.15 metres, respectively. Camel can run at a speed of up to 65km/h for a short distance and can sustain a speed of up to 40km/h for long distance.

Population

As per 18th Livestock census (2007), camel population in India is 0.52 million nos., which in consistently decreasing since 1992 census. During the inter-censual period from 2003 to 2007, the population of camel decreased by 18.25%. India shares about 2.2% of world camel population of 23.9 million nos. The camel population is maximum in the state of Rajasthan followed by Haryana, Gujarat, Uttar Pradesh and Madhya Pradesh.

Breeds

On the basis of utility, Indian camels are classified into two classes, viz. baggage and riding type. Region wise they can also be classified as hilly and plain camels. The plain camels are further classified into desert and riverine camels. Desert camels are of riding type while that of riverine and hilly camels are of baggage type. According to National Bureau of Animal Genetic Resources (NABGR), there are total eight recognized breeds of camel in India (Table 12.1) among them Bikaneri is the principal breed which constitutes more than 50% of total camel population. The chromosome numbers in camel is 37 pairs (2n=74).

Table 12.1 : Breeds of camels with their home tract.

Breed	Home tract
Bikaneri	Rajasthan
Jaisalmeri	Rajasthan
Joria	Rajasthan
Kutchi	Gujarat
Malvi	Madhya Pradesh
Marwari	Rajasthan
Mewari	Rajasthan
Mewati	Rajasthan & Haryana

Peculiarities of Camel

- The hump of camel is filled with fatty tissue and serves as a source of energy.

- Metabolism of 1 g of fat produces over 1 g of metabolic water which helps to survive them in the desert area where there is scarcity of water.
- Oval shaped red blood cells are much stable in order to withstand high osmotic variation without rupturing when drinking large amount of water (100 litres) at a time.
- Camels are able to withstand changes in body temperature. Their temperature ranges from 34 °C (93 °F) at night and up to 41 °C (106 °F) during the day, and only above this threshold will they begin to sweat.
- Normal pulse and respiration rate are 30-45 and 5-12 per minutes, respectively.
- Thick coat (skin) of camels reflects sunlight, and also insulates it from the intense heat radiated from desert sand. Water loss through evaporation is also less in camel.
- Camel hair blankets are warmer than the woolen one. Finest hair is obtained from young camel in the wild state.
- Camel's mouth is very sturdy which helps to chew thorny desert plants.
- Long eyelashes and ear hairs, together with sealable nostrils, form a barrier against sand.
- The kidneys and intestines of a camel are very efficient at retaining water.
- Urine comes out as thick syrup, and their feces are so dry that they can fuel fires.
- Thick foot pad of camel helps to walk on loose hot sands easily.
- Camels can survive for long period of about two weeks on dry food as they have usually low metabolic rates.
- Camels along with Llama, Vicuna and Alpacas are called pseudo-ruminants as they lack one stomach chamber (omasum) out of four chambers in ruminants. Reticulum contains glandular cells.
- There is no gall bladder like horse.
- There are four mammary glands in the udder of camel.
- Skin over the udder is blackish.

- Front quarter of udder is comparatively larger than hind quarter.
- Wolf teeth are sometimes present in camel, while the upper incisors are sometimes absent in females, but when present they are smaller than in males. The dental formula of camel is I=2/6, C=1/1, W=1/1, PM=6/4, M=6/6. Total number of permanent teeth is 34.

Feeding and Watering

Dry matter requirement of an adult camel is about 1.8 – 2% of the body weight. Most of the camels when they are not in work can thrive solely on grazing and browsing trees, shrubs and bushes. They graze for 6-8 hours a day to meet the requirements. The size of the hump can be considered as indicator of good nourishment. With ample availability of feed, the size of hump increases. If there is no sufficient grazing available, extra ration in the form of green and dry fodders can be offered to them. It is better to provide chaffed fodder and dry fodders must be given to them even sufficient grazing is available. When camels are at heavy work and sufficient grazing is not available, they must be fed grain ration. For camels, gram is considered as best feed. Grains after crashing or soaking in water should be given to them. A daily ration of 1 kg crushed gram, 1 kg crushed barley, 9-18 kg green fodder, 7 kg straw and about 50 g salt can be considered adequate for stall fed camels at moderate type of work.

Camels have exceptional ability of conserving water. They can go for weeks or sometimes for months without water if sufficient green succulent fodder is available. Camels can withstand up to 20% of body loss due to continuous dehydration. Type of feed, climate and working condition of camel are deciding factors for quantity of water requirement. Normally camels are watered twice a week in the field conditions, but should be given regularly once or twice daily. Camel requires 18-36 litres of water, if watered daily. They can consume large amount of water in a short period of time, sometimes up to 135 litres.

Housing

Sheds for camel should be simple with open space and should provide shelter against sun, rain and drought. A three walled enclosure open on one side with thatch is sufficient. The walls may be made of *kachcha* brick. Open side should not face the direction of prevailing winds. The cement concrete floor is not required but floor should be given slope with proper drainage. The manger should be put at the height of 3-3.5 feet.

Reproduction

Camels are seasonal breeders. The breeding season starts from mid November to March during winter. Their reproduction is different as compared to other livestock. Camels are induced ovulators. On an average 23 days follicular cycle is exhibited by adult female during breeding season. Ovulation is induced by mating and occurs after 36-48 hr post-coitus. Both male and female comes into sexual heat. In male it is called as Rut or Musth. Rutting season varies from 3-5 months depending on plane of nutrition. During the musth condition the male camel emits an offensive odour due to dark reddish secretion from the poll gland or occipital gland present on the back of the neck, it is rudimentary in female. A male in rut extrudes soft palate from its mouth (pink bladder like) and produce typical sound. The ratio of breeding stud to females is kept at 1:30-35 in a breeding herd. Female matures at the age of 3 to 4 years and male at 4-5 years. Male throws a black pigment from his pole gland which helps him to attract a female. Mating takes place while sitting and mating time varies from 5-15 min. Gestation period in camel is 390 days. About 96% pregnancies occur in the left horn. The average birth weight is 35-40 kg. Normally she camel can be bred until 20 years of age and some respond even up to 30 years of their age.

Pregnancy diagnosis in camels can be done by different methods. The most important is the cocking of tail. After about three weeks of service, the camel which has conceived develops tendency to raise the tail upward whenever approached by a male or handled by man. If this behaviour is not shown by the female and tail hangs in the pendulous position, she is considered as empty or non-pregnant. Some other methods of pregnancy diagnosis are rectal palpation for the examination of corpus luteum and slip of the foetal membrane, serum progesterone test (more than 1ng/ml- pregnant), pH of the urine (alkaline- pregnant) etc.

After the confirmation of pregnancy care has to be taken to avoid abortion. Extensive running by pregnant camel must be avoided. It should be assured that the camels are getting balanced nutrition with plenty of water. As the foetus grows, feed quantity should be increased by 25%. Camel used as pack animal must be given rest and avoid carrying extra load and exertion. Plenty of greens and dry fodder is necessary. Concentrates with a crushed gram 2 kg, crushed barley 2 kg, green fodder 15-20 kg, dry fodder 7 kg and salt 30 g should be given. Excitement and over exertion may lead to the abortion. In the last trimesters of pregnancy the female must be given maximum rest. In few camels heat during pregnancy is expressed and in such cases mating must be avoided.

Restlessness, swelling of vulva, frequent urination, off feed and camel finds a corner or dark place and cleans it with the help of four legs are some of the normal signs of parturition. Generally parturition occurs in sitting position. The fore limbs of young ones appear first followed by head. The duration of labour is pronounced and more. Naval Cord is generally broken itself due to licking of young one. Placenta is expelled soon after parturition. Camel calf stands on its feet in 6-8 hours after birth. One calf per parturition is common in she camels.

Utility of Camels

In different states of the country depending upon the soil, climate and rainfall, camels are put to a variety of uses. Camels are potential suppliers of commercial products such as milk, meat, hair, hides and dung apart from acting as a work animal.

1. **Work:** Camels can be put to work when they are 4 years of age. For long duration of work, camel can produce one horse power of energy. Camels can carry load equivalent to 40% of their weight depending upon distance traveled and speed for load carrying operation. Camel can cover 100 - 120 km per day with a speed of 25-30 km per hour. Camel can carry 200-250 kg of weight. In view of all above characteristics camel can be used for (i) pulling carts (ii) driving oil mills (iii) carrying water from deep wells (iv) ploughing fields (v) crushing Sugar cane.
2. **Milk:** Camel has great genetic potentiality to yield milk. Lactation yield in camel is generally affected by climate, breed, age, feeding management and health. Average Indian camel yields 3.5 to 10 kg of milk with a peak yield of 20 kg a day in some exceptional animals. Camel milk has high vitamin C content and comparatively longer shelf-life due to low bacterial count compared to cow milk. Camel milk is affected by grazing and its taste depends upon availability of pastures. Milk is salty and contains insulin like protein (52 units/liter of milk), which is useful in diabetic patients (Agarwal *et al.*, 2003). Daily consumption of 500 ml camel milk by a diabetic patient may help to reduce the dose of insulin by about 30% (Agarwal *et*

al., 2002). Fat globules in camel milk are smaller in size and contain comparatively more unsaturated fatty acids. The milk of camel has a medicinal value and useful is Jaundice, Tuberculosis, Asthma etc. Camel milk is useful in making of Mawa and Khoa but due to presence of more salt and lesser fat it is not suitable for preparation of curd and ghee.

3. **Meat:** Camel meat is not popular in India, but it is used for human consumption in different countries. Camel can produce animal protein at lower cost in arid zone. Many studies revealed that camel meat is similar to beef meat. Dressing percentage varies from 55% to 65%. Camel carcass consists of 53-57% meat, 4-8% fat and 16-38% bones. The optimum economical age for slaughter of camel is 2-3 years. Meat quality deteriorates with increase in age.
4. **Hair, leather and dung:** Indian camel produces 1-1.5 kg coarse hair per year. Hair are used for making ropes, rough clothing, paint brush and sometimes items for tourist interest. Camel leather is useful for making shoes and toys. It can also be used for making suitcase as well as for large skin containers (*kuppas*) used for storing oils and ghee. Camel dung is good source of organic manure and can also be used as fuel after drying it.

Health Care

The camels are generally raised and utilized in harsh arid and semi-arid conditions. Camels suffer from all those diseases from which other animals commonly suffer. However, unlike other animals camel carry on without outward symptoms of illness, until exhausted by disease. The most common ailments which occur in camel are pica (deficiency of phosphorus), luxation of patella, chronic ulcers, lower jaw fracture (due to biting by male during rutting season) and some parasitic diseases like Surra.

Commonly Occurring Diseases in Camel are

- **Bacterial diseases**: Actinomycosis, Anthrax, Black quarter, Haemorhagic septicemia, Mastitis, Pneumonia, Tetanus and Tuberculosis.
- **Viral diseases:** Rabies, Rinderpest and Pox.
- **Parasitic diseases:** Bot, Bronchitis, Colic, Gid, Filariosis, Coccidiosis, Hydatidiosis, Mange and Surra.

- **Miscellaneous diseases:** Asphyxia, Blindness, Dehydration, Debility, Epilepsy, Five days convulsive sickness in male camel, Impaction, Poisoning, Mad staggers, Satyriasis, Snake bite.

Important Points Regarding General Management of Camel

- Weaning is done at the age of six months.
- Castration is done between the ages of 4 to 6 years.
- Camels should never be used for work during the last 2 months of pregnancy period and first 3 weeks after parturition.
- By the time camels are 2 years old, the nose peg and rope should be fixed.
- Best age for training a camel for riding is 3 years.
- Young camels can be initiated to regular work by the time they are 6 years old.
- Clipping should be done in camels with long hairs to keep them cool and healthy.
- The camel if clipped in March or April, yield on an average 0.90 - 1.35 kg of hair, but the yield in cold countries may be as high as 5.4 kg.
- Colostrum must be fed to calves for first 7-10 days.
- For preventing the attack by skin parasites, the clipped skin of camel should be smeared with mustard oil, after 48 hours of mustard oil application, the body is smeared with mud and allowed to remain for 3 days till it gets automatically sloughed off.

Chapter - 13

Avian Production

Md. Manzarul Islam

Poultry sector in India has grown from a backyard activity into a major commercial activity in four and half decades. Poultry is one of the fastest growing subsectors of agricultural sector in India. Within the poultry sector, broiler production is growing faster than egg production and about 66.7% of the total output (in value term) comes from the poultry meat sector and only 33.3% from egg production (FAO, 2008). India ranks fifth in chicken and sixth in duck population. Presently, India is world's third largest egg producer and the fifth major producer of chicken meat (FAO, 2010). Poultry is today the major source of meat in India. As per an estimate about 2 million people are employed, fully or partly, along the poultry value chain and that an increase in annual per caput availability of one egg or 50 gm or poultry meat generates about 20,000 to 25,000 full time jobs (Ciamarra & Otte, 2009). Distribution of chicken and duck population as well as poultry meat and egg production among different regions, countries of the world and among different Indian states has been already discussed in chapter one of this book.

Important Economic Traits of Poultry

i. **Egg Production:** In domestic fowl the most important economic trait is egg production. Generally the commercial layer lays till its death starting from 20 weeks of age. Commercial layers are uneconomical for rearing after first laying cycle (after 72 weeks) as the rate of lay declines below the economic level. Present day commercial layers give as high as 310-315 eggs in first laying cycle. After 6-8 weeks of first egg is laid the peak production in each cycle is reached. After attaining peak production there is declining trend in production level. The time taken for attaining the peak and decline thereafter varies from breed to breed. Egg production is generally expressed in terms of either hen housed or hen day. Hen housed egg production is calculated by dividing the total number of eggs produced by the numbers of hens housed in the laying pen initially. Hen day egg production is calculated by taking the mortality into account i.e., by dividing the total number of eggs produced by the numbers of hen survived. In this system it also accounted that a particular layer died after performing how many days. Egg production in chicken is affected by genetic factors such as age at first egg, persistency, intensity or rate of lay, broodiness and pause as well as environmental factors like season, disease, feeding, and duration & intensity of light.

ii. **Egg weight and egg quality**: Next to egg production the important economic trait is egg weight. The egg weight is generally regarded same as the egg size as both are related to each other. The egg weight is affected by many genetic factors like rate of lay and age & body weight at first egg. Smaller size egg is produced by birds that mature early with low body weight. Larger size egg is produced by heavy breeds as compared to light weight breeds. Smaller size egg is generally produced by the strain of birds which lay at higher rate with more number of eggs. The variation in size of egg also depends on the order of lay within the clutch. The egg which is laid first in the clutch is larger in size as compared to that of laid in later part of the clutch. As high ambient temperature results in low feed intake by the bird, the size of egg laid by that particular bird is adversely affected. As the age of bird increases, the size of egg also increases simultaneously. At the start of lay, bird produces egg with only 75% of its maximum weight. As compared to deep litter system of housing, the larger size egg is produced in cage layer house. Although the bird cannot produce the same number of eggs in second cycle after moulting, the size of eggs certainly increases. The optimum weight of chicken egg varies from 50-55g with an average of 53g. The weight of duck egg varies from 65-70g while that of turkey varies from 80-85g. The weight of quail egg is much smaller and varies from 7-14g with as average of 10g.

Colour, shape, texture and shell thickness in the external part of egg and quality of albumen, yolk and presence of blood and/or meat spot are important for judging the quality of egg. Breaking strength of egg is an economically important trait as it has to be transported for long distance and can be measured by shell thickness. Season is having effect on shell thickness as it decreases in summer. Thickness of shell also decreases in some diseases conditions. Yolk index and Haugh unit are indicative of yolk and albumen quality, respectively.

iii. **Growth:** Growth is an important economic trait in relation to meat type birds. In case of broilers, the growth rate is fairly high up to approximately 12 weeks of age and then slows down. While in case of quail and turkey the growth rate is high up to 5 and 16 weeks of age, respectively and then slows down. Fast growth rate helps to reduce the cost of production by saving labour, time and feed.

iv. **Feed consumption and feed efficiency:** In broilers the ratio of feed consumption to weight gain is called feed efficiency. For egg production the bird with small body are considered to be most efficient as they consume less feed. In case of layers the feed efficiency is measured as amount of feed consumed in kg to produce a dozen of egg or kg of egg mass. Category wise feed consumption by birds is shown in Table 13.1.

Table 13.1 : Feed consumption by different category of birds.

Age (Weeks)	Category	Approximate required quantity (Kg)
0-4	Broiler	1-1.25
5-7	Broiler	1.7-1.9
0-8	Chick	2
9-20	Grower	5.5
21-72	Layer	40(110g/day/bird)

Source: ICAR (2002).

v. **Fertility and hatchability:** Fertility and hatchability is expressed in percentage in relation to total egg set. Sometimes the hatchability is also expressed in percentage as proportion of fertile egg set. Inbreeding leads to decrease while out breeding leads to increase in fertility and hatchability. As the fertility and hatchability are reproductive traits, they are lowly heritable. Better fertility and hatchability can be seen in flocks which are in high rate of lay. The hatchability is low in eggs with abnormal shape, extremely large or small size and the eggs stored at very low or high temperature. Inbreeding and bacterial contamination tend to decrease hatchability while high rate of lay increases the hatchability.

vi. **Plumage characteristics:** For sexing of day old chicks and for identifying breeds and varieties within the breed, plumage characteristics (colours) are very important trait. White (White leghorn), black (Jersey black giant), red (Red leghorn) and blue (Blue Andalusian) are solid plumage colours, few other colours are also encountered sometimes.

vii. **Comb type:** Single, rose, pea, walnut, duplex and breda are six types of comb found in birds.

New Coloured Feathered Birds Developed

The new coloured feathered birds developed in public and private sectors for meat and egg production for rural poultry are-

- Nandanam Chicken-1 & Nandanam-Broiler II (TANUVAS, Chennai)
- Giriraja (UAS, Bangalore), Indian Broiler developed at Bangalore under AICRP on Poultry (IBB-83)
- Swarnadhara (KVAFSU), Coloured broilers-Rajali under AICRP on Poultry, Bangalore
- Vanaraja, Gramapriya and Krishi-bro (PDP, Hyderabad)
- Krishna-J (Jabalpur farm varsity)
- Grama Laxmi (Kerala Agricultural University) and Kalinga Brown (Bhuvaneshwar- based poultry organization – CPDO)
- Development of scavenging chicken at CARI, Izatnagar with native fowl base
- CARI NIRBHEEK
- CARI SHYAMA

- UPCARI
- HITCARI

Reproduction in Fowl

Fowl reproduction is characterized by ovoviviparity. After a very short period of embryonic development, the offspring leaves the body of the mother. There is marked difference in the reproductive physiology of poultry when it is compared to that of mammals. After being fertilized at infundibulum (part of reproductive tract), they are supplied with nutrients and surrounded by a shell and finally expelled from the body of birds while in case of mammals, the fertilized egg remains in uterus until birth.

Male reproductive system: The male reproductive system is composed of two testes, each having an epididymis and vas deferens which leads to the copulatory organ or papillae. Unlike most of the other domestic animals where testes are located in a sac called scrotum, the testes remain in abdominal cavity along the backbone in birds. Left testis is usually larger than the right one. In case of birds the accessory reproductive organs like seminal vesicles, prostrate, cowper's gland and urethral gland are absent. The spermatozoa are produced in the seminiferous tubules present in the testis as slender tubes. The spermatozoa are than carried out of the testes by the seminal fluid, which is also produced in the testes. The vas deferens conveys the spermatozoa and seminal fluid from the testes to the cloaca. The copulatory organ is located on the median ventral portion of one of the transverse fold of the cloaca. The sperms are introduced by the papillae into the oviduct of female at the time of mating. Chicken normally produces about 1ml of whitish semen per ejaculate; the turkey produces about 0.5ml of yellowish to brownish semen per ejaculate. The sperm concentration in case of turkey is almost double than that of chickens. In case of turkey the number of sperm per ejaculate varies from 1.75 to 3 billion numbers. Thus a chicken also produces same numbers of sperms as the ejaculate volume is double than that of turkey.

For breeding of meat producing turkeys, male type duck and egg type chicken the artificial insemination can be used. About 80-100 million spermatozoa per insemination are necessary to maintain good fertility.

Female reproductive system: The reproductive system of female consists of ovaries and oviduct. The oviduct has five parts viz., infundibulum, magnum, isthmus, uterus and oviduct. Out of the two ovaries (left and right) at the time of chick's birth, the left one develops and become functional while the right one remain as rudimentary ovary. Very rarely it is seen that both the ovaries are functional. At the time of hatching, the female chicks left ovary contains up to approximately 3600 to 4000 tiny ova from which full size yolk may develop when the hen matures. When the pullet matures or comes into egg production, some of the ova develop to mature yolks. Ovulation Inducing Hormone (OIH) which is similar to Luteinizing Hormone (LH) is responsible for ovulation. Ovulation occurs after 30 minute of oviposition in turkey and chickens. The fertilization takes place in the infundibulum.

Formation of Eggs

Approximately about 25 hours of time is required from ovulation to laying. The details of egg formation are given in Table 13.2.

Table 13.2 : Parts of oviduct and their contribution to egg formation.

Parts of oviduct	Time taken for developing egg to pass through (hours)	Major contribution in egg formation
Infundibulum	0.25	Engulf the yolk after ovulation
Magnum	3	Thick white and practically all white is secreted
Isthmus	1.25	Two shell membrane plus some salt and water are added.
Uterus	21	Some water and minerals are added to egg white. Shell, shell pigment and cuticle are also secreted.
Vagina	-	Egg passes through it without spending much time. Function is not clear, some believe that it secretes cuticle.

Source : Singh (2009).

Structure of Egg

The egg consists of shell, shell membrane, albumen and yolk. The yolk is produced by the functional left ovary during the final 10 days before release; remaining portion of the egg originates in oviduct.

i. ***Shell:*** The outermost covering of egg is called shell which consists of about 11% of the total egg weight. It is mainly composed of calcium carbonate (94% of shell weight). On the entire surface of the shell, minute pores are present which are partially sealed by keratin. The number of pores varies from 6000-8000. Only few of the pores (hardly 12-20) may permit bacterial penetration within the egg under specific circumstances. Carbon dioxide and moisture from the eggs are lost through the pores present on the shell surface. There are three basic units of shell structure viz., outer cuticle made of keratin, middle spongy/calcareous layer and inner mammary layer.

ii. ***Shell membrane:*** There are two shell membranes, outer thick and inner thin. The outer shell membrane is attached to the shell. Both the shell membranes are usually attached to each other except at the broad end of the egg forming an air cell. Both the layer of shell membrane together measures 0.01-0.02 mm in thickness. Shell membrane acts as an effective barrier against bacterial invasion. At the time of storage of egg as there is loss of moisture and shrinkage of egg content, air cell continues to increase in size.

iii. ***Albumen:*** The white portion of egg is called albumen which constitutes about 58% of total egg weight. There are four layers of albumen viz., outer thin, outer thick, inner thin and inner thick or chalaziferous layer. Thick content is comparatively more than thin one. Inner thick albumen or chalaziferous layer surrounds the vitelline membrane of the yolk and continues as a cord like structure called chalazae which holds the yolk in position.

iv. ***Yolk:*** The yellowish portion of egg is called yolk. About 31% of total egg weight is constituted by yolk. Vitelline membrane covers the yolk and holds the fluid mass. There are 7-9 concentric rings of light and dark yellow materials under the vitelline membrane. A cone like structure called Latebra is there in the yolk which holds the germinal disc in place.

Scavenging System of Management for Raising of Chicks, Scavenger Feed Base of Village

The scavenging system is the oldest method for raising of chick followed in village conditions where there is no shortage of land. This is the system which provides sufficient space to the birds where they can have the required amount of feed in the form of herbage, seeds and insect. Care must be taken to protect the birds from predator animals and many infectious and parasitic diseases. This system is not popular nowadays due to popularization of intensive methods.

In scavenging type of raising system, clean, roomy and airy shelter should be provided for night time. Both fixed and mobile type of housing may be constructed. If the house is mobile type, it can be rotated on the range. For about 20 free-range chickens one night shelter can be constructed and can be attached to any existing structure, such as the farmer's outhouse, kitchen or dwelling. In areas where the rainy season is prolonged and rains heavily, the floor can be raised with an optimum roof overhang especially at the entrance. The floor can be made up of solid earth or raised bamboo platform. The walls of the building can be made of mud or bamboo, and the windows and door of bamboo slats. The space requirement for birds is 1 square meter for 3-4 birds.

The total amount of food products available to all scavenging birds in a given area is called scavenging feed base. Number of households present in the area, types of food crops grown and their crop cultivating and crop processing methods are some of the deciding factors for availability of scavenging feeds in that area. Periods of fallow land or flooding, cultivation, harvesting and processing are some of the seasonal fluctuations in the scavenging feed base of village.

The scavenging feed includes termites, snails, worms, insects, grains lost during sowing, harvesting by-products, grass, fodder tree leaves and non-traditional feed materials. Scavenging birds can only harvest the scavenging feed base. Due to allelomimetic behavior several types of poultry scavenging together can make more effective use of scavenging feed base in village. Rearing of poultry under the free-range and backyard systems depends to a large degree on the quality of the feed available from scavenging feed base.

Low Input Technology: Backyard and Semi Intensive Unit of Various Sizes

As the Indian farmers are having less land holding capacity and most of the farmers are having limited capital and resource, the development and adoption of low input technology can uplift their livelihood. As a low input technology Backyard and Semi-intensive units can be started by poor farmers.

Backyard poultry farming: There are many indigenous breeds which are suitable for backyard system of poultry rearing. The need of poultry keeper's will decide the choosing of right breed for backyard flock. Chicken breeds with milder temperaments are better suited for families with small children who would interact with the birds. For the shelter and protection of birds from predators chicken coops must be provided. Coops must be big enough for the chickens to live comfortably by allowing them room to flap their wings, walk about, and build comfortable nests. Chicken coops must also provide warmth through insulation, heat lamps, or extra bedding and nesting material. A good coop will provide birds with shade and shelter, freedom from draughts and protection from rain. Feeders and waterer should also be provided for the birds. Feeders should be easily cleaned and replenished, and if possible, positioned to prevent access by wild birds. Chicks will usually be fed on crumbles, which are high in protein and energy. The pullet will be consuming from 80 to 100g of feed per day by the time the bird is ready to lay. About 250 to 300 ml of water a day will be required by a grown hen. To ensure the health of birds, they should be checked regularly. Regularly check up for parasites such as lice or mites should be done and isolated for treatment. Poultry droppings should be removed regularly and clean litter or bedding should be added. Egg collection should be done regularly to prevent breakage by birds. Further, the remaining egg in the coop will encourage chickens to go broody. Artificial light used in the coop will encourage hens to lay for longer in winter season. Eggs can be stored for up to 6 weeks if refrigerated.

Semi-intensive unit of poultry rearing: Combination of extensive and intensive systems of poultry rearing where birds are confined to a certain area with access to shelter is called semi-intensive system of poultry rearing. Birds in this system are reared in two ways i.e., half-way in houses and half-way on ground or range. Semi-intensive production system is sometimes also referred to as backyard production system. Flock sizes in semi-intensive flocks ranges from 50 to 200 birds and it is adopted where the amount of free space available is limited, but it is necessary to allow the birds 20-30 square yards per bird in open area. The whole space is divided giving a run on either side of the house of 10-15 square yards per bird. During the day, birds are confined in an enclosed outside area and housed at night. Feed and water are available in the house to avoid wastage by rain, wind and predators. The deep litter system may be the most common housing system in the semi-intensive systems of poultry rearing. The birds are fed 1 or 2 times a day with grains.

Mixed Farming and Poultry Raising

Poultry raising along with other components such as crops, livestock, aquaculture and agro forestry is called mixed farming. It is also known as Integrated Farming. Minimum use of external inputs by enhancing the recycling of materials within the system through a process of value addition is to be considered in mixed farming. In mixed farming, between farming families and their natural resources a symbiotic relationship is formed.

Chicken or duck cages can be constructed on the top of the pond. The dropping of birds will be utilized by the fishes as potential source of feed. The waste of about 1500 birds is sufficient for rearing of fish in one hectare of land (production up to 10 MT/ha). In comparision to other livestock, poultry wastes contain more nutrients.

Poultry can also be integrated with crop and agro-forestry for good returns. In this type of farming cows, goats and poultry comprising of chicken, ducks, turkeys etc are used. Ducks can be reared along with paddy cultivation and thereby fertilize the field with its dropping.

In mixed farming the involvement of women is much more and family labour also plays important role. The mixed faming is special because it provides balanced diet to all the members of integrated farming. Round the year profit is one more specialty of mixed farming system.

The Advantages of Mixed Farming are

- Income throughout the year
- Eco friendly
- Reduced risk of crop failure
- No dependency on input from outside
- More employment generation
- Self dependency

Concept of Self Local Market

In rural areas, poultry products in the form of live birds or eggs hardly need any organized market as the people prefer egg and meat from the indigenous birds rather than from the exotic birds. The birds and eggs are sold and purchased exclusively for local household consumption. The price is also not fixed and it is finalized by mutual negotiation. The specialty of self local market is that the products are fresh and unadulterated as it is sold by shortest channel of marketing i.e., producer to consumer.

Poultry Farming in Hilly Areas

In the hilly areas the farmers are having small land holding. They use to raise local/indigenous breeds of poultry as there is provision of scavenging in the hilly tract. For the families of hilly area, either landless or land holder, the poultry provides economical and lively security to them. In the tribal communities the chickens are traditionally reared. Almost every farmer in the village rears few poultry alone or with other livestock and are mostly reared by scavenging system. In very cold season, the birds are allowed to scavenge in the hotter part of the day and then are closed in small house with extra supplemented feed. The manures produced by the birds can also be used as fertilizer for agricultural fields. They spent very little money in rearing and birds or livestock become part of the family. Crosses of breeds like Rhode Island Red, White Leghorn, New Hampshire are also sometimes reared in hilly areas.

The hatchability is one of the biggest disadvantages of hill farming, for every 300 meter increase in altitude there is decrease in 10% hatchability. It has also been recorded that there is decrease in hatchability up to 30% if the birds are reared at 2000 meters. The broilers reared in hilly area use to have reduced growth rate and higher mortality rate due to lower oxygen level as the broilers require high oxygen for metabolism.

Limitations in Hill Farming

- Small land holding
- Soil erosion & slopes
- High mortality
- Low water quality
- Difficulty in marketing
- Natural hazards

Organic Poultry Farming

The farming where the birds are produced by feeding with organic feed, which is produced without synthetic pesticides, drugs, antibiotics or genetically modified crops is called organic farming. The ration which contains meat by-product such as meat meal or bone meal should not be given to birds in organic farming. The eggs produced by such birds are called organic eggs. The birds must be reared in deep litter system with sufficient access to free-range. In organic farming there should not be routine prophylactic drug treatments. But the birds which are actually ill are treated. It should always be tried that the birds do not get ill and the breeding strategy should be made accordingly.

The aim of organic farming should be to produce food of high quality in sufficient quantity. The fertility of soil should be maintained and increased for a long time. By organic farming the pollutions of all forms can be minimized. 100% certified organic feeds, 1.75 square feet space per bird, supply of clean water, natural light, adequate ventilation to prevent ammonia build-up, egg storage at 60 ^{0} F are some of the mandatory requirements of organic farming.

Brooding and Rearing Practices used for Chicken, Duck, Quail, Turkey and Guinea Fowl

Brooder Management

Brooding: During the early part of life i.e., up to 3-5 weeks, the young chicks are given special care which is called brooding. Brooding is helpful in preventing cold shock as the environmental temperature is lesser than brooder temperature. Body temperature regulation is not developed in newly hatched chicks as the thermo regulatory centre is non-functional in early part of the life. Another probable reason for this may be lack of feathers. Additional heat may be provided during hatching to 3-5 weeks for the obvious reason. The brooder house should be located at minimum distance of 100 m from other poultry houses. Mainly there are two types of brooding i.e, Natural and Artificial brooding. In case of natural brooding, broody hens are used for first 3-4 weeks of chick's age. Under artificial brooding, again there are two methods i.e., hot room brooding and cold room brooding. In case of hot room brooding, the entire room is heated by central heating devices while in case of cold room brooding; only the area under brooder canopy is heated. In most of the tropical countries the later one is mostly used. The cold room brooding can be done either by floor brooding or by battery brooding (Figure 13.1).

i. ***Floor brooding:*** The ideal floor brooder house should be draft-free, rain-proof and protected against predators. For adequate environment there must be proper ventilation. Floor brooding can be done with the help of canopy or sometimes without canopy by infrared bulbs. In canopy brooding; a hover is hanged from the roof which is fitted with bulbs. The canopy helps in trapping hot air which is required by the chicks. A hover of 1.8 m diameter can hold about 500 chicks. During the first week of brooding a chick guards is placed 60 - 75 cm from the edge of hover to prevent the moving of baby chicks away from the source of heat. Generally after one week the chick guard is not needed. In case of infrared brooding, the bulbs are hung at least 25-30 cm above the litter. There are many advantages of using infrared for brooding such as germicidal effect of infrared light, triggers synthesis of vitamin-D, lasts for longer time and reduces cannibalism.

ii. ***Battery brooding:*** In commercial farms the brooding is mainly done by battery brooders. The advantages of using battery brooder in comparison to floor brooder are- less mortality, free from coccidiosis or other litter borne diseases, less labour requirement, better feed efficiency etc. In battery brooder the cages may consist of 4 or 5 tiered. Heating is done by electric heater.

Figure 13.1 : Brooder

Preparation before Chick's Arrival

Before keeping the chicks in brooder house, everything should be ready. Old litter should be removed from the house if the floor brooding is chosen. For the first 3 days paper should be spread over the litter so that chicks can take proper feed instead of litter material. The house should be properly scrubbed, cleaned and disinfected with any disinfectant solution at the recommended concentration at least one week before arrival of the chicks. All the equipments should be cleaned properly. To avoid insect threat insecticide should be spread. For controlling of ticks and mite, malathion spray/ blow lamping or bath can be used. After cleaning the house new litter should be spreaded. The lights should be on before 24 hour of arrival of chicks to maintain the proper heating. As the poultry diseases are highly contagious, All-in-all-out system helps in the control and prevention of diseases. At the entrance of each building, foot bath of recommended disinfectant should be kept which will help in preventing the spread of diseases.

Brooding Temperature

During the first week of brooding, temperature should be 95 °F (35 °C) which may be reduced by 5 °F per week during each successive week till 70°F (21.1°C). About 24 hours before arrival of chicks, the brooder should be switch on to maintain the required temperature. A thermometer should be kept in brooder house to control the temperature. Indication provided by the chicks is clue that whether they are getting desired amount of heat or not. The chicks try to get closure to the source of heat and huddle down under the brooder, when the temperature is not up to the required level. On the other hand when the temperature is too high the chicks will get away from the source of heat and may even pant. The chicks will be formed evenly scattered when the temperature is right. After the chicks are about 3 weeks old, brooders are not necessary.

Floor, Water and Feeder Space Requirement

A chick should be provided floor space of 0.05 sq m at the start, this should be increased by 0.05 sq m after every four weeks until the pullet are about 20 weeks of age. For broiler female and male a floor space of 0.01 sq m and 0.15 sq m, respectively be provided till 8 week of age. For 1000 chicks there should be a provision of 50 linear cm of water space for first two weeks and has to be increased to 152-190 linear cm at 6-8 weeks. Waterer height should be maintained at 2.5 cm above the back height of chicks to reduce the spoilage. Cleaning of waterer should be done daily. From 0 – 8 weeks of age the requirement of feeder varies from 2.5 – 3.0 cm per bird.

Grower Management

After brooding till sexual maturity a bird is called grower. At about 8 weeks of age the birds are shifted from brooder house to grower house. There are many types of floors in grower house viz., litter, part slat, part wire, all slat, all wire etc. Floor, feeder and water space recommended for a grower are 950-2350 cm sq, 7.5-12.5 cm and 2-2.5 cm, respectively. Before transferring the growers into layer house, the cockerels which remain in the flock due to sexing error (2%) should be identified and separated from pullets. The feed of grower birds should contain 16% crude protein and 2500kcal/kg ME. About 5-6 kg feed is required for egg type chicken during growing period from 9-20 weeks. Temperature, humidity, age, dietary constitution, activity and air movement influences water consumption. To keep the birds free from parasitic diseases, deworming is very much essential at least once bimonthly. In a grower house, decrease lightning should be followed. Growing pullets should never be exposed to increased lighting schedule. At the age between 12-16 weeks debeaking is recommended. In order to encourage the birds to lay, the laying nests should be placed during the last month before laying. For ducks and turkeys the nest should be placed at ground level while in case of chickens multi-tier nests are preferable. The nests should be closed at night. For transfer of pullets to laying house, "all-in, all-out" practice should be followed.

Layer Management

The female birds from 20 weeks onward which are sexually mature and producing eggs called layers. Laying starts at the age of 20 weeks and continues economically till 72-75 weeks. After one year of economical laying, birds are sold as spent hen. At the age of 18-20 weeks the flock should be transferred from grower to layer house. 1-2 days prior to housing the females, males should be placed in the laying quarters if it is breeding flock. The requirement of floor space is 0.23 - 0.28 sq m. Feeder space and water space should be 10 cm and 2.5 cm per bird, respectively in floor house. The floor requirement for commercial cage operation is 465 sq cm per bird. Five hanging feeders with 50 cm diameter of 20-25 kg capacity are required for 100 layers. Water space of 2.5 linear cm per bird should be provided. Feeders should be filled more than 1/3rd full. One laying nest for every four pullets is necessary. Lighting of layer house is very essential for optimum egg production. The lighting should be increased gradually from 21 weeks onwards till it reaching 16-17 hours per day and maintained at that level thereafter. It has been found that in comparison to irregular light, correct lighting boost up egg production by 5-10 %. There is no advantage in increasing the light over 17 hours per day. Duration of light period should not be decreased during the laying period.

Male Management

The management of breeder male remains almost same as that of layer management except to ensure proper fertility; the male breeding diet should be fortified with extra calcium, manganese and Vitamin E.

Brooding and Rearing Practices used for Ducks

Mainly the duck production remains in the hand of poor's. The east and south Indian states like Tripura, Assam, West Bengal, Tamil Nadu, Andhra Pradesh, Karnataka and Orissa posses most of the duck population. Duck rearing is more economical than chicken, as duck eggs are 15-20 gm heavier than the chicken eggs. Further ducks supplement a part of their food requirement by foraging. Unlike chickens ducks lay eggs in second and third year also economically. Ducks are resistant to many diseases as they are hardy in nature. Duck lay hatching eggs with optimum size of about 70-75 g as compared to chicken eggs which is just 50-58 g. Fumigation of eggs is done only for 20 minutes after collection as duck eggs are more porous. The incubation period is 28 days for duck eggs. But 35 days is required for incubation of Muscovy duck egg. During incubation period, duck eggs require less temperature and more humidity compared to chicken eggs. Turning of eggs during incubation should be at angle of 180° as against 90° for chicken eggs. During incubation period daily 30 minutes of cooling during incubation period is needed in case of ducks.

At the age of 20 – 22 week egg type duck starts laying. One drake for each 5 to 6 ducks is required for optimum fertility. Most commonly used breed for commercial eggs production are Khaki Campbel and Indian runner. For meat production White pekin, Aylesburry and Muscovy are well known breeds due to their fast growth rate which attains a body weight of about 2.2-2.5 kg in about 7 weeks. Mule ducks are sterile progeny when Muscovy ducks are crossed with other ducks produce.

Rearing Practices used for Turkey

The native place of turkey is North America. Turkey are reared only to produced meat as the meat production can be compared with other livestock species but they produce egg very uneconomically when compared with chicken, quail or duck. At the age of 30 weeks the first egg is produced. When there is presence of adequate light the turkey hens lay as much as 100 eggs during first 24 weeks from the age at sexual maturity. Turkey egg is approximately 85 g which is 1.5 times larger than chicken eggs. Turkey eggs are as palatable and nutritious as chicken eggs. The incubation period is 28 days for hatching of turkey eggs. Turkey meat is famous as festive food for special occasions like thanks giving and Christmas. For meat purpose a turkey poults weighing around 4 kg seem to be much more in demand. Turkey achieves a body weight of about 4 kg at the age of 12-14 weeks. Approximately at 20-24 weeks of age when they achieve a body weight of 8-9 kg are sold as roasters. The general management conditions remain more or less similar to that of chickens. Double hover space is required by turkey poults as compare to chicken. Disease resistance is more in case of turkeys when compared to chickens. More protein, mineral and vitamins are required in the diet of turkey. Turkey meat is costlier than chicken meat. Margin of profit is less as the cost of production is more than chicken.

Rearing Practices for Guinea Fowl

Meat of guinea fowl is slightly darker than that of chicken but it is tender. Guinea fowls survive well in tropical climate as well as temperate due to their hardiness. Guinea fowls are resistant to some of the diseases which are fatal to chicken. The young one of guinea fowls is popularly known as keets. Incubation period is 28 days. At around 36-38 weeks of age indigenous guinea fowl start laying eggs and lay about 50 - 100 eggs during the year. Improved varieties lay from 180 - 200 eggs as they mature early, approximately 28 weeks of age. The eggs of guinea fowl are smaller in size than that of chicken eggs. Fowls attain about 1.2 kg body weight at 13 weeks of age under good management conditions.

Rearing Practices used for Japanese Quail

Unlike other species the male is lighter then the female in Japanese quail. Egg production starts as early as 5 weeks of age due to their high prolificacy. By about 50 days of age birds attain full production level. They produce as much as 250 eggs per year under favourable condition. Eggs are multi coloured, ranging from dark brown, blue and white. Weight of the egg is approximately 10g (6.7-13.8) which is about 7% of the hen's body weight. It needs extreme care while collecting eggs as they are very thin shelled. The incubation period is 17-18 days. Up to 3 weeks of age 27% protein with 2800 kcal ME/Kg of feed is recommended for optimum growth. The laying quail should be fed on a ration containing 22% protein and 2900 Kcal ME/Kg. Diseases of quail are quite similar to that of chickens but are reported are to be resistant from Ranikhet disease, ascariasis and chicken coccidiosis.

Incubation and Hatching Norms of Different Species of Poultry

i. **Selection of hatching eggs:** Hatching egg should have optimum size for proper hatching as the extra large and very small eggs do not hatch well. The standard weight of chicken egg should be 56g (54-58g). In case of duck and turkey the weight of hatching egg should be 65-70g and 80-85g, respectively. The shape of the egg should be ovoid, other than this such as round or excessive long egg does not hatch well. Darker shelled egg hatch well than light shelled egg. Hatchability is also determined by the thickness of shell, which should be between 0.33 and 0.35 mm for best result. Egg with tremulous air cell decreases hatchability. The eggs should also be free from any blood or meat spot for better hatchability.

ii. **Care of hatching eggs:** For clean egg production the nest box should be cleaned regularly and the eggs should be removed as early as possible. Three or four times a day egg should be collected to prevent undue loss due to breakage or cracks. Eggs should be washed with lukewarm water before storing the physiological zero temperature which is 21.1 °C. An ideal temperature for storing hatching egg is 10 °C (7.2-15.5 °C). A relative humidity of 75-80% should be maintained at storing room. Length of storage should not be much longer as the hatchability starts decreasing after one week of storage and reaches zero by 4th week. The eggs should be stored with broad end up and turning may be beneficial when stored for more than 7 days. At the time of transport of hatching eggs, care should be taken that the eggs are not shaken too much.

iii. **Fumigation:** Fumigation is the process of destroying microorganisms in the hatchery room or from the surface of the eggs which may cause many diseases like pullorum, paratyphoid, infectious coryza, fowl pox, ranikhet disease etc. Fumigation can be done by using formaldehyde gas. For fumigation, 40ml of 40% formalin is mixed with 20g of potassium permanganate. This will produce enough gas to fumigate the area of 2.83m^3 if allowed to react for 30 minutes. This is called 1x fumigation. Similarly, 2x, 3x or 5x concentration are also used depending upon the requirement.

iv. **Incubation and hatching of eggs:** Eggs are placed in certain environment where they get required temperature, humidity, turning etc. for certain period of time is called incubation and the production of baby chicks from egg is called hatching. Two methods of hatching of egg are: (1) Natural method and (2) Artificial method. Broody hens were used (natural method) in the earlier days for hatching of eggs, still in village areas the eggs are hatched by broody hens. *Desi* hens are considered as ideal for brooding of eggs naturally. About 8-13 eggs can be placed under one hen depending on body size of the hen. For large scale production of baby chicks, this method of hatching is highly unsatisfactory as it will be time taking and output of chicks are very slow. There will be less production of chicks at a time and the market demand cannot be fulfilled satisfactorily. Nowadays incubators (artificial method) are used more efficiently for hatching of eggs as it provides similar environments as supplied by broody hens.

Frequent collection of fertile eggs should be done to prevent spoilage by high environmental temperature especially in hot and humid conditions. 50 to 70 °F is regarded as optimum for holding the fertile eggs before incubation. Relative humidity of storage chamber should be maintained at 70 to 80 % as less humidity promotes loss of water from eggs.

Temperature, humidity, gaseous environment and turning of eggs are some important physical factor necessary for successful incubation of eggs which results in to proper hatching. The normal incubation temperature usually varies from 99.5 to 99.75 °F (37.5 to 37.7 °C) for first 18 days and 97 to 99 °F (36.1 to 37.2 °C) for last three days. Hatchability is always affected by humidity in the incubator. Temperature inside the incubator is measured by dry and wet bulb thermometers. For the first 18 days the required humidity for chicken egg is 60% and during last three days the humidity should be 70%. Pipped egg and soggy chicks are the result of low and high humidity, respectively. Oxygen is required by the developing embryo for their metabolism. For best hatching results the optimum level of oxygen should be present in the incubator is 21%. 1% decrease in oxygen leads to 5% decrease in hatchability while 1% increase in oxygen leads to 1% decrease in hatchability. In high attitude hatchability reduce due to reduction in concentration of oxygen. CO_2 concentration inside the incubator should not exceed 0.5%. Hatchability is decreased by higher concentration of CO_2. Hatchability becomes zero when CO_2 level reaches 5%. Turning of eggs is another important requirement for proper hatching. Turning of eggs in the incubator improves hatchability by preventing the adhesion of egg mass with the shell. Turning of eggs should be at least for four times during a day when turning is done by hands. In case of automatic turning of eggs at least eight times or more turning during 24 hours is needed. The eggs trays are turned through an angle of 90°. No turning is required after 18 days. It is better to load the fertile eggs with broad end up in the incubator.

For testing the actual condition of egg they are candled during 5 to 7 days and on 18th day of incubation. The first candling is important for removing unfertile eggs from the incubator while the second one is done to remove dead germs. Only the eggs with live embryo are transferred to the hatcher on 18th day of incubation.

The incubator should be properly cleaned before setting eggs into it. The incubator is disinfected and fumigated to kill disease causing organisms. Before storing and after transfer of eggs from incubator to the hatcher the hatching eggs should be cleaned and fumigated. This reduces the incidence of spread of disease. Fumigation can be done with formaldehyde gas using 40 ml of 40 % commercial formalin and 20 gm of $KMnO_4$ for an area of 2.83 m^3 inside the incubator or hatcher. $KMnO_4$ may be placed in a glass container and formalin poured over it. Never the reverse mixing is done.

After hatching of eggs, the sexing of chicks is important. For sexing Japanese or vent method of sexing is commonly used. Rudimentary raised structure seen in the cloaca of male chicks. Proper skill and experience is necessary as the sexing of chicks involve identification of very small structure at day old stage.

Management for Proper Incubation and Hatching of Eggs

- The hatchery should be situated at least 500 ft (153 m) from poultry houses.
- Medium sized and brown shelled eggs hatched better.
- Ovoid shape is better for high hatchability.
- Minimum egg weight should be 52 to 56.7 gm for increase hatchability in case of chickens.
- Sex can't be predetermined until hatching.
- Eggs losses 11.5 % of its weight in first 18 to 19 days (0.6 % / day).
- Small eggs will lose more weight than large eggs.
- Female hatch 3 hours before male.
- There is no effect of sound on hatchability of eggs.

Incubation Periods in Different Species of Poultry

- Ostrich : 42 days
- Muscovy duck : 35 days

- Duck : 28 days
- Turkey : 28 days
- Geese : 28 days
- Chicken : 21 days
- Quail : 17 days
- Pigeon : 17 days

v. **Sanitation and hatchery hygiene:** Sanitation and hatchery hygiene is an important aspect of hatchery management as the hatchery is regarded as greatest potential source from where disease can spread. There are two types of diseases found in hatchery; one is egg transmittable and other is transmitted by contact. Egg surface, boxes of eggs can convey infection from the poultry farms. For preventing transfer of disease the hatchery should not be located near poultry farm. There should be distance of at least 500 ft between hatchery and other poultry houses. Only the nest clean eggs should be collected from the layer house. Incubators should be located in separate houses and the gate should be signed with "No Entry". Fumigation of eggs must be done which are entering the hatchery. In the hatchery always one way flow from egg room to incubators to hatchery to chick holding room should be maintained and no reverse flow is allowed. The common chemical disinfectants used to maintain sanitation in the hatchery are Quaternary Ammonium Compounds, Hypochlorites, Iodophores, Phenolic compounds, Formaldehyde etc.

vi. **Sexing of chicks:** In commercial operations the day old chicks are separated either by examining the rudimentary copulatory organ (vent method) or by auto sexing which depends upon the sex linked external characters. Vent method was originally developed in Japan and is most popular method of sexing chicks. Vent sexing is done by use of a special instrument, *"proctoscope"*, it is possible to see the testes of male chicks through the intestinal wall of day-old chicks. The disadvantage of this method is that it needs skill and there is chance of injury while examining chicks. In auto sexing, plumage colour or growth of an organ or any other character linked with sex chromosome are taken into account to determine the sex.

vii. **Packing and dispatch of day old chicks:** The day old chicks are packed in boxes of 56x46x15cm size which is partitioned in four compartments. The boxes are made of corrugated card boards for single use and for reuse the plastic boxes are also available in the market. The plastic boxes should be properly washed and fumigated before reuse. Season and distance to be travelled by chicks are the factors which decide the numbers of chicks to be transported in one box. In winter and summer season the number of chicks to be transported are 100 and 80, respectively.

Poultry Housing

Proper housing is needed for protection of birds from adverse climatic conditions, parasites, predators, thieves and to avoid the development and spread of infectious and contagious diseases. It is also important for ensuring scientific feeding in a controlled manner and for providing proper micro-climatic conditions in the near vicinity of bird. A good house ensures easy and economic operation in a poultry rearing along with comfort and protection. An ideal house should have environmental conditions suitable for poultry production i.e., temperature- 70-75 °F, relative humidity- 30-60%, ammonia level less than 25 ppm, litter moisture- 20-25% and air flow- 10-30metres/minute.

Location of Poultry Houses

1. Site should be elevated, not low-lying so that stagnation of water can be prevented during rainy season or flooding.
2. There should be ample space for future expansion.
3. It should be nearer to main road for easy transport.
4. There must be availability of ample quantity of clean, wholesome and palatable water.
5. It should be nearer to the city/town for easy marketing of poultry products.
6. It should be away from other broiler farms in the area to prevent competition in the market.
7. There must be availability of electricity for easy operation in poultry rearing.
8. It must be of adequate size to accommodate required number of birds.

9. It should not be too near or far away from the farmer's house.
10. Fertile and well drained soil is desirable.

Important Points Regarding Construction of Poultry Houses

1. The long axis of the house should be in east-west direction to prevent the direct sunlight on the birds.
2. The size of the house should be finalized depending on the number of birds to be reared. Generally in deep litter system of poultry rearing the space requirement of 1 and 2 sq ft for broiler and layers is needed, respectively.
3. The distance between the houses of same age group of birds should be about 20 m while the young stock house should be at least 45 to 100 m away from adult stock house.
4. Depending on number of birds to be reared and availability of land, the length of house can be any. But the house should have a width not more than 9 m in order to allow ample ventilation and aeration at the mid-portion.
5. The height of the shed should be 2.4 to 3 m from foundation to roof line.
6. The different types of floor in use are all litter, all slat, slat and litter, wire and litter and sloping wire floor.
7. In areas where summer season is prolonged, the roof should be painted with reflecting type of paints such as aluminium paint.
8. An overhang of about 0.9 m is beneficial as it will prevent rain splashing inside the house.
9. The roof can be made up of thatched, tiled, asbestos or concrete depending upon the cost involvement.
10. The position of light should be fixed at about 2 m above the ground level. The interval between two bulbs should be 3 m.

Different Types of Poultry Houses

Following are the different types of poultry houses-

- **Hatchery:** In this house the hatching of fertile eggs takes place with the help of different type setters and hatchers. After hatching the chicks are transferred to brooders.

- **Brooder/chick house**: This house is used to brood and rear chicks from 0 to 8 weeks of age. Generally the chicks which are to be reared till layer stage are kept in this house.
- **Grower house**: Growing birds from 9 to 20 weeks of age are kept in this house.
- **Brooder cum grower house**: The birds to be reared till layer stage from 0 to 18 weeks are kept in this house.
- **Layer house**: Birds more than 20 weeks of age are reared and are kept till 72 weeks of age i.e. up to completion of first laying cycle.
- **Broiler house**: Used for meat-type broiler birds which are reared up to 6 weeks of age.
- **Breeder house**: The house where both male and female breeders are maintained at appropriate sex ratio is called breeder house.
- **Environmentally controlled (EC) house**: The entire environment of the house is manipulated for providing optimum living conditions to the birds.

Systems of Poultry Housing

Following are the different types of poultry housing system

I. **Free range/extensive system:** Free range system of poultry housing is oldest among all and has been practiced over the years by small/marginal farmers where there is no limitation on availability of land. Under this system, birds are not confined and can search for food over a wide area, provided they are protected from predatory animals and various diseases.The birds may roost outside usually on trees planted in the field or runs and water trough may be provided there. Night houses in the form of wooden boxes may be provided to protect the birds from inclement weather and wild animals. Under backyard extensive systems birds are housed at night but allowed free-range during the daytime and usually scavenging is supplemented with grain feeding in the morning and evening.

II. **Folding unit system:** Recently this system of housing has been innovated. In this system the birds are confined to one small run and on daily basis the position of run is changed to provide bird's fresh land. Extra labour involvement, carrying food & water and egg collection are some of the disadvantages of folding unit system. The space requirement for one bird is 4 sq ft, out of which 1 sq ft is allotted for house and other 3 sq ft for run. The standard size of folding unit should be of 25 hens.

III. **Semi-intensive system:** This is a combination of the extensive and intensive systems and practiced where the amount of free space available is limited. In this system birds are confined to a certain area with access to shelter.

IV. **Intensive system:** In this system the birds are fully confined in house without any access to the land outside. This system is usually practiced in the areas where land is limited and costly. In this system production is higher. Among different types of intensive systems, cage system and deep litter system are the most common.

a. **Cage system:** For commercial layers this system of housing is very well suited as it is most intensive system of housing. This method is beneficial for the farmers who have very limited area of land. Four bird cages are ideal for commercial layers. The cage should be made with suitable specifications like length (45cm), height at back (38cm), height at front (42cm) and width (38cm). The floor of the cage is made up of strong galvanized wire (mesh size-2.5x5cm) with a slope to facilitate rolling of egg to the tray. Floor of the cage should be about 1m higher than the ground level. Feed and water are supplied in long continuous trough attached to the cages. Water trough is placed above the feeding trough. Among different types of cages the California cage system is most popular nowadays as there is convenience of feeding, watering, egg collection and droppings removal.

b. **Deep litter system:** The system of housing where large floor is covered by litter material like saw dust, chopped wheat straw, paddy husk, groundnut shell, leaves etc. and birds are reared in a large group is called deep litter system of housing. The litter material should be provided with a standard thickness of 5cm which varies according to season. The thickness of litter should be thin and thick in summer and winter season, respectively for the obvious reason. The litter should be stirred at frequent interval (preferably twice a week) for avoiding caking. The wet litters should be replaced by dry one immediately. The moisture content of deep litter should be 20-25%. The spaces for floor, feeding and watering are given in Table 13.3 and 13.4.

Figure 13.2.1 : Cage system of rearing.

Figure 13.2.2 : Deep litter system of rearing.

Table 13.3 : Minimum floor space requirement per bird

Age (Weeks)	Floor space per bird (cm^2)	
	Light bird	Heavy bird
0-8	700	950
9-18	950	1350
19-22	1400	1950
23 and above	1950	2300 to 2800

Source : Sastry and Thomas (2005).

Litter management: Litter should be managed in such a way that it should show the qualities like- 1) absorb moisture from the dropping quickly, 2) dries up rapidly, 3) no formation of cake, 4) free from mould, 5) must be non toxic, 6) must be cheap and locally available and 7) must be biodegradable.

Table 13.4 : Minimum feeding and watering space requirement.

Age (Weeks)	Feeding space (linear cm/100 chicks)	Watering space per 100 birds	
		Channel type trough (cm)	Fountain type (lit)
0 to 2	250	60	4
3 to 6	450	150	10
7 & above	1000	250	20 to 25

Source: Sastry and Thomas (2005).

Groundnut hulls, sawdust, chopped straw, bagasse, paddy husks are commonly used as litter materials. Height of the litter should be 5 cm. In winter the depth should be increased to double i.e., about 10 cm. It is very essential to keep the litter material as dry as possible. As the droppings of bird contains water, the moisture levels in litter material increases daily. There will be formation of cake if the moisture in litter material is too high and in the other hand there will be formation of dust if the moisture level is very low. There will be formation of excess ammonia if the moisture level in litter is beyond 25 percent. If the ammonia levels in the air exceed 25 ppm, the birds will be subjected to various stresses. Irritation of eyes and nasal membranes will lead to conjunctivitis, as well as poor feed intake and growth rate. To reduce the ammonia level, there should be addition of super phosphate at 2 kg per 10 sqm area.

The dropping from chicken synthesizes Riboflavin (Vitamin B_2) and Cyanocobalamine (Vitamin B_{12}) by the chemical and bacterial actions when it is mixed with litter. Niacin, phosphorus, potassium, magnesium, sodium and calcium are also present in built-up litter. Birds kept in deep litter often peck and eats a small quantity of litter material which contains above nutrients. This phenomenon is known as recycling. The deep litter bird do not show deficiency symptom even if the feed lack in some nutrient as the birds get it from litter. Recycling of nutrient is completely absent in cage system.

Poultry House Equipments

Poultry house equipments should be simple, portable, cheaper, easy to clean & disinfect and durable. Further, it must provide comfort to the birds and increase the work efficiency of the poultry man. Some of them are

- **Feeders:** Proper type of feeder is essential to avoid feed wastage. It should not be too shallow or too dip and be filled easily. Trough, pots and pans of suitable size can be used as feeder (Figure 13.3.1).

Figure 13.3.1 : Different types of feeder.

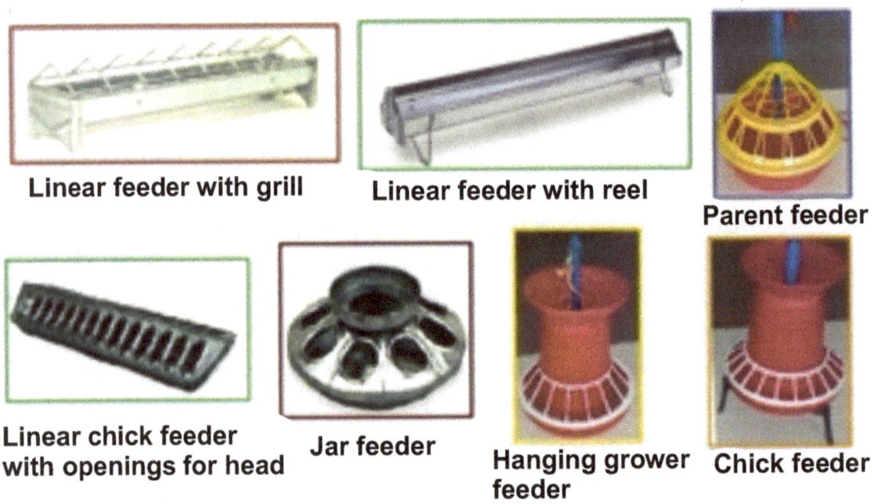

- **Watering devices:** Ample amount of clean and fresh water should be made available to the birds at all times. An ideal watering device should be large enough for a whole day supply, keeps the water clean & cool, shouldn't rust or break easily, topple over readily and is easy to clean and refill, does not allow the birds to spoil the water and is relatively cheaper. It may be fountain, trough and permanent channel type. Some of the waterers have been shown in Figure13.3.2.

Figure 13.3.2 : Different types of watering devices.

- **Brooder guards:** These are solid or wire mesh placed around the source of heat to prevent the chick from going away from the source of heat. Solid materials are used in cold while wire mesh in hot weather condition.
- **Perches or roosts:** Perches or roosts besides fulfilling the natural instinct also help to keep the bird's feet and plumage clean. The perch space varies with the size of the bird. Generally wooden perches are used however two inches thick split bamboos or other poles can also be used as roosts. All perches should be of equal level and placed at a distance of 35 cm from each other. They can be placed at about 25 cm from rear wall and at a height of about 30-40 cm from the floor.
- **Nest boxes:** Each pen of laying birds should be provided with nest boxes for laying eggs (Figure 13.3.3). Nest should be adequate in size, well ventilated, dark and easily cleaned & disinfected. It may be individual, community, roll away and trapnests.
- **Grit and shell containers:** Shells and grits are provided to the poultry separately and for this purpose ordinary wall hoppers made either of wood or metal can be used.

Figure 13.3.3 : Poultry house equipments.

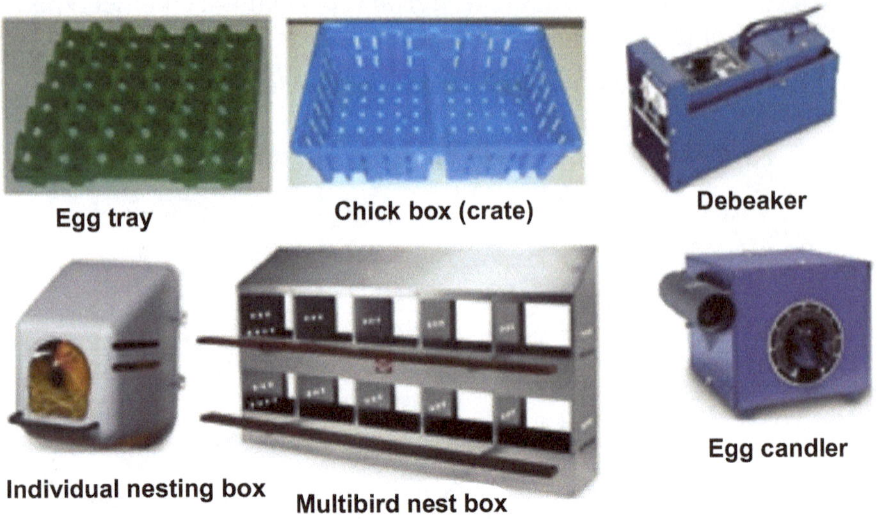

Apart from these, equipments like broody coop, egg candling & grading machines, egg trays, chick boxes, chick sexing machine, debeaker, cleaning and sterilizing equipments etc may be required at a poultry farm.

Summer Management of Poultry

The most difficult and stressful period for the birds is summer. The adult birds suffer more in hot weather than in the cold weather. The bird gives maximum egg production at 55°F and this temperature is regarded as ideal. Egg production starts declining as the temperature starts rising. With the increase in temperature, the feed consumption starts declining and so production declines. The upper lethal temperature of birds is 117°F.

Reduced feed intake, increased water intake, poor feed conversion efficiency, drop in egg production, reduction in egg size, increased cracked eggs, poor shell quality, panting, prostration and finally mortality are the effects of heat stress in birds.

Management Points to be Followed for Preventing Birds from Heat Stress are

1. East-West orientation of house to avoid direct sunlight.
2. Sufficient distance between two houses (20-30 meter) for proper ventilation and air flow.

3. Construction of house under tree shed.
4. Roofing at higher level (2.6-3.3 meter from the floor) and should be properly insulated by using straw/thatch or white washing of top of the roof with aluminium paint.
5. Increase in floor space by 10-15% to reduce overcrowding.
6. Sprinkler on the roof top can reduce shed temperature by 5-10°F.
7. Foggers may be installed in the house but operated only when the humidity is below 70% because it increases the humidity of house.
8. Exhaust fans can be used.
9. Green surrounding round about the sheds helps in reducing heat stress.
10. Availability of clean drinking water all the time.
11. Feeding of birds during cooler parts of the day.
12. Provision of wet mash during summer days.
13. Feeds with marginal increase in proteins, minerals & vitamins and decrease in metabolizable energy.
14. There should be provision of glucose water (8g glucose with 2 g electrolytes) or 50 g of cane sugar molasses in 1 liter of water.
15. Vitamin-C @ 10 mg/bird/day as an anti stress medication.

Winter Management of Poultry

Winter season is less problematic compared to summer season. Following are some of the management points to be adopted for preventing loss due to winter stress-

1. To prevent chilliness gunny bags should be fixed on the wire netting of the sidewalls, specially during night time.
2. During the foggy days light can be provided at day time.
3. Proper cross ventilation must be maintained.
4. Feed of poultry should contain high energy with low level of proteins.
5. Slight increase in Vitamin C, A and B complex to prevent illness.

6. Frequent feeding should be done instead of once or twice feeding a day.
7. Lukewarm water should be provided for drinking purpose.
8. Thickness of litter should be increased to give warmth to birds.
9. Dampness of litter is reduced by frequent stirring.
10. Proper preventive measures must be taken to control the outbreak of disease during winter season.

Feeding and Watering of Birds

Digestive system and digestion in chicken: Unlike ruminants, the digestive system of chicken consists of simple stomach. Teeth are absent and chewing does not take place. In chickens the rate of metabolism is higher than other farm animals. The feed is picked up by horny beak and tongue, the ingested feed goes to the crop which is distended pouch in the oesophagus. The main stomach is divided in to proventriculus and gizzard. The proventriculus is rich in secretory cells and producing HCl and pepsin while the gizzard helps in grinding of feeds with the help of thick muscular wall. Part of small intestine i.e., duodenum and ileum are similar but shorter in chickens when compared to other farm animals. At the junction of ileum and colon (large intestine) there is presence of two blind sacs called ceaca. Both the ceaca functions as fermentation chamber. The last part of the digestive system i.e., cloaca has large storage chamber for urine and faeces, the coprodium. Two main important accessory organs help in digestion are pancreas and liver.

Different types of enzymes and acids are secreted from the parts of digestive system of chicken's and help in digestion are- Mouth (Amylase), Proventriculus (Pepsin, HCl), Gizzard (Koilin), Duodenum (Secretin, Enteropeptidase, Pancreozymin), Pancreas (Amylase, Trypsin, Chymotrypsin, Carboxypeptidase, Lipase) and Liver (Bile).

All disaccharides except lactose are digested in the digestive system of chickens. Lack of enzyme lactase prevents digestion of lactose in chickens. Chickens mostly derive their energy from starch as they cannot digest complex carbohydrates like cellulose, hemicelluloses and lignin. Carbohydrates in the form of monosaccharides are absorbed into the blood of portal venous system from the intestine. Because of acidic reaction the fat digestion is almost negligible in proventriculus and gizzard. Pancreatic lipase and bile present in the intestine are responsible for digestion of fat. The fats are digested in the path way from triglycerides to 1, 2-diglycerides to 2-monoglycerides and finally to free fatty acids and glycerols. The end products of protein digestion are amino acids which are readily absorbed from the intestine.

Feeding norms for different species of poultry: Since major expenditure (60-70 %) in poultry rising is feed cost, the feeding constitutes the fundamental and major management concern in poultry production. One of the key factors for successful poultry production is efficiency in feeding. More than 40 nutrients are required by the poultry which can be arranged in 6 classes' viz., water, proteins, carbohydrates, fats, minerals and vitamins. The principle sources of energy are carbohydrates and fats. Fats yield 2.25 times more energy than carbohydrates as they are concentrated form of energy. The major sources of essential fatty acids are linoleic acid, linolenic acid and arachidonic acid. The required amino acids by the body are supplied by protein. The essential amino acids for poultry are: arginine, glycine, histidine, lucine, isolucine, lysine, methionine, cystine, phenylalanine, threonine, tryptophan and valine. Out of all, arginine, lysine, methionine. cystine and tryptophan are critical amino acids in the diet of poultry. Minerals and vitamins regulate several essential metabolic processes in the body but do not supply energy. Calcium, Phosphorus, Sodium, Copper, Iodine, Iron, Manganese and Zinc are critical minerals needed in poultry ration. The most important vitamins needed for poultry are Vitamin- A, D_3, B_2 and K although almost all the vitamins are nutritionally essential in poultry ration.

Classification of Poultry Feeds

The feeds of poultry can be classified into following groups-

1. **Energy rich feeds:** The chief source of energy in poultry ration is provided by cereal grains and their by-products, fats and certain other carbohydrate feed stuffs. Maize, jowar, barley and bajra are some of the feeds generally used as carbohydrate source in feed.

2. **Protein rich feed:** Animal and plant source are the two types of protein rich feeds. Fish meal, blood meal and meat meal are some of the major source of animal proteins while plant protein source includes groundnut oil cake, coconut oil cake and soybean meal.

3. **Fat rich feed:** Fat of pig (Lard) and fat of cattle/buffalo (Tallow) are some of the common source of fat to be incorporated in poultry feed. Sometimes vegetable oils can also be used as source of fat.

4. **Minerals:** The minerals are available in larger proportion in animal protein as compared to vegetable protein. Lime stone, oyster shell, bone meal and dicalcium phosphate are some of the major source of minerals.

5. **Vitamins:** Vitamins are available in higher proportion in green leafy fodders such as lucerne, berseem and groundnut plant meals.

Feed Ingredients in Poultry Ration

Maize, barley, oat, wheat, wheat bran, rice, rice polish, sorghum, GNC, fish meal, lime stone etc. are conventional poultry feeds while tapioca meal, dried poultry waste, molasses, mustard oil cake, soyabean meal, sesame meal, sunflower seed meal, cotton seed cake, linseed meal: liver residue meal, silk worm pupae meal, hatchery by-product meal, feather meal, meat and bone meal are some of the unconventional feed ingredients for poultry ration.

Nutrients Requirement for Different Age Groups

The nutrient requirement of poultry depends on type and age of poultry. Further the requirement varies on the basis of purpose for which the birds are raised. Table 13.5 represents the BIS (1992) recommendations for most important nutrients requirement by different age group of poultry.

Table 13.5 : BIS specifications for nutrients requirement by poultry.

Nutrients	Layer starter	Grower feed	Layer feed	Breeder feed	Broiler starter	Broiler finisher
Crude Protein (%)	20	16	18	18	23	20
Crude Fibre (%)	7	8	8	8	6	6
Ash (%)	4	4	4	4	3	3
Salt (%)	0.6	0.6	0.6	0.6	0.6	0.6
Calcium (%)	1	1	3	3	1.2	1.2
Phosphorus (%)	0.5	0.5	0.5	0.5	0.5	0.5
Metabolizable energy (Kcal/Kg)	2600	2500	2600	2600	2800	2900

Feed Formulation

Feed formulation can be done with the help of computer by least cost method using linear programming. The steps to be followed for formulation of poultry feed are-

1. Calculation of nutrient required in ration.
2. Selection of ingredients on the basis of cost and their nutritive value.
3. Toxic principles present should take care with regard to its maximum inclusion level.
4. Selected ingredients are added to formula to provide nutrient per 100 kg.
5. 5, 10 and 8-10% of the total feed should be reserved for additive and supplements, animal protein and cereal by-products, respectively.
6. To meet the desired ME level, energy source to be included.
7. Checking of Ca, P, Na, crude fibre and limiting amino acid in the ration.
8. Finally the total ration should be checked for different nutrient to be required.

Feed Restriction

Feed restriction is mostly applied in broiler breeding stock but can also be used for layer breeding stock. For getting good egg production with higher fertility rate, the layer should be kept in such a condition that there is no fat accumulation in the body of birds. For this the weight of the bird is controlled by feed restriction. For feed restriction during the growth period of 9-22 weeks the birds are offered 70-75% of *ad libitum* feeding while at laying period the birds are offered 90% of *ad libitum* feeding. There are many advantages of feed restriction such as better egg size, uniformity in flock, better livability, less problem of fertility and hatchability, reduces mating difficulties etc.

Water Requirement in Poultry

Poultry needs constant supply of clean drinking water as the water retaining capacity is very limited. Birds use to drink water freely and frequently. On an average a bird takes about 250ml of water per day. In large scale poultry production, automatic waterer is used. Type of feed, size of the birds and environment plays important role in consumption of water by birds. 20-30 litres of water is consumed per day on an average. A pH of 6 to 8 is preferred for broiler production, but birds can tolerate a pH range of 4 to 8. A pH range of greater than 8 could cause reduced water consumption. The strength of hardness, calcium, chloride, copper and iron in drinking water of poultry is allowed to the limit of 110, 500, 250, 0.6 and 0.03 ppm, respectively. Bird's health is not affected at iron level as high as 660 ppm, but it will have detrimental effects on water lines and fogging systems.

Poultry Health Care Management

Poultry farming is nowadays taking the shape of industry. The diseases affecting poultry are important constraints for its development. For getting continuous production of meat and egg to the optimum level, it is important to keep the birds in healthy conditions. Bacteria, virus, fungus and parasites leads to different disease conditions in poultry. Table 13.6 represents the diseases of poultry with causative agent, important symptoms and their preventive measures.

Table 13.6 : Diseases of poultry with their causative agent, symptoms and prevention.

Name of disease	Causative agent	Important symptoms	Preventive and control measures
Bacterial diseases			
Pullorum disease	*Salmonella pullorum*	Drooping of wings, anorexia, respiratory disturbances, profuse whitish diarrhea with pasting of the vent region	Proper hygiene and management, regular screening and elimination of reactors
Fowl typhoid	*Salmonella gallinarum*	Drop in egg production, watery to yellowish diarrhea, purple discolouration of the comb and wattle	Smililar to pullorum disease,Vaccination with killed vaccine
Colibacillosis	*Escherichia coli*	Respiratory disturbances, loss of appetite, diarrhea and weakness	Proper hygiene and management

Contd.

Infectious coryza	*Haemophilus paragallinarum*	Depression, submucoid nasal discharge, conjunctivitis, facial oedema, swollen wattle, coughing and gurgling	Vaccination simultaneously with killed ranikhet disease, treatment with antibiotics
Fowl cholera	*Pasteurella multocida*	Anorexia, marked depression, mucus discharge from the mouth and foetid diarrhoea	Antibiotic treatment after sensitivity test, clean and disinfection of building and equipments
Viral diseases			
Avian influenza	*Orthomy -xoviridae family*	High mortality, rapid drop in production, fever, sinusitis, respiratory signs, excessive lacrimation, oedema etc.	Secondary spread of disease in avoided by destroying incontact and sick birds
Ranikhet disease	*Paramyxovirus*	Four forms of disease are- Doyl,s, Beach,s Beaudette,s form and Hitchner,s form. Important symptoms are greenish blood stained diarrhoes, tremor, torticollis, paralysis of legs and wings, lower egg production etc.	Proper hygiene and management, live vaccines- lentogenic and mesofenic vaccines.
Infectious bursal disease (IBD)	*RNA Virus*	Ruffled feather, watery diarrhea, soiled vent feather, depression and anorexia.	Proper hygiene and management practices, live vaccines are given in drinking water during 3rd week of age
Fowl pox	*Fowl pox virus*	Mild or focal cutaneous lesions, nodular proliferative legions are seen on comb, wattle, legs, feet, cloaca and around the eye.	Recovered birds attain immunity for life long, live pigeon pox or fowl pox vaccines are given, tissue culture vaccine is also given

Contd.

Marek's Disease	*Herpes virus*	There are four forms of disease are- classical form (paralysis and thickness of sciatic nerve, torticollis), visceral form (lymphoid nodular tumours may affect the gonads, liver, lung, kidney, spleen etc.), ocular form (blindness-fish eye or pearly eye or gray eye) and cutaneous form (nodular lesions up to 1 cm in diameter)	Vaccine derived from all three serotypes are available.
Parasitic diseases			
Coccidiosis	*Eimeria species*	Haemorrhagic lesions in the intestine, soft mucoid and pink coloured faeces, dysentery is usually seen	Proper hygiene and management, vaccination with all the species of *Eimeria*
Mycotoxicosis			
Aflatoxicosis	*Aspergillus flavus*	Effect depends on types of birds affected, Ducklings are highly susceptible followed by turkey poults, adult ducks, adult turkey, chicken and quails.	Prevention of growth of fungus in feeds and stores, addives can be used to reduce the toxicity

The diseases caused due to deficiency of vitamins and minerals are given in Table 13.7.

Table 13.7 : Diseases of poultry caused due to deficiencies of vitamins and minerals.

Vitamins/Mineral	Deficiency symptoms
Vitamins	
A	Nutritional roup
Thiamine (B_1)	Polyneuritis
Riboflavin (B_2)	Curled toe paralysis
Pyridoxine	Poor growth, convulsions
Pantothenic acid	Crustations at corners of mouth and feet, fatty liver
Biotin	Crustations at the bottom of feet
Folic acid	Poor growth & feather development, sitting on hock (perosis)
Cyanocobalamine (B_{12})	Very poor growth
D_3	curved legs, leg weakness, thin shelled eggs, reduced egg production
Tocopherol (E)	Crazy chick disease, muscular weakness
Phyloquinone (K)	Slow blood clotting (Haemorrhages)
Minerals	
Calcium	Soft bones, poor growth, thin shelled/shell-less eggs
Phosphorous	Poor growth, soft bones
Magnesium	Muscle spasms, slow growth, lethargy, loss of appetite
Sodium and Potassium	Poor growth of cardiac and respiratory muscles
Manganese	Enlargement of joints, staggering gait
Zinc	Poor growth and feathering, shortening of leg bones
Iron & Copper	Anaemia
Iodine	Impaired body response, lowered activity
Selenium	Muscular dystrophy, poor immune response
Cobalt	Poor growth

Vaccination Schedule for Poultry

Vaccines are the weakened organisms, dead organisms or individual proteins that are introduced into human/animal body to give an immune response that mimics a natural infection without causing disease. The vaccine is a prophylactic measure and available against most of the poultry diseases. Vaccination schedule for broilers and layers is given the Table 13.8 and 13.9, respectively.

Table 13.8 : Vaccination schedule for broilers.

Age	Disease	Type of vaccine	Route of vaccination
Day old	Marek,s disease	Attenuated Marek,s disease vaccine	S/C (In hatchery itself)
7-8 days	Ranikhet disease (RD)	F-strain	Intra nasal/ Intra ocular
14-15 days	IBD	Live vaccine	In drinking water

Table 13.9 : Vaccination schedule for layers.

Age	Disease	Type of vaccine	Route of vaccination
Up to 15 days same as broilers			
5-6 weeks	RD	F-strain	Oral drops
7-8 weeks	Pigeon pox	Live virus vaccine	S/C, Feather follicles
9-10 weeks	RD	R_2 B strain	S/C or I/M
15-16 weeks	Fowl pox	Live virus vaccine	I/M
16-18 weeks	RD	R_2 B strain	S/C or I/M

Precautions During Vaccination

1. Always fresh vaccines should be used and two different vaccines should never be mixed to save time.
2. Cold chain should be strictly maintained while transporting the vaccines.
3. Only healthy birds should be vaccinated.
4. Vaccination should never be done when there is outbreak of disease.
5. Vaccination causes stress in the flock so anti-stress medicines like vitamins, antibiotics etc should be used three days before and after the vaccination.
6. All the birds of a flock should be vaccinated at one time.
7. Vaccination should be done in the cooler part of the day, preferably in morning or evening time.
8. Proper vaccination record should be maintained.

Poultry Bio-Security

In poultry farming, bio-security simply means, safety from transmissible infectious diseases, parasites and pests and that embodies all of the measures that can or should be taken to prevent virus, bacteria, fungi, protozoa, parasites, insects, rodents and wild birds from entering or surviving and infecting or endangering the well being of the poultry flock.

Bio-security is a principle which stresses isolation of healthy birds in order to protect them from any microbial infection i.e. bacterial, viral, fungal and parasitic. Isolation of healthy birds is just as important as quarantine of diseased poultry flocks. This helps in getting more weight from broilers, low feed conversion and diseased number of birds condemned at slaughter.

Poultry Judging

Judging of commercial layers for egg production: For improving the efficiency of egg production, periodical judging or culling in commercial farms is needed. Judging of a good layer is based on certain extrinsic characters and related measurements expected to indicate a laying bird. Accordingly, layers can be graded based on the following score card (Table 13.10).

Table 13.10 : Score card for judging a layer.

Sr. No.	Phenotypic appearance	Points
1	Health and vigour of bird	25
2	Body conformation and abdominal capacity	25
3	Moulting and plumage	15
4	Pigmentation	10
5	Condition of head, wattle, vent and others	25
	Total	**100**

Judging of Broilers for Meat Production

Broilers are judged on the basis of phenotypic appearances which are having importance with regard to meat production. Table 13.11 shows the points to be given for different phenotypic appearances for judging of broilers.

Table 13.11 : Score card for judging a broiler.

Sr. No.	Phenotypic appearance	Points
1	General appearance	10
2	Health and vigour	15
3	Breast	30
4	Back and neck	15
5	Leg and thigh	15
6	Feathering	15
	Total	**100**

Culling of Layers

The layers are judged for culling on the basis of different characters which are discussed in Table 13.12.

Table 13.12 : Judging for poor production in chickens.

Characters	Good layer	Poor layer
i. Eyes and expression	Bright and alert	Sunken and dull
ii.Comb	Large, red, full, glossy, warm	Small, pale, scaly, cold
iii. Wattle and ear lobes	Prominent, soft, smooth	Inconspicuous, rough, dry
iv. Feathering	Tight and close feathering, tail and wings well carried up	Loose and scattered feathering, tail and wings are droopy
v.Plumage	Initially bright and dark coloured but faint and dull in latter stage	Always bright and dark coloured
vi. Vent	Large, dilated, moist	Small, round, dry
vii. Pubic bones	Wider apart (usually about 3 fingers)	Close together (1 to 2 fingers)
viii. Abdominal cavity	A depth of 4-5 fingers from the end of the keel to pubic bones	A depth of 2-3 fingers
ix. Skin	Soft, thin, loose, silky	Rough, thick, tight and under laid with fat

Moulting and Depigmentation

Moulting is an important condition in layer bird which helps in deciding good or poor layers. Moulting is a natural physiological process for the birds to renew old feathers at the end of first year of laying. Moulting process starts from the head and feathers fall in the order of head, neck, breast, body, wings and finally tail. The time of start of moulting and time taken in whole process is the deciding factor for good or poor layer. Early moulters are usually poor layers and they take more time to complete the moulting process (24 weeks or so) and as a result they give less egg during one laying cycle because often they stop laying during moulting process. Good layers start moulting late and complete the process in about 1-2 months and continue to lay eggs.

The yellow carotenoid pigments (xanthophylls) coming from feed are deposited in the different body parts of pullets. As the pullet starts laying the pigments are transferred from the body to the eggs. The depigmentation starts according to blood circulation in different parts of the body. The bleaching of the body is indicative of laying capacity of the layer. The order of depigmentation is vent, eye ring and ear lobe, beak, shank (bottom, front, entire).

Vices in Poultry and its Remedial Measures

The habits which are not normal and affect the economics of the farming is called vices. Many birds use to suffer from certain vices which become quite extensive right from the beginning and their eradication becomes a big problem. The vice generally causes huge loss to the poultry farmer. Cannibalism, Egg eating, Egg hiding, Pica etc. are some of the important vices seen in case of poultry.

- **Cannibalism:** The habit of attacking by birds to the flock mate and eating their flesh is called cannibalism. In laying birds, vent picking is a common vice. Presence of wound in the body of otherwise healthy birds is an indicative of cannibalism. Overcrowding, genetic predisposition, haemorrhages in the external genitalia, protein deficiency, deficiency of salt and minerals are some of the important causes of cannibalism. Debeaking is the way to avoid cannibalism which can be done either manually or mechanically with the help of electric debeaker. In debeaking, one-third of the upper beak and tip of the lower beak is to be trimmed. Birds involved in cannibalism and those birds which are wounded must be segregated immediately. Overcrowding should be reduced to required level. Birds should be provided balanced ration with required vitamins and minerals.
- **Egg eating:** Birds sometime develop the tendency to eat their own eggs. Presence of cracked eggs, soft egg shell, lack of

sufficient bedding material and presence of eggs for longer period in the pens are some influential factors for this vice. Isolation of affected birds, increase in quantity of lime and protein in diet, keeping birds in cage, debeaking and frequent egg collection can reduce the effect of this vice to a considerable limit.

- **Egg hiding:** This vice is generally not seen in case of domestic fowl, but its seen in jungle fowl as it is a maternal instinct in these birds. Birds use to hide the eggs in field and bushes etc. This vice can be prevented by restricting the movement of birds. Laying area should be inside the house to protect eggs from hiding.
- **Pica:** It is a vice caused due to deficiency of phosphorus in the diet. Unusual materials like feathers, litter material, threads, etc taken by the affected birds. This vice can be cured by providing good care and balanced diet to the birds.

Poultry Farm Records

In any business the record keeping is a crucial activity which serves the important source for monitoring and evaluation. The records should be simple, easy to understand, without repetition; but must provide all the needed information. To study the production performance and check whether the firm is in profit or loss condition, record keeping a very important. Two types of records are production/ technical records and financial records. Feed and feed ingredient register, feed additives and medicines register and purchase/ expenditure register, egg production register, egg turn out register and attendance register of labours are some of the important registers maintains in the poultry farm.

Marketing of Poultry and Poultry Products

There is no well organized market in India for marketing of poultry and its product. Producer to consumers, producers via retailers to consumers, producers via assemblers to consumer and co-operative societies/ egg or poultry meat and other product markets organizations are few distributor and marketing channels in India. Relatively shorter channels should be selected for egg and raw meat distribution.

Only 10% of the broilers produced in the country are processed in the dressing plant as the poultry processing sector is highly underdeveloped in India. About 90% remaining broilers are either sold alive or freshly dressed. Traders and commission agents handle the marketing of eggs. The price of egg is fixed by National Egg Co-ordination Committee (NECC) all over India. To assist the farmers in the marketing of egg, number of co-operatives societies has also been established in several states. Procurement of eggs, grading, processing, packing, storage, transport and distribution are some of the tasks involved in egg marketing.

Some Organized Channels of Marketing are

- Producer - Consumer
- Producer - Collector–Assembly Merchant - Wholesaler - Retailer - Consumer
- Producer - Wholesaler -Retailer - Consumer
- Producer - Collector- Assembly Merchant - Consumer
- Producer - Retailer - Consumer

Important Terminologies

AGMARK : Short form of Agricultural Marketing.

Axial feathers: Soft feathers in the middle of the wing.

Blind check: Egg with crack in shell, detected only by candling.

Bloom: A layer of protective coating on external surface of the egg. It is also called cuticle.

Broiler/Fryer: These are immature young chicken of either sex which is soft, pliable and tender of about 6 weeks of age suitable for meat purpose.

Broodiness: Desire of hen to sit on eggs.

Candling: It is the process of visual examination of an intact egg against a light for its internal quality.

Chick: Young one of chicken.

Clutch: Number of eggs laid on consecutive days. It is generally 8-13 eggs.

Cock: Male of chicken.

Croiler: It is a colour feathered broiler.

Culling: It is the process of elimination of unproductive or undesirable birds from the flock.

Debeaking: It is the process of cutting of the part of upper beak of birds.

Down feathers: Initial hairy covering of the body of chicks.

Drake: Male of duck.

Dubbing: Removal of comb and wattle.

Duck: Female of duck.

Ducklings: Young one of duck.

Fluff feathers: Soft downy feathers around the thigh.

Fumigation: It is a method of disinfection using disinfecting gases especially Formaldehyde gas.

Gander: Male of Goose.

Goose: Female of goose.

Gosling: Young one of Goose.

Grit: These are the hard insoluble materials fed to birds which provide a grinding material in the gizzard.

Hen: Female of chicken.

Incubation: It is the process of providing suitable warmth, humidity and turning movement to eggs which facilitate growth and development of hatching of an egg and finally hatching takes place.

Keel bone: Breast bone of birds.

Keet: Young one of Guinea fowl.

Layers: Mature female fowl from 21-72 weeks which has been kept for egg-laying purpose.

Moulting: It is the process of renewing the feathers at the end of laying cycle.

Oviposition: Process of laying (expelling) of a fully developed egg through the cloaca is called oviposition.

Pause: Any cessation of egg laying for a period of 7days or more is called pause.

Pipping: The act of pecking the shell and shell membranes by the chick in order to break the shell and hatch out of the shell is called pipping.

Poult: Young one of Turkey.

Rales: Abnormal respiratory sounds.

Sickle feathers: Long curved feathers of male tail.

Signet: Young one of Swan.

Snood: A fleshy protuberance hanging from the base of the beak in Turkeys.

Spur: A projection of horny substances on the shank of males and sometimes of female.

Squab: Young one of Pigeon.

Stub: Short, partly grown feathers.

Torticollis: Twisted or wry neck.

Vaccine: A preparation of micro-organisms (killed or living) which when given to the bird produces or increases immunity against specific disease.

Wattles: The flap of skin under the chin of a chicken or turkey.

Chapter - 14

Climatology and Animal Production

Virendra Kumar Singh

The biotic component of earth is continuously and significantly affected by the abiotic components. Physical factors of the environment (temperature, humidity, rainfall, wind, solar radiation, light, atmospheric pressure, etc) surrounding the animal have influence on the production, reproduction and welfare of the animal. Livestock performance is best when environmental factors are optimum which is difficult to achieve due its variable nature. To this complex livestock-climate interaction, animal responds favorably by undergoing 'adaptation' or unfavorably by being under 'stress'. Type and magnitude of stress due to physical environment depends on climatic factors of a particular area, animal adaptability, breed characteristics and managemental interventions by humans to minimize the losses. To understand the physical environment of livestock frequently used terms are mentioned below

- **Climatology**: The scientific study of the climate of a place or region on the basis of weather records accumulated over a long period of time.
- **Bioclimatology:** The study of climate in relation to living organisms.
- **Meteorology**: The study of an interdisciplinary science involving integrated system of atmosphere, land and ocean for observation, understanding and prediction of weather.

- **Climate:** The long term (about 30 years) average of meteorological variables in a given region.
- **Microclimate**: The climatic condition directly surrounding the animal.
- **Macroclimate**: Climatic condition of a large geographical area like state or country e.g. climate of countries near equator is tropical.
- **Mesoclimate**: Climate of an area which is different from the climate of a larger area which smaller area is a part of. e.g.: Climate of state Jammu & Kashmir is somewhat cooler but its capital city Jammu is comparatively warmer.
- **Weather:** Short-term day to day fluctuations in meteorological variables (as distinguished from climate which is long-term manifestation of weather) *viz.* daily/weekly weather reports and forecasts.
- **Air temperature or ambient temperature**: Temperature of the air surrounding a body or an object irrespective of the maximum or minimum temperature of the day.
- **Maximum and minimum temperature:** In a given time duration, highest air temperature is known as maximum temperature and lowest air temperature is known as minimum temperature. Maximum (mostly observed during afternoon/ late afternoon) and minimum temperature (mostly observed during morning/early morning) are commonly measured for a period of 24 hours.
- **Humidity:** It is the amount of water vapor present in the atmosphere. It is expressed in two ways-*Absolute humidity* which is actual mass of water contained in a unit volume of moist air and *Relative humidity* which is the ratio of the actual vapor pressure of the given air to the saturation vapor pressure (or ratio of actual water vapor content to the maximum water vapor content which can be present in the given air) at a given temperature.
- **Dew point (dew-point temperature)**: The temperature to which a volume of air should be cooled to saturate it keeping pressure and water-vapor content constant. When this temperature is below 0 °C, it is called the *frost point.*
- **Stevenson screen**: Stevenson screen (instrument shelter or thermometer screen) is a part of weather station which protects

meteorological instruments especially temperature and humidity measuring instruments (kept in it for recording) from rain, snow, winds and animals. Stevenson screen maintains uniform environment and air circulates freely inside. It is painted white to reflect heat radiation. Guidelines for its construction and exposure are specified by the World Meteorological Organization.

- **Greenhouse effect & global warming**: The greenhouse effect is a process by which thermal radiation from a planetary surface is absorbed by atmospheric greenhouse gases (water vapor, carbon dioxide, methane, nitrous oxide, ozone) and is again radiated in all directions. Since part of this radiation goes back towards the surface and the lower atmosphere, it results in an elevation of the average surface temperature above what it would be in the absence of the gases. Green house gases differ in their heat trapping capacity which is known as Global-warming potential (GWP) i.e. relative measure of how much heat a greenhouse gas traps in the atmosphere. It compares the amount of heat trapped by a certain mass of the gas to the amount of heat trapped by a similar mass of carbon dioxide (GWP of CO_2 =1, GWP of Methane = 20 times that of CO_2, GWP of Nitrous Oxide = 300 times of CO_2). GWP of chlorofluorocarbons, hydrofluorocarbons, hydrochlorofluoro carbons, perfluorocarbons and sulfur hexafluoride are very high. Global warming primarily causes rise in global temperatures apart from melting glaciers and rising sea levels. It also increases the frequency of extreme weather events. However, in livestock the effects of global warming may be loss of habitat and food, altered hibernation, breeding, and migration patterns, increased stress and emergence of new diseases.

Classification of World Climate

For classification of climate 'Koppen' (German climatologist) system or its modified form is most commonly used which is based on monthly rainfall and temperatures as described below:

1. Average monthly temperature of the warmest month
2. Average monthly temperature of the coldest month
3. Average thermal amplitude between the coldest and warmest months

4. Number of months with temperature exceeding 10°C
5. Winter and summer rains

According to this system a three letter code is allotted to different climatic zones as follows : 1st letter from A,B,C,D,E; 2nd letter from S,W,T,F,M, s, w and 3rd letter from a, b, c, d, h, k. For e.g. *'Tropical rainforest* which is *hot, rainy through all seasons' is coded* as 'Af' and *'Humid continental which is severe winter; dry winter; warm summer'* is coded as Dwb. Nowadays Koppen-Geiger climate classification is frequently used as follows:

Table 14.1.1 : Koppen-Geiger climate classification.

Main climates	Precipitation	Temperature
A : Equatorial	W: desert	h : hot arid
B :Arid	S : steppe	k :cold arid
C :Warm temperate	f :fully humid	a :hot summer
D : Snow	s :summer dry	b :warm summer
E :Polar	w :winter dry	c :cool summer
	m :monsoonal	d : extremely continental
		F : polar frost
		T :polar tundra

Source : Kottek *et al.* (2006).

Climatic Regions of India

Indian climate may be broadly defined as tropical monsoon type which comprises 4 main seasons 'winter' (January-February), 'hot summer' (March-May), 'rainy south-western monsoon' (June-September) and 'north-eastern monsoon' (October-December). Specific variations are seen in higher altitude regions like Kashmir valley which has typical temperate climate and Ladakh, Lahul and Spiti, which have a typical cold-arid desert climate. Indian agro-climate is classified on the basis of moisture index (Thornthwait), thermal index (Thornthwait) and under National Agricultural Research Project (NARP). India has 15 agro-climatic zones and 21 agro-ecological regions mentioned in chapter 1 of this book.

Table: 14.1.2 : Commonly used instruments in meteorological/climatic study.

Climatic variable	Derivatives measured	Instruments	Units
1.Atmospheric temperature	Maximum temperature, minimum temperature, average temperature.	Mercury-in-glass thermometer (Maximum thermometers), Liquid-in-glass thermometers (Minimum thermometers), bimetallic thermographs, electrical thermometers semiconductor thermometers, thermocouples, Six's thermometer (combined maximum and minimum).	Degree Celsius (°C), degree Fahrenheit (°F), Kelvin (K).
2.Soil temperature	– –	Soil thermometer (mercury-in-glass thermometer).	Smaller depths (bent stem thermometer; 90° bent and 120° bent), deep soil thermometer.
3.Atmospheric humidity	Absolute humidity. Relative humidity. Specific humidity. Dry bulb, wet bulb, dew point temperatures.	Hygrometers (hair hygrometer, electrical resistive and capacitive hygrometers), Psychrometer (whirling sling psychrometer, assmann aspirated Psychrometer, screen psychrometer, heated psychrometer), Dew cell (lithium chloride heated condensation hygrometer).	Absolute humidity – grams (water vapor) / kg (total air). Relative humidity- %
4.Atmospheric pressure	– –	Mercury barometer, Aneroid barometer, Digital piezoresistive barometers, cylindrical resonator barometers, reading electronic barometers, Bourdon-tube barometers.	Pascal (Pa) , newton per square metre), mm of Hg.
5.Radiation	Direct solar radiation.	Absolute pyrheliometer Pyrheliometer	Radiant energy (Q) in joule =W s, radiant flux (Φ) in W

Contd.

		Spectral pyrheliometer	
		Sun photometer	
	Global (solar) radiation, diffuse sky (solar) radiation, reflected solar radiation.	Pyranometer, Spectral Pyranometer, Net pyranometer	
	Total radiation.	Pyrradiometer	
	upward long-wave radiation, downward long-wave radiation .	Pyrgeometer	
6.Precipitation (Rainfall)	– –	Rain gauge, udometer.	Millimeter
7.Drop size distribution and velocity of falling hydrometeors.	– –	Disdrometer	
8.Wind	Wind speed	Anemometer	m/sec or knots
	Wind direction	Wind vane	South, north, east, & west
	Wind speed and direction (upper atmosphere).	Theodolite balloon	Same as for wind speed & direction
	Air velocity in mines, cooling power of the environment, personal comfort of workers.	Kata thermometer	watt/m^2
	Wind profile by scattering of sound waves by atmospheric turbulence.	SODAR (Sonic Detection And Ranging)	– –
9.Sunshine	Solar radiation.	Campbell-stokes sunshine recorder (burn method), pyranometer, actinometer, pyrheliometer, solarimeter.	"hours per day" or "daily sunshine hours"

Contd.

10.Evaporation	Evaporation, Evapotranspiration.	Atmometers , lysimeter (plants), open pan evaporimeter, hook gauge evaporimeter.	Millimeters/day
11.Snowfall	Snow cover extent,Snow cover depth.	Direct snow depth measurement, direct snow water equivalent, snow pillows, radioisotope snowgauges, natural gamma radiation.	Extent: km^2, Depth: cm
12.Light	Luminous flux	– –	lumen
	Luminous intensity	Lux meter.	candela
	Luminance	Luminance meter, colorimeter.	Candela/$metre^2$
	Illuminance	Lux meter.	Lux
13.Cloud	Cover	Ceilometers.	Cover-oktas, height-ft, km
	Height		
14.Altitude, direction, and velocity of clouds		Nephoscope.	-------
15.Visibility	Suspended particulates in a liquid or gas colloid.	Nephelometer, transmissometer.	-------

Table 11.1.3 : Classification of Indian Climate based on evapotranspiration.

Climate type	Region and state
1. Arid	West Rajasthan
2. Semi-arid	Northern and eastern Rajasthan, Gujarat, Marathwada, Mysore, Rayalaseema, Punjab, Delhi, Western U.P.
3. Sub-humid	Central and eastern UP. Western and Central MP, Vidarbha, Bihar, Sub-mountain tract of UP, HP, West Bengal and Nilgiri.
4. Humid	Konkan, Kerala, coastal Madras, Assam, Orissa, West Bengal, Eastern MP and Coastal Andhra Pradesh.

Physical Environment and its Interaction with Animal

Meteorological variables like ambient temperature (minimum, maximum and average), humidity (absolute and relative), wind (speed and direction), solar radiation (duration and intensity), rainfall, atmospheric pressure, etc influences the energy exchange process between animal and its surrounding environment which results in either gain or loss of heat by the animal. Within certain limits the animal can tolerate energy loss or gain from its environment but beyond these limits (excess hot or excess cold) normal energy exchange process is disturbed. So, it is important to understand the mechanism of heat flow through the gradient between inner core of the animal's body, outer body surface of animal and its environment. Conduction, convection, radiation and evaporation are the processes through which energy is being continuously exchanged between an animal and its environment. For the survival of an animal in a given environment, over a longer duration of time net energy exchange (heat gain from all channels-heat loss from all channels) must be zero.

1. **Conduction:** In this process heat is transferred through a medium without movement of any particle/molecule of the medium i.e. heat transfer from a warmer to a cooler surface through direct contact. Heat transfer occurs through legs or body parts which touch the ground when lying down or when standing in water etc.
2. **Convection:** In this process, heat is transferred by movement of particle/molecule of the medium from a warm to a cooler object E.g. heat is lost when particles of surrounding layer of warm air next to the skin (air trapped by hair follicles) is replaced with cooler air. Air speed increases the effects of convective cooling, so animals will try to stand where there is a breeze or under fans.
3. **Radiation:** In this process heat radiates in the form of electromagnetic waves from a warmer object to cooler environment through space and heat flow by this process depends on temperature and surface of the radiating body. This explains the fact that heat gained or lost by the cow depends on the difference in temperature between the cow and the surroundings. Providing shade to the cow from direct sunlight reduces solar radiation exposure to half or lesser. Cows with black coat absorb more solar radiation than cows with lighter colored coats.

4. **Evaporation:** In this process heat is lost by the evaporation of water from the skin (sweating) or respiratory tract (panting). The animals which have higher capacity for sweating has lower capacity of panting. Sweating capacity of the domesticated livestock species are in the order of horse = camel > cattle > sheep = goat > pig > dog = cat > bird and reverse order is for their panting capacity implying that one who has more efficiency for sweating will have less efficiency of panting. Evaporation can only take place when humidity or water vapor content of surrounding air is less and it decreases with increasing level of humidity. There is no evaporation when relative humidity is 100%.

Based on the mechanism and effect of heat loss, it is categorized as sensible and insensible. *Sensible heat loss* increases the temperature of surrounding air which occurs by radiation, conduction or convection. *Insensible heat loss* (latent heat loss) does not increase the temperature of the surrounding air. If surrounding air temperature is less than core body temperature, then there is sensible heat loss (conduction, convection, radiation or combination of any of these) but if surrounding air temperature is higher than core body temperature then there is insensible heat loss (evaporative or latent heat loss from skin and respiratory tract).

Heat Load in Animals

A healthy animal normally tries to maintain the core body temperature within the optimal limits by balancing the metabolic and the absorbed environmental heat. If the sum of metabolic heat produced internally by the animal and the heat gained from the external environment begins to exceed that lost, the heat load develops in the body.

Heat	+	**Heat**	–	**Heat**	=	**Heat Load**
(Metabolic process)		*(Gained from external environment)*		*(Lost to external environment)*		

$$\mathbf{M - E \pm F \pm C_d \pm C_v \pm R = 0}, \text{ hence } \mathbf{M = E \pm F \pm C_d \pm C_v \pm R = 0}$$

Where **M:** metabolic heat production, **E:** heat loss from skin and respiratory passages by evaporation, **F:** heat loss or gained while bringing ingested food and/or water to body temperature, $\mathbf{C_d}$: heat lost or gained by direct contacts between the body and surrounding surfaces, $\mathbf{C_v}$: heat lost or gained by convection due to contact between the air and skin and/or lining of the respiratory passages, **R:** heat lost or gained by radiation.

Calculation of Heat Load

It is important to estimate the existing heat load or predict the risk of its occurrence in order to initiate steps to alleviate or prevent the adverse effect. In cows, estimation of heat load may be easily understood by following indications/hints but this approach can also be used for other species also. *Firstly,* the hot weather forecast with immediate previous exposure to cool weather (abrupt change) is likely to cause heat stress. Adverse effects of only high temperature are less severe than the combined effects of high temperature and high humidity. *Secondly,* the score of temperature humidity index and its correlation with thermoregulatory limits of cattle may hint any risk of heat stress in future. *Thirdly,* when the night temperature is not as low as expected and next day is predicted to be hot then a situation of heat stress may arise next day.

In a herd if in most of animals rectal temperatures and respiratory rates are higher and dry matter intake and milk production are lower than normal, then it means that either animals are suffering from heat stress or are likely to suffer after some time depending on the extent of increase or decrease.

A measure of heat tolerance or adaptability is quite useful to assess the animal's ability to cope with heat stress or to anticipate management practices for the herd to reduce heat stress. Heat tolerance or adaptability can be measured by following methods:

1. **Iberia heat tolerance test** (Rhoad, 1944): This test is mostly done for cattle which basically determines increase in rectal temperature above 101°F (38.3°C). It is performed by exposing the animals in the open area to the conditions of a calm clear day with temperature of shade between 85°F and 95°F (29°and 35°C) and measuring the rectal temperature at 10.00 and 15.00 hours on three consecutive days. Following formula is used to find out the heat tolerance

$$\text{Coefficient of Heat Tolerance} = 100 - 10\,(\text{BT} - 101)$$

BT = average body temperature based on rectal temperature readings (at 10.00 and 15.00 hours on three different bright sunshine days) with the average air temperature 85°F - 95°F . (Cattle held in open area during test period, with access to water, but having no shade); 101

= normal body temperature of cattle; 10 = factor to convert degrees deviation in body temperature from the normal to a unit basis; 100 = ideal value for perfectly maintaining temperature at 101° F.

The result is expressed as a percentage of maximal efficiency in maintaining rectal temperature at 101°F and hence the animal with the least rise in rectal temperature (and the lowest respiratory rate) is considered as most heat tolerant. If two or more animals have same coefficient of heat tolerance then the animal with higher respiration is less heat tolerant. Iberia heat tolerance test is useful, reliable and easy to perform test to differentiate heat tolerances of various breeds of cattle. Limitation of this test are that it does not account for environmental variations and variations in average rectal temperature value of 101° F due to age, breed, level of feeding, level of production, etc which sometimes may lead to falsely low figures for young animals and high figures for animals whose normal body temperature is below 101° F .

2. **Benezra's coefficient of adaptability (BCA)** (Benezra, 1954): This is based on body temperature and respiration rate responses after exposure for 7 hours on 3 consecutive days.

BCA = Body Temperature (in °C)/38.33 + Respiration rate per minute/23

If BCA = 2: Maximum Adaptability, If BCA > 2: Lower Adaptability

The limitation of this coefficient is that it gives equal weightage to body temperature and respiration rate which is not practically possible as 10% increase in body temperature may lead to death but respiration rate will still be only slightly high.

3. **Dairy search index** (Bonsma, 1949): It takes into consideration the rectal temperature, pulse rate and respiration rate. Here, rectal temperature, respiration rate and pulse rate of the animals is recorded at 07:00 and 15:30 hrs daily. Observations in the morning are taken before offering any feed and water and in the afternoon 30 minutes after animals return from grazing (to nullify effect of grazing activity). Adlib water is offered.

$$DI = 0.5\ X_1/X + 0.3\ Y_1/Y + 0.2\ Z_1/Z$$

Where X_1, Y_1 and Z_1 are temperature, pulse rate and respiration rate after exposure and X, Y and Z are the same parameters before exposure respectively.

measurement of parameters (under field conditions) which are indicative of an animal's water loss through various channels, namely: respiratory rate (index of moisture loss from the respiratory tract), loss of moisture (from selected skin areas); nitrogen concentration of the urine (index to water expenditure through the kidneys); moisture content of the feces (index to water expenditure through the excreta). Procedure involves measuring these parameters for certain time duration at regular interval while making water available throughout at surrounding temperature. Even though it emphasizes an important phenomenon of water loss and clear breed differences, its disadvantages are that it lacks any formula, complicated nature of test and non accurate measurement of water loss.

5. **R-Value/Heat tolerance test of Lee** (Phillips & Lee, 1948): It is also considered as laboratory equivalent of Iberia heat tolerance test. Its procedure consists of exposure of the animal to a series of atmospheric conditions at different known temperatures and humidity. Simultaneously, rectal temperature is measured regularly and an expression of its mean behavior, called R-value is calculated. The R-value denotes the area between the x-axis and the curve in a time-rectal temperature plot. From R-value for the different combinations of temperature and humidity, the values of the constants a, b and c were calculated and used in following formula:

$$\boxed{\text{Log } R = aT + bH + c}$$

Where T is temperature and H is humidity, values of a, b and c obtained upon two different animals indicate the relative rise pattern of rectal temperature in response to atmospheric temperature for a given humidity.

Advantage of this test is that it is done in controlled environment and an integrated animal response and similar procedures can be adopted with respect to respiratory rate, pulse rate, etc. However disadvantage is that animal showing too rapid rise in body temperature has to be removed from the test environment before the 7 hours of exposure.

6. **Lines of equal effect** (Barrada, 1957): In this the lines of equal effect are drawn empirically by recording separately rectal temperature, respiratory rate and respiratory volume of cattle with reference to different combinations of wet and dry bulb

temperature in a temperature-vapor pressure diagram. Lines change their slope as one passes from cooler conditions to hot conditions in the same way as those obtained for humans. It is advantageous in terms that it is very useful concept for the description of the combined effect of temperature and humidity on physiological responses but the only disadvantage is that this procedure is too elaborate to be used as a test of heat tolerance.

7. **Dowling "Cooling efficiency" test** (Dowling, 1956): Since animal loses thermoregulatory control in excessive solar radiation, this test does not consider body temperature as a meaningful criterion of heat adaptability. Hence, it proposes that heat tolerance test is based only on an animal's capacity for dissipating excess body heat without the complication of a simultaneous heat load from solar radiation. The procedure involves exercising the animal on a warm day for some time until the rectal temperature reaches above normal and then allowing animal to stand in shade to cool. The rate of cooling is followed by regularly measuring rectal temperature which leads to interpretation that the animal which shows the maximum rate of decrease of body temperature is considered to have the highest cooling efficiency. Its drawbacks include exclusion of solar radiation effect in an area of intense solar exposure and chances of thermal imbalance (hyperthermia) in animal being tested.

8. **Predicted heat tolerance from calf total moisture vaporization rates** (Yeck & Kibler, 1958): This test is done for calves and its result is expressed as ratio of total evaporation (cutaneous and respiratory) in an environment of 27°C / total evaporation in an environment of 10°C. According to this higher ratio means greater heat tolerance. This test takes into consideration the evaporative cooling as thermoregulatory means of cattle but an accurate measurement of total evaporation is doubtful.

9. **Average relative deviations (ARD)** (Marai and Habeeb, 1998): It is based on calculation of the average relative deviations (positive or negative) from normal in thermal, water and/or nitrogen imbalances of the animals (or in all traits measured) caused by exposure to hot climates. It can be mathematically expressed as:

$$\text{Adaptability (\%)} = [100 \text{ “ ARD}] \times 100$$

10. **Tunica dartos indices (TDI)** (El-Darawany, 1999 and Marai *et al.*, 2006): It is used to measure the ability of the ram to tolerate increased ambient temperatures. It is based on the fact that scrotum autoregulates its temperature through the tunica dartos muscle and distance between the testes and the abdominal wall indicates the magnitude of vascular heat exchange which can be estimated by using following equations:

TDIa = ((Max. SCL - Min. SCL)/(Max. SCL - TL))*100
TDIb = ((RT - SST)/RT)*100
TDIc = (((RT - SST)/RT) ((Max. SCL - Min. SCL)/ (Max. SCL - TL)))*100

Where, RT = average rectal temperature in summer; SST = average scrotal skin temperature in summer; Max. SCL = average maximum scrotal length in summer; Min. SCL = average minimum scrotal length in winter; TL = average testis length during the winter and summer seasons.

11. **Temperature-humidity indices:** In order to quantify the heat load in animals, mathematical expressions to calculate thermal or heat load indices are developed which are based on associations between climatic variables (Temperature, humidity, solar radiation, air movement etc) and responses (physiological and behavioural) of the animal. Though, there are many indices which are available but the most basic and easily applicable is the temperature-humidity index (THI) which suits best to dairy cattle physiology. Thorns formula for calculating THI is as follows:

$$THI = 0.72 (T_{Db} + T_{Wb}) + 40.6$$

Where T_{Db} is Dry Bulb Temperature (in ° C) and T_{Wb} is Wet Bulb Temperature

NRC (1971)

Relative Humidity

TEMPERATURE	30%	35%	40%	45%	50%	55%	60%	65%	70%	75%	80%	85%
100°	84	85	86	87	88	90	91	92	93	94	95	97
98°	83	84	85	86	87	88	89	90	91	93	94	95
96°	81	82	83	85	86	87	88	89	90	91	92	93
94°	80	81	82	83	84	85	86	87	88	89	90	91
92°	79	80	81	82	83	84	85	85	86	87	88	89
90°	78	79	79	80	81	82	83	84	85	86	86	87
88°	76	77	78	79	80	81	81	82	83	84	85	86
86°	75	76	77	78	78	79	80	81	81	82	83	84
84°	74	75	75	76	77	78	78	79	80	80	81	82
82°	73	73	74	75	75	76	77	77	78	79	79	80
80°	72	72	73	73	74	75	75	76	76	77	78	78
78°	70	71	71	72	73	73	74	74	75	75	76	76
76°	69	70	70	71	71	72	72	73	73	74	72	75

THI = Tdbf - (0.55 - (0.55 x (RH / 100))) x (Tdbf – 58)

Normal <74	Alert 75-78	Danger 79-83	Emergency >84

(based on Thorn, 1959).

A representative temperature humidity index (THI) chart which can be interpreted as follows: Normal <74; Alert 75-78; Danger 79-83; Emergency >84 (Associated Livestock Weather Safety Index i.e. LWSI; LCI, 1970). However the interpretation of THI score needs to be standardised for specific climatic conditions in the region and more specifically for a specific breed because a cold climate adapted cattle may show signs of heat stress at a THI score of 84 whereas a hot climate adapted cattle may not show any signs of heat stress. Equation given below is also an example of various climatic factors that affect the animal's thermal environment:

$$ETE = K_1{*}DBT + K_2{*}WBT + K_3{*}AM + K_4{*}RD + K_5{*}EI$$

Where ETE, *Effective Thermal Environment;* K_1 – K_5: *factors;* DBT: *dry-bulb temperature* (C); WBT: *wet-bulb temperature* (C); AM: *air movement;* RD: *radiation;* EI: *energy intake.*

(Yamamoto, 1983)

12. **Other tests :** Bonsma (1949) described many tests such as Felting test, Walking tests and Water deprivation tests, which cannot be categorized as heat tolerance tests but they give significant information on an animal's general capacity to

survive in a hot climate under field conditions. In felting test, sample hair from animal is damped and rubbed in palms which is interpreted subsequently. Felting test is indirect test for animal's heat tolerance, simple and applicable to very young animals. Walking tests" and "Water deprivation tests" is done by allowing animals to walk long distances without water over certain periods of time, in a semi-arid region.

Apart from qualitative and quantitative estimation of heat stress, it is also very important to identify animals with high risk of suffering heat stress by categorizing them as animals that have experienced a weaning, handling or transportation stress, animals who are grazing/ have grazed infested pastures, animals with a history of illness (especially pre-existing respiratory disorders), animals with black or dark color coat, heavy breeds (especially which calve during hot season), old animals, underfed/undernourished animals and high producing animals.

Morphological, Anatomical, Behavioural and Physiological Adaptation in Harsh Climate

Most of the domestic livetsock are homeotherms (animals that maintain the constant body temperature). They can thrive well in a range of ambient temperature without any changes in basal metabolism. In this process of resisting the ill effects of harsh climate, animal undergoes *adaptation* which means an adjustment in natural or animal systems to a new or changing environment. Adaptation to climate change refers to 'adjustment in natural or human systems in response to actual or expected climatic stimuli or their effects, which moderates harm or exploits beneficial opportunities' (IPCC, 2001). Adaptation may be biological, genetic or physiological. Biological adaptation refers to 'morphological, anatomical, physiological, biochemical and behavioral characteristics of animal which favor comfortable survival in a specific environment'; Genetic adaptation refers to 'heritable alteration {selection by nature (evolution) or man} characteristics of animal favoring comfortable survival of a population in a specific environment' and Physiological adaptation refers to 'capacity and process of adjustment of animal to itself, other biotic factors and to its external physical environment'.

Depending on the species, breed, age, sex and physiological/ pathological status, the animal differs in its ability to adapt against the climatic variable/variables. Hence, adaptability/adaptive capacity is defined as 'ability of a system to adjust to climate change (including climate variability and extremes) to moderate potential damages, to take advantage of opportunities, or to cope with the consequences' (IPCC, 2001) . Adaptive changes in animal are categorized as acclimation or acclimatization. Acclimation refers to 'the adaptive changes in animal in response to single climatic variable which are generally seen in experimental conditions like of climatic chamber and Acclimatization refers to 'the long term adaptive physiological changes in animal to climatic variations which results in increased tolerance to continuous or repeated exposure to complex climatic stressors which are generally seen in natural and field conditions'.

Scientist Hans Selye gave the term *General Adaptation Syndrome* which states that the physiological response of livestock to stress is divided into three phases i.e. *alarm, resistance* and *exhaustion.* During first phase of *alarm,* sympathetic response (commonly called "fight-or-flight" response) leads to rapid release of glucose for energy through release of catecholamines (adrenaline/epinephrine or noradrenaline/norepinephrine) and cortisol from adrenal glands (stimulation of hypothalamic-pituitary-adrenal axis). Second phase is of *adaptation or resistance,* where parasympathetic response restores normalcy however the hormones released during first phase remain high. If animal is not able to adapt or develop the resistance then the third phase of *exhaustion* occurs wherein the body reserves and hormones from the adrenal gland are inadequate, hence animal does not recover from the stressor leading to failure of the homeostatic mechanisms and finally death.

Out of morphological, anatomical, behavioural and physiological adaptation, morphological adaptation is generally permanent and evolutionary in nature. Such adaption occurs to minimize the adverse impact of heat or cold and altered quality and quantity of feed and water available due to such climate. Based on adaptive features, following ecological or eco-geographic rules were concluded:

1. **Allen's rule:** It states that in the same species of endotherms, body extremities like limbs or appendages (ear, tail, bill etc) of animals are relatively shorter in cooler climates than in warmer climates.

2. **Bergmann's rule:** It states that within a broadly distributed taxonomic group, the body size of animals living in cold climates is larger than the animals of the same species living in warm climates.

3. **Foster's rule (Island rule):** It states that members of a species become smaller or bigger depending on the resources available in the environment.

4. **Golger's rule:** Within the same species, mammalian and avian breeds of warm and humid regions have more melanin pigmentation than those in the cooler and drier regions. Thus animals in the arid regions are characterized by an accumulation of yellow and reddish brown phaeomelanin pigmentation; whereas in cold climates this pigment is reduced.
5. **Rapoport's rule (Hypothesis):** It states that latitudinal ranges of plants and animals are generally smaller at lower than at high latitudes.
6. **Thorson's rule:** It states that benthic marine invertebrates at low latitudes tend to produce large numbers of eggs developing to pelagic and widely-dispersing larvae, whereas at high latitudes such organisms tend to produce fewer and larger lecithotrophic (yolk-feeding) eggs and larger offspring, often by viviparity or ovoviviparities, which are often brooded.
7. **Wilson's rule:** It states that climate of an area influences the insulation characteristics (thickness of adipose tissue or hair/ fleece length) of an animal.

Salient Morphological, Anatomical and Functional Adaptation in Domestic Animals

Animal species through a gradual process try to adapt morphologically, anatomically and physiologically against the harsh environment. Some of these adaptations have been cited below (Hafez, 1968)

- To combat intense solar radiation camels possess long limbs and gazelle have developed short but reflecting coats.
- Camels are well adapted for deserts for thorny vegetation by the virtue of thick skin, hard tissue around mouth; thick mouth lined with long papillae and for water scarcity by virtue of increased drinking capacity, hump (for pseudo water storage), conservation of metabolic water ability to survive dehydration.
- In order to survive in low atmospheric oxygen concentration at high altitudes, llamas and alpacas have increased O_2 carrying capacity of blood by virtue of increased number of RBC, better ability to transfer O_2 from capillary blood to tissue cells at lower partial pressure and high efficiency in extracting nutrients from feeds.

- High temperature has led to summer hair shedding in ungulates; altered sheep wool characteristics i.e. loose, coarse in Awassi and fine, dense in Merino; increased heat loss from increased surface area of skin folds in cattle (brahman), prominent dewlap and hump in zebu cattle, higher number of sweat glands in zebu cattle and disproportionate long ears as compared to body in donkey.
- Skin of buffaloes are dark pigmented and sparsely haired as an adaptive response to high humidity.
- Low temperature adaptation is seen as long hair intermixed with fine hair is seen in cattle (Scottish highland); minimum exposed extremities in yak (Tibetan), good grazing behaviour in musk-ox (Arctic) , thick subcutaneous tissue in arctic species and abundant brown fat in neonate of many mammals.
- Due to climate variability, feed availability becomes seasonal which has further led to digestive system transformation to ruminant stomach ungulates and development of depots of reserve adipose tissue in humps of camel, in fat-tail sheep (Awassi, Kudri, Masai, Karakul), in fat-rump of sheep (Somali, Sudan blackhead) and in the rumen of Antelopes.

Why is Zebu Cattle More Heat Tolerant than European Cattle?

Zebu breed of cattle are more thermo-tolerant than the European breeds due to lower heat production and higher efficiency of heat loss. Low metabolic rate due to reduced growth rate and milk yield also play important role. As compared to vascular passages involving capillary networks, *Bos indicus* posses high density of arteriovenous anastomoses which offers lower tissue resistance to heat flow from the body core to the skin and facilitates increased blood flow to the skin during heat stress (Hales *et al., 1978*). Smooth and shiny hair coats of zebu cattle also help to reduce heat exchange via radiation (Hutchinson and Brown, 1969). Zebu cattle also possess light coloured hair (for better reflection of incident radiation) and many sweat glands to augment heat loss by increased sweating rate and its evaporation.

Behavioural and physiological responses in harsh climate: Mechanisms by which an animal maintains a constant internal environment are behavioural responses, chemical balance, feed consumption, body water balance, thermoregulation, pH control, osmotic and electrolyte regulation, circulatory balance of cardiovascular activities, metabolic process and blood pressure. Maintenance of homeostasis is mainly under the control of the central nervous system (CNS) and endocrine system. In order to escape the effect of heat and cold, homoeothermic animals undergo hibernation in cold (winter) and estivation during hot (summer) season. *Hibernation* is 'cessation of coordinated locomotor activity along with reduction in body temperature, total metabolism, heart beat and respiration rate during winter. The animal returns to normal state from hibernation on its own without any external energy'. *Estivation* refers to 'torpid sleeping state during the summer season wherein reduced metabolic rate prolongs the period of energy reserve consumption and water is conserved. Domestic animals do not show any of these phenomena. In general, term *Torpor* means 'a state of decreased physiological activity in an animal, usually by a reduced body temperature and rate of metabolism'. Torpor is used to enable animals to survive periods of reduced food availability.

Body Temperature Regulation

Body temperature is regulated by hypothalamus (part of brain) which acts as a thermostat to maintain temperature as close as possible to a normal set-point. The hypothalamus receives input from internal and external thermoreceptors and it activates physiologic and behavioral activities that influence heat production, heat loss and heat gain. Based on the thermoregulatory behavior and physiology, animals are classified into following categories:

Term	Description	Examples
Homeotherm/ temperature regulators	An Animal that maintains its body temperature at a constant level, usually above that of the environment, by its metabolic activity.	Mammals and birds.
Endotherm	An animal that can generate internal heat; a warm-blooded animal.	
Poikilotherm/ temperature conformers	An Animal that cannot regulate its body temperature (except by behavioral means) which varies with the temperature of its surroundings.	Fish and reptiles.
Ectotherm	An animal that is dependent on external sources of body heat.	
Stenotherm	An animal capable of living or growing only within a limited range of temperature.	Marine or soil dwelling animal.
Eurytherm	An animal that can function at a wide range of different body temperatures.	Cat, Dog, Tiger.

Thermoregulation occurs in a given range of temperature beyond which the thermoregulatory processes change in intensity and type which is well supported/manifested by responses of the animal. To understand these limits and their effects, various zones can be studied as follows:

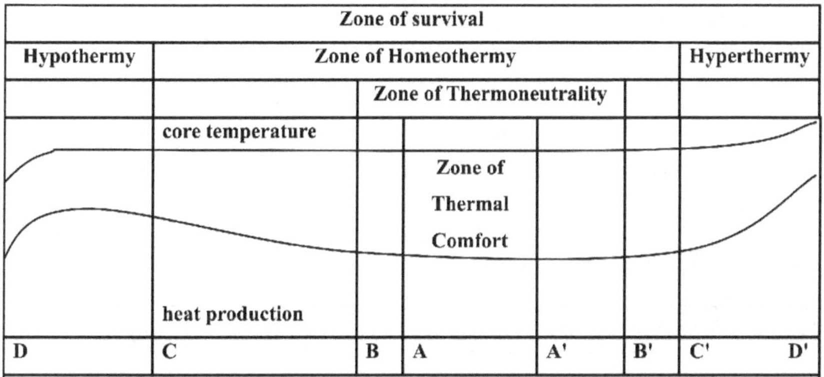

Limits of thermoregulation and related zones (adapted from Bianca 1968)
*Above diagram is indicative in nature

- **Zone of Thermal comfort (A - A′):** Little sensation of heat or cold, and a steady state core temperature can be maintained without any thermoregulatory responses;
- **Zone of Thermo neutrality (B-B′):** Core body temperature maintained by vasoconstriction and piloerection (if below thermal comfort zone) or vasodilation, sweating and panting (if above thermal comfort zone). Lower limit is *Lower critical temperature* (B) and upper limit is *Upper critical temperature* (B);
- **Zone of Homeothermy (C - C′):** Mammal maintains a constant, or steady state core temperature by using thermoregulatory responses as well as increasing or decreasing metabolic rate (if needed);
- **Zone of Survival (D - D′):** Environmental temperature range within which adult mammals can survive. Beyond the lower limit (*lower lethal temperature*-D) and upper limit (*upper lethal temperature*-D′) death occurs due to hypothermia and hyperthermia, respectively. Lower lethal temperature for most mammals is 15-20°C (about 20°C below the normal core temperature) and upper lethal temperature is about 42-45°C (3-6°C above the normal core temperature). Limit of upper lethal temperature is more critical as it is very near to normal core temperature compared to lower lethal temperature.

Hyperthermia and Fever

Hyperthermia (also called as 'heat stroke' or 'sun stroke') refers to any increase in body temperature above the normal range (mostly without change in hypothalamic set-point) as in cases of heat stroke, exercise-induced hyperthermia, malignant hyperthermia, seizure, etc. Body temperature reaches near 106° F.

Fever refers to any increase in body temperature where thermoregulatory processes act at a higher hypothalamic set-point possibly caused due to locally synthesized prostaglandin E_2, pyrogens like drugs, toxins, and viral or bacterial products (e.g. endotoxin), etc. Body temperature may be 103-106° F.

Table 14.2.1 : Difference between Hyperthermia and fever.

Features for comparison	Hyperthermia	Fever
Hypothalamic set point	Same	Raised
Thermoregulatory process	Absent	Present
Body temperature	106° F (approximate)	103-106° F (approximate)
Caused due to	Heat stroke, exercise, drug etc	Pyrogens like infectious microorganisms etc
Treatment	Antipyretic drugs like paracetamol/ acetaminophen are not useful, management is crucial.	Antipyretic durgs like paracetamol/ acetaminophen are useful, medication is crucial.

Hyperthermia is more stressful than fever because in fever due to rise in hypothalamic temperature set-point, thermoregulatory process help in controlling the rise of body core temperature which is not the case in hyperthermia. Accurate differential diagnosis should be done for fever and hyperthermia because falsely prescribing antipyretics for hyperthermia may lead to ignoring of managemental measures to be taken for hyperthermia. Hyperthermia should be treated on an emergency basis by restoring the blood volume through rapidly administering cool Ringer-Lactate solution and spraying and sprinkling water droplets on skin followed by directing warm (not cold) air flow to facilitate evaporative heat loss.

Animal Defences Against Harsh Climate

In hot weather, heat stress may occur either because heat is transferred from surrounding environment to the animal body or transfer of heat is restricted from animal to the surrounding environment. Heat stress may be a combined effect of high ambient temperature, humidity and solar radiation.

Defences against heat are brought about by following ways:

1. Increase in heat loss is facilitated by behavioural means (estivation), decreasing body insulation (increase blood flow to skin, season induced hair shedding and decreased adipose layer thickness) and increasing evaporation. Behavioural means includes seeking shade, wallowing & rooting, body licking, preferring succulent feeds, stretching body, anorexia, dispersed groups, increased water intake, reduced locomotor activities, etc in mammals and gular flutter (open-mouthed panting in birds), wing extension, avoid feet-soil contact, water splashing, moistening incubated eggs, dispersed groups etc in birds.

2. Reduced heat gain is done by increasing reflectance of smooth/ glossy summer coat.

3. Decreased heat production is brought about by reducing BMR (reduced thyroid activity, reduced feed intake) and by reducing physical activity.

Cold weather either facilitates transfer of heat from animal to surrounding environment or reduces the chances of heat gain from external natural resources. Cold stress is a combined effect of low ambient temperature, air movement and sometimes rain.

Defences against cold are brought about by following ways:

1. Decreased heat loss is done by behavioural means (hibernation) and by increasing body insulation. Behavioural means include huddling, tucked up posture, seeking warmer microclimate, increased locomotor activity, nest building, etc in mammals and fluffed feathers, tucked in head, torpidity, incubation and broodiness in birds. Increased body insulation is facilitated by tissue insulation (cutaneous/subcutaneous vasoconstriction and increased adipose layer thickness), hair coat insulation (increased thickness of hair coat), air insulation (piloerection, increasing thickness of boundary layer/air).

2. Increased heat gain is done by reduced reflectance through dry and rough hair coat.
3. Increased heat production is done by non-shivering thermogenesis/increased BMR (Increased thyroid activity, Increased feed intake) and shivering thermogenesis i.e. increased muscular activity (Bianca, 1968).

General Effects/Signs of Heat Stress

Domestic animal generally show similar signs of heat stress which may vary depending on animal factors as well as severity of heat stress. Followings are the common signs of heat stress in cow in ascending order of severity:

- *Major signs of heat stress*: Respiration rate - 80 to120/minute - moderate heat stress, 120 to 160/minute - moderately severe heat stress; > 160 breaths/minute severe heat stress; seeking shade, refusing to lie down; reduced feed intake; crowding around water trough; body splashing, agitation and restlessness; reduced or no rumen contractions; grouping to seek shade from herd mates; open mouthed panting/gasping, excessive drooling of saliva; recumbency; collapse, convulsions and coma; death (www.cattletoday.com).
- *Other signs of heat stress include* : restlessness; panting (open-mouthed breathing); reduced rates of GIT motility; lethargy, decreased activity; incidences of premature and small calves; reduced weight gains in growing animals; frequent laminitis; increased incidences of milk fever, uterine prolapse, mammary gland infections, uterine infections; udder edema; prominent decline in the intake and rumination in lactating cows, increased sweating, water intake and rectal temperature; reduced heart rate and drop in daily milk production (www.lhu.emu.ee).

Effect of Climate on Animal Production

Direct effects of harsh weather and extreme climatic variables on animal are seen on health, growth and reproduction and indirect effects are seen as alteration in the quality and quantity of feed, its cost and frequency of incidences as well as distribution pattern of livestock diseases and pests.

Growth; production of milk, egg & wool; reproduction and feed intake are most affected due to adverse climate. Heat stress causes decrease in feed intake (esp. dry matter) and metabolic rate in order to maintain thermal balance by reducing heat load. Effects are severe when there is water deprivation, nutritional imbalance or nutritional deficiency. Heat stress is severe for pig as they lack functional sweat glands. Poultry cannot tolerate heat stress and are most affected. Digestibility of feedstuffs are increased during mild heat stress but it is not of much use as during hot season the forages in pasture is low in nutrient and high in fiber/lignin. However during cold stress dry matter as well as voluntary feed intake along with metabolic rate is increased. Cold stress lowers the digestibility and leads to diversion of nutrients for generating heat. Stress due to heat as well as cold affects the immune status of the animal. Heat stress apart from suppressing humoral and cellular immune responses in poultry, in general negatively affects immune system as seen by altered levels of serum immunoglobulin, lymphocyte proliferation, cytokine secretion, NK cells, T-cell population, etc.

Effect of Climate on Animal Growth

Heat stress negatively affects growth of livestock which is attributed to lower feed intake and higher expenditure of energy to combat stress. Maintenance requirement of the animals increases both in heat and cold stress. Reduced daily weight gain occurs due to reduced feed intake in heat stress and diversion of feed nutrient for heat generation during cold stress. During heat stress decreased nitrogen retention occurs due to reduced protein digestibility and increased protein breakdown for energy. Nitrogen is lost through sweat which further reduces nitrogen retention in the growing animal. It is generally noticed that smaller breeds grow better than larger breeds due to large body surface area per unit mass of the body. A major economical impact of heat stress on growth is late attainment of puberty in farm animals.

Effect of Climate on Milk, Meat and Egg Production

Generally during heat stress maintenance requirement increases but feed intake, milk production, daily weight gain, egg production, egg shell thickness etc decreases. Heat stress (high temperature-humidity index) affects milk yield significantly in high yielders (more in exotic and crossbred cattle than Indian breeds). Heat stress effects are seen in the composition of milk as reduced total milk proteins (percentage is unaffected), altered fatty acid content, decreased fatty acids (C_6 –C_{14}) in milk fat, decreased fat, slight reduction in lactose and reduced SNF. Indirect effects of heat stress on milk yield is mediated through decreased thyrotrophic hormone, decreased growth hormone and increased prolactin hormone and on milk composition through increased epinephrine, decreased aldosterone hormone and altered substrate supply of ingredients used in milk synthesis.

Meat production also is affected especially in case of poultry due to reduction in feed intake, poor growth characteristics, diminished carcass weight, low muscle protein, lowered initial muscle pH (leading to pale, soft and exudative i.e. PSE meat), depleted muscle glycogen content (leading to dark, firm and dry i.e. DFD meat) and enhanced drip losses.

Egg production is severely affected by high temperature-humidity which can be seen as effects on egg size, clutch size, increased hen mortality, hormonal imbalance, poor quality of egg shell and lowered rate of body weight gain in poultry (causing delay in puberty).

During cold stress, maintenance requirement and feed intake increases but milk production, daily weight gain, egg production, etc decreases. However, egg shell thickness remains unaffected. Due to increased maintenance requirement most of the nutrients are diverted for heat generation leading to decline in milk production and the decline in amount of fat and lactose. However, milk becomes richer in long chain fatty acid as compared to short chain fatty acid during cold stress. Meat production due to the cold stress is not much affected as compared to heat stress.

Effect of Climate on Wool Production

Extreme or fluctuating temperatures adversely affect the wool growth in sheep depending on its physiological status (for e.g. diversion of nutrients during reproduction and lactation phases), breed, nutrition, sex, shearing, etc. Being seasonal breeders, physiology of sheep is profoundly influenced by duration of light/dark exposure and consequently the wool production is affected which is evident by increased wool growth during summers and decreased wool growth in winters. Providing heat during winters may slightly increase wool growth and cooling during summers may slightly decrease wool growth but it is widely believed that altered wool production is more due to fluctuation in seasonally related photoperiods rather than temperature.

Effect of Climate on Animal Reproduction

Generally during heat stress puberty is delayed. In females, heat stress affects ovulation, oocyte maturation, follicular development and sexual behavior probably due to changes in the level of hormones such as GnRH, LH, FSH and estradiol. There is decrease in duration and intensity of estrous cycle followed by impaired fertilization and low conception rate. Owing to the heat stress, oocytes may show nuclear damage and molecular changes which affect the embryonic development leading to poor quality embryo or embryonic mortality. Exposure to heat stress during 2nd or 3rd trimester of pregnancy may affect the gestation period and birth characteristics of offspring. In males during heat stress process of spermatogenesis is likely to be disturbed. The volume and quality of summer semen is generally good except at very high ambient temperatures. Semen/sperm characteristics like sperm concentration, sperm motility, sperm viability, sperm morphology and acrosomal integrity are affected at high temperatures. Capacitation and fertilizing ability may be lowered during high temperature and humidity accompanied with lowered testosterone levels. Fertility and hatchability also show a declining trend in poultry during heat stress.

Effect of Photoperiodism on Animals

Season as well as day length influences numerous biological and behavioral changes in animals. Temperature changes along with varying photoperiod affect mainly reproduction (mostly in poultry), production performance (egg, milk), colour of fur and feathers, migration, hibernation onset, sexual behavior, etc. Reproductive responses to photoperiod are determined largely by the extent of photoresponsiveness, nature of the photoperiodic signal, nutritional and social environment. (Wildeus, 1995; Walkden-Brown and Restall, 1996). Pineal gland upon stimulation by darkness secretes melatonin hormone which regulates the onset of breeding season in many domestic animals. Goat and sheep responds to decreased daylength/photoperiod and horse and cat responds to increased daylength/photoperiod. So any change in climate which affects the light exposure and ultraviolet exposure during daytime is likely to alter the breeding pattern in the domestic animals. In cattle, long day light exposure increases circulating prolactin as well as Insulin like Growth Factor-I which possibly mediates increase in milk yield.

In case of poultry, the photoreceptors located in hypothalamus get stimulated by light and further stimulate the hypothalamic-pituitary gonadal axis. GnRH-I after being secreted, facilitates the secretion of LH and FSH which further stimulates follicle development and steroid production in female as well as sperm and androgen production in male. Light has photostimulatory effect on the poultry making it reproductively active. After setting of approximately 12 hour photosensitive phase at dawn, photostimulation will occur even if the light period is not continuous. Generally poultry is influenced by the duration of light exposure of previous day and depending on current exposure it perceives it as increased photoperiod (stimulatory) or decreased photoperiod (inhibitory).

General Measures to Reduce Heat Stress/Summer Management

In order to reduce heat stress in farm animals various steps can be taken. These steps are classified into following categories

1. Feeding and watering management
2. Housing management
3. Milking management
4. Health management
5. Breeding management and Biotechnological approaches

1. **Feeding and watering management**: Even though different feedstuffs have same energy level yet they may produce different amounts of body heat known as heat increment (HI). Fats produce lowest HI, followed by carbohydrates and then proteins. Carbohydrates like cellulose have higher HI than more soluble carbohydrates such as starch and sugar. As the HI increases in order from fats and oils to oilseeds, grains, concentrates, roughages, forages they also become associated with decreasing digestibility. So summer ration/feeding strategy should include maximum amount of low HI diets which are locally available and cheap feed resources for e.g.: highly digestible, low fiber, low protein and high energy feed. However, dietary inclusion of fibers, partial replacement of grains by fat, oil and molasses in feed are essential for reducing the problems of acidosis. Since protein quality is important during hot weather so rumen degradable protein should be avoided to improve its availability. Inclusion of caseinated iodine is beneficial.

During heat stress body water and electrolytes (such as K^+ and Na^+) are lost through evaporative cooling via sweating and panting. Excess loss of K^+ is compensated by urinary Na^+ excretion to maintain the ratio of sodium and potassium ions. Panting also leads to loss of CO_2 leading to respiratory alkalosis which is compensated by increasing urinary excretion of HCO_3^-. Hence, aim should be to replenish important electrolytes such as Na^+, K^+, Mg^+ and HCO_3^-. In order to achieve dietary electrolyte balance, homeostasis, appropriate osmoregulation of body fluids, appetite stimulation, normal skeletal development etc, addition of salt and electrolytes to the drinking water and feed is essential (www.rci.com.au). Rumen stabilizing and digestion aids like yeast cultures, fungal products and buffers should be included. Feed supplementation of niacin, enzymatic antioxidants like superoxide dismutase, catalase, glutathione peroxidase and non-enzymatic antioxidant like vitamins C, A and E, proteins like albumin, transferrin, peptides like glutathione (GSH) along with Zinc and Chromium is essential for combating heat stress.

Feeding frequency should be higher but bulk feeding at a time and sudden changes in ration composition or its regime should be avoided. Feeding or grazing should be preferably done during morning or evening time to avoid day heat. Animal should not be allowed to walk long distances while grazing. Care should be taken to make the feed more acceptable and palatable by physical modification, adding molasses, etc. Water intake can increase 50% above normal levels (especially in lactating animals) during heat stress so *ad lib* cool and clean drinking water should be available at all times.

2. **Housing management:** Physical protection from hot environment especially direct sunrays can be done by natural as well as artificial shade (temporary or permanent). Trees are an excellent natural shade which blocks solar radiation and cools the surrounding air due to evaporation of moisture from leaf surface. Criteria like orientation, floor space, height, ventilation, roof and wall construction material, feeding and water facilities, waste disposal system should be considered carefully in permanent shades for heat stress management. Shade orientation of east-west direction is recommended for tied animals and north-south orientation for cows which are free to move. Light or white color reflects maximum solar

radiation so it should be used for painting walls and roofs on the outer side. Low stocking density and sufficient air movement will reduce the heat load in animals. Temporary shades should be constructed in the grazing field with drinking water facilities. Wallowing ponds should be constructed in the vicinity of grazing field to avoid animal walking long distances. Air-conditioning is expensive so evaporative cooling pad, foggers, misters, sprayers, sprinkler and fan cooling systems should be used in arid regions of low humidity and high temperature,.

3. **Milking management:** Heat stress should be reduced in the milking parlor by physical modification of the parlor, sprinkling water on animals during afternoon hours to avoid heat stress for evening milking, provision of drinking water near the exit of milking parlor, reduce milking time, milking followed by feeding during cooler times of day (early morning and late evening), reduce animal density during milking in the parlor, avoid excess handling of animals prior to milking, avoid animal walking long distance to reach milking parlor, etc.

4. **Health management:** Heat stress during summer along with high humidity may compromise the physiology especially the immune system of animal to fight the infectious diseases resulting in sick animals. High temperature and humidity favors the survival and breeding of flies, ticks, fleas, mosquito, etc leading to increased vector borne diseases. Hence, preventive measures should be taken which includes regular cleaning and disinfection of animal sheds, avoiding water logging, closure of cracks and crevices and increasing sunlight exposure and ventilation in animal sheds. Regular deworming of all animals prevents parasitic infestation. Preventive measures should be taken to avoid heat/sun stroke and emergency preparedness to deal with such conditions should be made.

5. **Breeding management and biotechnological approaches:** Reproductive management is of prime importance during heat stress and hence efforts should be made to improve detection of oestrus and reduce incidence of anoestrus by drug therapy, offering feed supplements, hormonal therapy, etc apart from following timed artificial insemination to overcome losses due

to oestrus detection failure. Selection of native local breeds which are adapted in the local climate should be preferred. Breeding strategy should be focused on selective breeding through controlled mating to develop breeds maintaining the balance between heat resistance and productivity. Problems of low fertility, silent or no oestrus, low conception rate due to heat stress can also be solved by using embryo transfer technology. Genomic and proteomic approaches are recently being utilized to identify and understand roles of genes and molecules involved in thermo-tolerance and thermo-sensitivity (e.g. heat shock proteins esp. HSP 70). These studies with the help of techniques like 'Gene knockout' will help in developing breeds which are heat resistant.

Chapter - 15

Organic Livestock Farming

Triveni Dutt and B.H.M. Patel

Globally livestock sector is changing rapidly in response to globalization and increasing demand for quality foods of animal origin. Animal husbandry is the backbone of rural economy of our country. The value of output from livestock sector is about 29 percent of the total value of output (GDP) of the agriculture and allied sectors. The livestock sector has been contributing over 5% to total GDP. India ranks first in milk production which accounts for more than 15 percent of total world output and 40 percent of Asia's total production. It continues to grow at about 4 percent per annum far exceeding the global average of 1-2 percent. Further, the manure from animals provides a good source of organic matter for improving soil fertility and crop yields. The *gobar* gas from the dung is used as fuel for domestic purposes. The surplus fodder and agricultural by-products are gainfully utilized for feeding the animals. Almost all draught power for farm operations and transportation is supplied by bullocks. Indian animal husbandry practices are based on minimal use of off-farm inputs and on management practices that restore, maintain and enhance ecological harmony. This principle also holds good for organic farming. The primary goal of organic agriculture is to optimize the health and productivity of interdependent communities of soil life, plants, animals and people (NOSB, 1995). Organic agriculture is an ecological production management system that promotes and enhances biodiversity, biological cycles and soil biological activity.

History of Organic Livestock Farming

Although, organic farming is not new to India, the first scientific approach to organic farming can be quoted back to the Vedas of the "Later Vedic Period" 1000 BC to 600 BC. The essence is to live in partnership with, rather than exploit the nature. The "Vrikshayurveda" (Science of plants}, the Krishishastra (Science of agriculture), and the "Mrigayurveda" (Animal Science) are the main works. Here agriculture was not developed just as a production system but as a culture. Animal Husbandry was an integral part of farming practice. Therefore, organic farming is a common sense farming which requires mind set tuning only. It is all about our realization with the nature rather rediscovering the traditional wisdom. The most notable characteristic of organic farming is the emphasis on the production process rather than a product itself.

Status of Organic Animal Rearing in India and World

Organic livestock farming is an emerging area of livestock production and gained significance recently because of many adverse side effects due to chemical residues and food toxicity of animal origin in conventional livestock farming. Of the total organic land in the world, Oceania, Europe and Latin America contributes nearly 39%, 23% and 19%, respectively. Thus, organic meat has highest market in Europe. Similarly in Denmark nearly 40% of total milk production is organic. However, Asian countries are still in efforts to balance between the food demand by poor people and quality food demand by elite families.

Land for organic farming in India is about 1, 50,790 ha (0.1% of total agricultural land) and exports 35 organic products worth USD 21 million. Major organic products include- rice, tea, coffee, spices, honey, medicinal plants, cashew, fruits and vegetables. Negligible attention has been paid to organic livestock farming despite the fact that its traditional animal husbandry system is ideally suitable for conversion to organic animal husbandry system.

In real sense, organic farming is already in vogue by default. Our traditional agriculture and animal rearing system could be considered as organic farming. Villagers in the hilly region today also use only organic manure for their agricultural land. However, the major demand for such organic products is restricted to developed countries, especially

European and North American countries. But in Asian and other developing countries, demand for such products is still restricted to few elite families and health conscious people. Thus organic farming in these countries is still in the stage of conception only. On the other side still we are not able to meet the minimum requirement of ordinary milk produced through conventional farming to this vast population. Thus, livestock farming in our country has to chalk out future road map with respect to how much organic farming is sustainable and can go hand in hand with conventional livestock farming. In such a scenario, information needed in the area of organic livestock farming has increased significantly.

Indian farmers are reluctant to take up organic farming because of the perceived high costs and increased risks over technology-assisted agriculture. There is also very little information on organic farming practices and efficiency available to the farmer. India ranks 33rd in terms of total land under organic cultivation and 88th position for agriculture land under organic crops to total farming area in the world. This only suggests how despite long-surviving traditional agricultural practices, India has not been able to shrug-off the technology-assisted agricultural practices in the post-Green Revolution era. The present scenario is attributed to increasing human population and pressure on agriculture and livestock sector to produce more and more food and livestock products.

Organic Livestock Farming

Organic livestock farming is an integrated system of farming based on ecological principles. It uses environment-friendly methods of crop and livestock production without use of synthetic fertilizers, growth hormones, growth-enhancing antibiotics, chemical drugs, synthetic pesticides or gene manipulation. The principle of organic livestock farming is to support conservation and promotion of local resources including species and breeds. Above all, organic livestock farming is a "sustainable, ecologically sound, economically viable, socially justified and humane" production system.

The following points are important challenges in organic livestock farming and same has been discussed below.

1. **Production of Organic Green Fodder and Concentrate:** Main specifications regarding organic livestock farming have to do with the avoidance of pesticides and nitrogenous fertilizers in feed and fodder production. The livestock should be fed 100%

organically grown feed. More than 50% of the feed should come from the farm unit itself, or should be produced in the region. However, for a transition period, the use of a limited proportion of conventional feed is authorized. The maximum percentage per year is 10% in case of herbivores and 20% for other species. Sufficient green fodder and *ad lib* clean and potable drinking water should be provided to the animals.

Getting organically grown green fodder and concentrate for animal feeding is a big challenge in organic livestock farming. To get organic feed and fodder, we have to use organic manure and organic pesticides. Arrangement of the Nitrogen-Phosphorus-Potash (NPK) by organic manure to agricultural land in large quantity is not an easy task. Therefore, we have to depend on livestock waste for the above purposes. Dung mixed with urine and straw after composting or vermi-composting can be a good source of NPK.

The modern methods of livestock waste recycling like bio-gas, compost, vermi-compost, or bio-fertilizers are not only useful as NPK but also help in minimizing environmental pollution. The nitrate content of organic fields is significantly lower. Organic livestock farming releases much less CO_2 than does conventional farming. Therefore, organic livestock farming provides suitable tools to minimize environmental pollution and nutrient losses on the farm level and seems to be more effective than conventional livestock farming.

In organic farming, *ad libitum* green fodder along with concentrate feeding is suitable system. Under this circumstance dairy cows may consume 40 to 50 kg green fodder. Chaff the roughage/green fodder before feeding to the dairy animals. Concentrate may be offered depending upon the stage of lactation and age. As we are going to maintain local breeds, which are generally low yielder require less quantity of concentrate and *ad lib* green fodder is sufficient for 5-6 lit of milk production. Further, offering additional 1-2 kg concentrate will meet the rest of the production. Total amount of ration to be fed to animal may offered in 3-4 times a day in equal intervals for better intake. To estimate the daily feed requirement, it is assumed that the animals consume about 2.5 to 3.0 percent of their body weight on dry matter basis.

Further, the use of synthetic growth-promoters, synthetic appetizers, preservatives, artificial colouring agents, synthetic amino acids, emulsifiers, solvent-extracted oil-cakes, urea, genetically engineered organisms or products are prohibited. However, seaweed powder, rock salt and natural sources vitamins, such as cod liver oil and yeast are the most preferred supplements, if any, are needed.

2. **Breeds and Breeding Strategy:** The principle of organic livestock farming is to support conservation and promotion of local resources including species and breeds. Appropriate choice of breeds is of utmost importance. Pure indigenous livestock breeds are preferred since they support the concept of organic farming *viz* low incidence of disease, better adaptability, low nutritional requirement, better fertility etc. Cross-breeding goes against the philosophy of organic production. Selective up-grading of the native breeds would be the only choice for improving the productivity of animals. In this regard, indigenous livestock resources and their breeding strategies have been listed below.

Indigenous cattle genetic resources and breeding strategy: India has 185 million of cattle which is 15.7 % of the world cattle population. Morphological configurations assist in adaption to tropical climate and disease resistance. There are 30 recognized breeds/synthetic strains of cattle in India. In addition to the defined breeds, vast cattle population (approx. 80%) falls under the 'non-descript' category. Majority of the local breeds considered uneconomical to keep due to low production. However, there is a great scope for the all indigenous breeds in organic farming as selective breeding (selection of genetically superior males and females of a particular breed for breeding purposes). It is followed for genetic improvement programmes in well defined indigenous dairy cattle breeds. Similarly, non-descript Indian cattle which constitute almost 75-80 % of total cattle wealth is to be upgraded through upgrading with improved indigenous cattle breeds. Since, this involves no involvement of latest assisted reproductive technologies; use of such low productive breeds for organic farming would be great boon to animal industry.

Country has 98 million of buffalo which is 55.4 % of the world buffalo population. There are 10 recognized breeds of buffalo found in various pats of the country. About 60% of the buffalo population is non-descript and rest 40% is descript classified into 10 breeds. Owing to its popularity among farmers population has increased by 8.9% over 1997. The genetic improvement of buffaloes is done by two ways - one is for the non-descript population and second is for the recognized breeds. Selective breeding of well defined buffalo breeds (Murrah, Bhadawari, Jaffrabadi, Surti, Nili Ravi) for milk production in the breeding tracts of these breeds and grading up of non-descript buffaloes with improved buffalo breeds.

Country has 125 million of goat which is 22.2 % of the world goat population. There are 20 recognized breeds of goat found in various parts of the country. Various breeds of goat found in different agro climatic conditions. Selective breeding of indigenous mutton and milch breeds (Black Bengal, Beetal, Jamunapari, Sirohi, Surti, Jhakrana, Malabari, Marwari, Ganjam, Kannaidu, and Osmanabadi) in their respective breeding tracts. Grading up of non-descript and dwarf breed of goats with superior indigenous breed (s) of recognized breeds (Jamunapari, Barbari, Beetal) for the improvement of milk as well as meat. India has 62 million of sheep found in various pats of the country. Similarly, selective breeding of indigenous sheep breeds in their respective breeding tracts and grading up of local non-descript breeds with improved local breeds for desired traits. Since, only few cross breeding programmes are successful in our country, going for any exotic breeds or evolved breeds is desirable in organic livestock farms.

The reproduction techniques should be natural whether breeding policy is selective or grading up. Artificial reproduction methods like Artificial Insemination (A.I.), Embryo Transfer Technology (E.T.T.), multiple ovulation and Embryo Transfer (M.O.E.T.) etc. are restricted. Hormonal heat treatment is not allowed. Use of genetically modified organisms (GMOs) also not allowed. When organic livestock are not available, certification programme shall allow brought-in conventional animals e.g.: calves up to 4 weeks of age that have received colostrums and have been fed on milk diet. Breeding stock may be brought-in from conventional farms but maximum replacement rate will be 10% only.

3. **Livestock Housing System:** Housing is generally recommended that animals are not caged, tethered or confined to buildings without adequate natural ventilation and light. Housing should be based on their physiological and behavioral needs. Animals should have sufficient area for grazing and housing should allow free movement and also maximum amount of fresh air and daylight. In organic systems the animals are allowed a larger housing area (including outdoor access) with obligatory straw bedding. Fortunately, in India animal are housed with sufficient space (Loose housing) in the rural areas and they are being taken for grazing during day time. Bedding in adequate quantity and dry litter is used as bedding material. Group penning is prescribed for all farm animals. The indoor area is supplemented by an outdoor area that must be at least 75% of the indoor area. The USDA organic rules do not establish any space requirements for livestock

living conditions, but state that a producer must accommodate the health and natural behavior of the animals. However, they also state that additional NOSB [National Organic Standards Board] recommendations are anticipated and public comment will be necessary for the development of space requirements. However, the European Union (EU) has suggested the floor space requirement for dairy animals.

Farm animals	EU-Regulation on Organic Livestock Farming
Dairy cows	
• Locomotion area per animal	: 6.0 m^2 indoors + 4.5 m^2 outdoors
• Floor characteristics	: lying space with litter (bedding)
• Husbandry practices	: keeping tethered is forbidden
Calves	
• Locomotion area per animal	: 1.5m^2 indoors + 1.2 m^2 outdoors
• Floor characteristics	: dry litter bedding
• Husbandry practices	: generally reared in group pens

The following minimum space requirements are based on Swedish experience, particularly experience with housing boars, pregnant sows, and sows with litters in the Swedish Thorstensson and Ljungstrom versions of deep-bedded pig housing.

1) Boars: 64 sq. ft. per individual (74 sq. ft. if no separate dunging area provided);
2) Sow and litter in pens: pens with manure gutter: 54 sq. ft. exclusive of gutter per sow and litter; pens without manure gutter: 64 sq. ft. per sow and litter;
3) Sow and litter in boxes: 48 sq. ft. per sow and litter;
4) Sow and litter in group lactation housing: 81 sq. ft. per sow and litter;
5) Gestating and mating sows (individual housing in crates is prohibited; the following requirement is for group-housed sows): 27 sq. ft. per sow (exclusive of feeding area);
6) Weaned, growing and finishing pigs: Space shall be provided to allow all pigs to lie down in full lateral recumbency at the same time.

The housing standards for pigs provide a table that lists the minimum bedded space allowance for growing pigs from 1.6 sq. ft. per

22-pound pig to 8.1 sq. ft. per 220-pound pig, with a minimum total floor space of 37.6 sq. ft. per adult sow. They state that pigs must always be provided with total floor space no less than 1.5 times the lying space (should at least be equal to the square of the length of the pig).

Floor space requirement for the different categories of animal are to be worked out for the organic farming in India. However, authors feel that ISI floor space requirement is even sufficient for the organic farming also.

4. **Disease Prevention:** Organic livestock farming strictly follows the common proverb "prevention is better than cure". Native breeds should be selected to avoid specific diseases or health problems, which prevail in conventional livestock farming. Livestock should be raised in such a manner that suits the requirements of the species in terms environment and available feed and fodder and promotes a good resistance against common diseases and infections. Utilization of nutritious and good quality feeds, which together with application of outdoor areas and grazing facilities strengthens the natural immune system of the animals. Securing adequate space allowance in order to avoid overcrowding and warding off its associated health problems. In such type of farming, prevention of any health problem is aimed at using alternate medicine which is free from any chemical residues such as homeopathy, herbal medicine, acupuncture, hot fomentation, etc. Prevention should be based on feeding plants and herbs that have reported potential for immune-modulation, antioxidants and rich in essential nutrients. These herbs help in increasing body immunity and maintain good health. Few commonly used herbs having these properties are Ashwaghanda, Aloevera, Turmeric, Tulsi, Giloi, Garlic, etc. Organic herbal medicine prepared for prevention of mastitis (Mastidip) by IVRI is available for use organic livestock owners. It is used as teat wash before and after milking which help in decreasing the microbial load in teats.

Organic standards classify the use of vaccines as 'restricted'. This means that their routine use is discouraged and is only permitted when it can be demonstrated by the organic farmer that a specific disease is endemic in the region or on the organic farm, or where their use is required by law, or in proven cases, where such a disease cannot be effectively controlled by other management practices. Organic certifiers require written verification from a veterinarian to confirm the presence or threat of disease infection prior to approving the application of the treatment. In addition, the vaccines must not contain genetically modified ingredients or by-products like recombinant and DNA vaccine. The use of vaccines under these circumstances will not prejudice certification and does not require quarantine procedures. Organic farming promotes use of native breeds which are disease resistant and therefore requirement of vaccines against common diseases can be minimized. Full records of all medication must be maintained.

The situation is somewhat different on farms with sheep, beef and pig enterprise, where parasites are likely to represent the biggest challenge. The control of internal parasites can be successfully achieved with management practices such as clean grazing systems, mixed stocking, rotational grazing and selection of resistant stock and breeds. A reduction in stocking rate will also have an effect in reducing the parasite burden of pastures. Drenching of particularly stressed and therefore more vulnerable animals such as sows with larger litters is considered to be acceptable. The situation is slightly more difficult with regard to the control of external parasites, whereas until now management practices are not so well developed and farmers have to resort to using conventional treatment on a more regular basis, or to comply with legal requirements.

5. **Treatment Using Alternative Medicines:** An important objective of organic livestock farming is the avoidance of reliance upon routine and/or prophylactic use of conventional veterinary medicines (antibiotics, anthelmintics etc.) to reduce chemical residues in the livestock products. Therefore, organic standards state that the adoption of good management practices is the preferred method to maintain livestock health in an organic system. However, they also recognize that sometimes this alone is insufficient to guard against diseases and illness. In such situations treatment of sick animals should be carried out immediately. Organic farming implicates use of non-allopathic medicines, natural medicines and methods, including homoeopathy, ayurvedic medicine, herbal drug therapy and acupuncture as they are assumed to be safe, organic and free from chemical residues.

The use of conventional veterinary medicines is allowed only in case of emergency when no other non-allopathic alternative is available. If permitted to use the withdrawal period should be twice or thrice of the normal recommended period. Prohibited treatment use is only acceptable in cases of emergency; however organic standards stress that such treatments (including medicines) should not be withheld where animal welfare concerns exist. If anthelmintics, antibiotics or any other chemical drugs are required, organic standards specify a number of management conditions which must be applied to treated livestock. Treated animal must be excluded for a period of time (equivalent to three times the legal withholding period of the substance in question, or a minimum of three weeks, whichever is longer) from certified organic land and separated from non-treated organic stock. Following this period, treated stock are permitted to mingle with organic (non-treated) stock and range on certified organic land, provided their identity should be traceable and proper record is maintained. If organic livestock are treated with anthelmintics and/or antibiotics a period of de-certification follows and the livestock and/or their products cannot be sold as 'certified organic'. The offspring of treated livestock, however, may attain organic certification status for milk and meat if managed in accordance with organic standards. This may occur provided the offspring's parent is managed in accordance with organic standards from the last trimester of pregnancy onwards, including whole lactation period. This means that any treatment applied to a breeder must occur prior to the start of the last trimester of pregnancy and full organic management must be sustained from this point onwards in order for the offspring's meat or egg products to be certified as organic. The use of anaesthetics does not result in the loss of certification status, but does require the treated livestock to be withheld from sale or slaughter for above recommended period. Keeping these points in mind, we can also exploit natural medicinal property of dung and urine for treatment of animals in production of organic livestock products. The most common affections in large herds are wound, cuts, burns and other skin ailments. In this circumstances, organic herbal preparation (Ollinall) prepared by IVRI is very effective for the above mentioned ailments.

6. **Livestock Welfare Concern:** Today animal welfare is very much concerned by the public. Further, animal welfare is again a primary concern within organic livestock production systems. Many questions have been raised on individual welfare of animals with respect to health care, as there is a prohibition on conventionally used veterinary medicines (except in emergencies) in organic livestock farming. Currently, all the leading animal welfare organizations support organic livestock farming as a means of alleviating the suffering of farm animals. However, the most common health problem in organic farms is parasitism. The parasitological studies

revealed a higher prevalence of parasites in organic herds, as could be expected. This was true for pigs, hens, sheep and probably also dairy cattle. Parasitic infestations can be regarded as a risk factor for animal welfare. But good management can bring down the infestations. Organic farming supports the animal welfare as animals are being raised with sufficient space in natural condition without using any chemical drugs.

Organic Certification Bodies

Basic requirements for organic livestock production involve certification of products right from origin of animals to marketing of their products. Products are produced, processed and packaged according to organic standards. Without certification, no products can be labeled as "organic". Certification guarantees not only the quality of product but also the quality of production. It involves certification of products right from origin of animals to marketing of their products. Broadly five international standards are available.

1) European Union (EU) Regulation (1804/1999)
2) Organic Food Products' Act (OFPA) Of U.S.A.
3) Draft Guidelines Of Codex / FAO / WHO
4) UKROFS of U.K.
5) IFOAM Basic Standards

Steps Taken by GOI to Promote Organic Livestock Farming

Considering the regional importance, Department of Commerce, Govt. of India has developed Indian National Standards for Organic Production (NSOP, 2000). The following certification bodies are fully accredited under the Indian National Program for Organic Production

i. ECOCERT International (based in France and Germany, branch office in Aurangabad)
ii. IMO India Pvt. Ltd. -Institute for Market Ecology (based in Switzerland, office in Bangalore)
iii. INDOCERT (based in India, office in Aluva, Kerala)
iv. LACON GmbH (based in Germany, office in Aluva, Kerala)
v. SGS India Pvt. Ltd. (based in Switzerland, offices in Delhi and other cities)

vi. SKAL International (based in the Netherlands, branch office in Mumbai)

National centre for organic farming was established in the year 2003 and government allotted Rs. 100 crores in 10th plan outlay for organic development. Further, a network project on organic farming was sanctioned by I.C.A.R. (2004-07) involving four I.C.A.R. institutes and nine SAUs. A pilot study on organic milk production system was undertaken at N.D.R.I., Karnal, and state government of Uttarakhand, Mizorum and Sikkim has been declared as organic states.

Risk Factors

Keeping the remunerative price of organic products, government has also declared the few states/districts as organic state/district (Uttarakhand, HP etc). Hilly states are generally considered to be ideal for organic farming due to traditional system of agriculture and animal husbandry. It has been generally observed that yield is generally low in organic farming compared to conventional method of farming. However, the economic losses due to lowered production can be compensated by higher price of these organic products. Similarly, when we use indigenous breeds for organic farming, total Indian milk production may affect further aggravating may further affect the food deficit in the country. Therefore, balance between the organic and conventional farming should be decided so as to meet the food requirement of our growing population. Organic farming may become economical when farmer get remunerative prices. Further, policy makers should also take active interest in building up of infrastructure, creation of separate 'green channels' for marketing of organic foods, announcement of premium prices for organic staple food crops, creation of demand by more awareness programs, subsidies for encouraging organic growers, more R & D investments on organic farming and finally cheap and quick certification process. Indeed, organic dairy farming requires greater time as it cannot depend on any mechanization which utilizes the chemicals. In case of organic farming, per unit area fodder production is comparatively less; this may require more land and man power to get the same amount of organic fodder.

Organic faming relies on use of alternative medicine for both prevention and treatment. However use of alternative medicine requires longer recovery period and thus affect production. Often these alternative therapies are not targeted against specific pathogens and therefore are ineffective against specific infectious pathogens. The major weaknesses of organic agriculture in the country are absence of linkages between the farmers and markets and absence of financial support from the governments. India has the potential to become a major organic producing country given the international demand for our farm products, different agro-climatic regions for the cultivation of a number of crops, the size of the domestic market and above all the long tradition of environment friendly farming and living. These risk factors can be mitigated by planning, implementing an appropriate policy, giving a prominent place to organic farming addressing the issues related to its coverage, financial support during the conversion period, creation of linkages among the farmers, processors, traders and consumers, inspection and certification of organic products. Further, increasing the public awareness of the benefits of organic agriculture along with the ill effects of the conventional system.

Conclusion

Indian farmers are reluctant to take up organic farming because of the perceived high costs and increased risks over technology-assisted agriculture. There is also very little information on organic farming practices and efficiency available to the farmer. Organic livestock farming is not a production method to solve all problems in livestock production. It is primarily a production method for a specific premium market with high requirements for the quality of the production process, demanding high management qualification. For the development of organic livestock farming it is important to ensure the availability of organic manure for the fodder production, which is the starting point of organic livestock farming. Further, organic livestock farming does not depend on extensive external inputs but it is an optimum internal input system. Therefore, it greatly helps a farmer to become self-sufficient in his/her requirements for agro-inputs and reduce his/her costs.

Bibliography

1. Acharya, R. M. (1988).Goat breeding and meat production. Goat meat production in Asia: proceedings of a workshop held in Tando Jam, Pakistan, Ottawa, Ont. IDRC, 268e.

2. Aganga, A.A. and Tshwenyane, S.O. (2003). Lucerne, Lablab and *Leucaena leucocephala* forages: production and utilization for livestock production. *Pakistan Journal of Nutrition*, **2** (2): 46-53.

3. Agrawal, R. P., Swami, S. C., Beniwal, R., Kocher, D. K. and Kothari, R. P. (2002). Effect of Camel milk on glycemic control, risk factors and diabetes quality of life in type-1 diabetes: a randomized prospective controlled study. *Int. J. Diab.*, **22:** 70-74.

4. Agrawal, R. P., Swami, S. C., Beniwal, R., Kocher, D. K., Sahani, M.S., Tuteja, F. C. and Ghorui, S. K. (2003). Effect of camel milk on glycemic control, lipid profile and diabetes quality of life in type- I diabetes: A randomized prospective controlled cross over study. *Indian Journal of Animal Science.* **73:** 1105-1110.

5. Ahuja, V. and Sen, A. (2005). Scope and Space for small scale poultry production in developing countries. W.P. No.2007-12-02. Indian Institute of Management, Ahmedabad-380 015, India.

6. Altman, P.L. and Dittmer, D.S. Biology data book, 2nd edn. Vol.1 1971, 606 pp; Vol.2, 1973, 1432 pp.Vol.3, 1974, 2123 pp. Bethesda, Maryland: Federation of American Societies for Experimental Biology.

7. Amanullah, M.M., Sekar, S. and Muthukrishnan, P. (2010). Prospect and potential of poultry manure. *Asian Journal of Plant Sciences*, 2010: 1-11.

8. Angus, R.C. and Barr, W.L. (1955). An appraisal of research literature dealing with loose and conventional dairy cattle housing: a review. *Journal of Dairy Science*, **38** (4): 391-406.

9. Arora, S. P. (1979). Dairy Handbook, NDRI, Karnal.

10. Baker, H.J., Lindsey, J.R. and Weisborth, S. H (ads.). (1980). The Laboratory rat, Vol. II, Research Application Academic Press, New York, U.S.A.

11. Ball, P. J. H. and Peters, A. R. (2004). Reproduction in Cattle. Third Edition, Blackwell Publishing Ltd, Oxford, UK.

12. Ballard, B. and Rockett, J. (2009). Restraint and handling for veterinary technicians and assistants. Delmar Cengage Learning. New York, USA.

13. Banerjee, G. C. (2000). A Text Book of Animal Husbandry. Oxford & IBH Publishing Co. Pvt. Ltd., New Delhi.

14. Banerjee, G.C. (1988). Feeds and principles of animal nutrition. Oxford & IBH Publication, Delhi.

15. Barrada, M.S. (1957). Responses of dairy cattle to hot environments with special emphasis on respiratory reactions. Ph. D. Thesis, Johns Hopkins Univ., Baltimore.

16. Battaglia, R.A. and Mayrose, V. B. (2007). Handbook of Livestock Management Techniques. Surjeet Publications. Delhi.

17. Benezra, M.V. (1954): A new index for measuring the adaptability of cattle to tropical conditions. *J. Anim. Sci.*, **13**: 1015.

18. Bhardwaj, K. R. and Rathore, D. S. (1990). CDRI-NLAC Animal profile. Central Drug Research Institute, Lucknow.

19. Bhardwaj, K. R., Purohit D. C. and Dhawan, B.N. (ads.) (1991). Laboratory animal ethics and technology. Central Drug Research Institute, Lucknow.

20. Bhat, P. N. (2010). Buffalo Production. Studium Press (India) Pvt. Ltd. New Delhi, India.

21. Bhat, P. N., Mohan, N.H., and Deo, S. (2010). Pig Production. Studium Press (India) Pvt. Ltd. New Delhi, India.

22. Bhatia, B. (1990). Adoption of farm mechanization in a developing economy, Daya Publishing House, Delhi.

23. Bianca, W. (1968). Thermoregulation, in Hafez, E. S. E., ed., Adaptation of domestic animals. Philadelphia, Lea and Febiger, pp. 97-118.

24. Bickert, W. G. (1982). Housing Systems. *J Dairy Sci.*, **65**:502-508.

25. Blech, M. (2004). Animal identification practices, Section 3. FAO & WHO.

26. Blokhuis, H.J. (1994). Intensive production units and welfare: domestic fowl. *Rev. sci. tech. Off. int. Epiz.*,**13** (1) : 67-78.

27. Boden, E. (2006). Black's veterinary dictionary. A & C Black Publishers Limited, 38 Soho Square, London.

28. Bonsma J. (1949). Breeding cattle for increased adaptability to tropical and subtropical environment. *Journal of Agricultural Sciences*. **39**:204-221.

29. Borellvon, E. and Sorensen, J. T. (2004). Organic livestock production in Europe: aims, rules, and trends with special emphasis on animal health and welfare. *Livestock Prod. Sci.*, **90**: 3-9.

30. Borghese, A. (2005). Buffalo production and research. REU Technical series 67. Food and Agriculture Organization of the United Nations, Rome, Italy.

31. Brady, S., Wolverton, D. R., Levis, D. G. and Brumm, M. C. (1993). Proper way to ear notch pigs. University of Nebraska-Lincoln. *http://digitalcommons.unl.edu/extensionhist/1125.*

32. Brooks, P. H., Russel, S. J. and Carpenter, J. L. (1984). Water intake of weaned piglets from three to seven weeks old. *Vet. Rec.,* **115**:513–515.

33. Broom, D. M. and Fraser, A. F. (2007). CAB International. Oxfordshire, UK.

34. Central poultry development organization. Duck Management Guide. Banglore, India. *http://www.cpdosrbng.kar.nic.in.*

35. Chander, M. and Mukherjee, R. (2005). Organic Animal Husbandry: Concept, Status & Possibilities in India-A Review. *Indian J. Anim. Sci.,* **75**(12): 1460-1469.

36. Cirelli, A. Equine Dentition, SP-00-08. University of Nevada, Reno, USA. *www.unce.unr.edu/publications/files/ag/2000/sp0008.pdf.*

37. Clough, G. (1976). The immediate environment of the laboratory animals. *In:* control of the animal house environment. Laboratory Animal Handbooks 7. (Sheehy, T.Mc.ed.) pp. 77-94 Laboratory Animals Ltd. London, U.K.

38. Conroy, C., Sparks, N., Chandrasekaran, D., Sharma, A., Shindey, D., Singh, L.R., Natarajan, A. and Anitha, K. (2005). Improving backyard poultry-keeping: a case study from India. Agricultural Research & Extension Network, Network Paper No. 146.

39. Copland, J.W. (1985). Goat production and research in the tropics: proceedings of a workshop held at the University of Queensland, Brisbane, Australia, February 1984. ACIAR Proceedings Series No. 7, 118p.

40. CPCSEA Guidelines for laboratory animal facility. (2002). Committee for the purpose of control and supervision on experiments on animals (CPCSEA), Chennai.

41. Daghir, N.J. (2001). Poultry production in hot climates. CAB International. Wallingford, UK.

42. Das, S. K. (2001). Prospect and potentiality for goat farming in north eastern region of India-A review. *Agric.Rev.,* **22**(3/4):228-233.

43. De Fanciscis, G. (1986). Buffalo for Meat Production. 3rd World Congress on Genetics Applied to Livestock Production. Paper 8. http://digitalcommons.unl.edu/wcgalp/8.

44. De, Sukumar. (1977). Outline of Dairy Technology. Oxford University Press.

45. Deka, R. and Wright, I. A. (2011). Training manual on smallholders' pig management. International Livestock Research Institute (ILRI) Asia-Pacific, New Delhi, India.

46. Deka, R., Thorpe W., Lapar, M. L. and Kumar A. (2007). Assam's Pig sub-sector: Current status, constraints and opportunities. Project report submitted to International Livestock Research Institute, New Delhi, India.

47. Devendra, C. (1980). Milk production in goats compared to buffalo and cattle in humid tropics. *J Dairy Sci.,* **63** (10):1755-1767.

48. Devendra, C. (1986). Feeding systems and nutrition of goats and sheep in the tropics. In Adeniji, K.O., and Kategeli, J.A. Proceedings of the Workshop on the Improvement of Small Ruminants in eastern and southern Africa. 18–22 August 1986. Nairobi Kenya. pp.91–110.

49. Devendra, C. (1988). Goat Meat Production in Asia. Proceedings of a Workshop held in Tando Jam, Pakistan.

50. Devendra, C. (1989). Shrubs and tree fodders for farm animals. In: Proceedings of a Workshop in Denpasar, Indonesia, 24-29 July 1989.

51. Devendra, C. (1991). Goat meat production potential in the developing countries. VII Annual Mexican Meeting on Goat Production in Monterrey, Mexico.

52. Devendra, C. (2007). Perspectives on animal production systems in Asia. *Livestock Science*, **106**: 1-18.

53. Devendra, C. and FAYLON, P.S. (1988). Sheep Production in Asia. Proceedings of the workshop on sheep production in Asia, Book series NO. 80/1989. PCARRD, Los Banos, Laguna, Philippines, April 18-23.

54. Devendra, C. and Liang, J. B. (2012). Conference summary of dairy goats in Asia: Current status multifunctional contribution to food security and potential improvements. *Small Ruminanats Research,* **108**:1-11.

55. Devendra, C. and McLeroy, G. B. (1990). Goat and sheep production in tropics. 1st edition. ELBS, Longman group, United Kingdom.

56. Dhanda, J. S. D., Taylor, G., Murray, P. J., Pegg, R. B. and Shand, P. J. (2003). Goat meat production: present status and future possibilities. *Asian-Aust. J. Anim. Sci.,***16** (12): 1842-1852.

57. Dickinson, F.N. and King, G. (1977). *Ann. Meet Am. Dairy. Sci.* Ass. Iowa.

58. Dikshit, A. K. and Birthal, P. S. (2010). Environmental value of dung in mixed crop-livestock systems. *Indian Journal of Animal Sciences,* **80 (7):** 679–682.

59. Dobson, H. and Kamonpatana, M. (1986). A review of female cattle reproduction with special reference to a comparison between buffaloes, cows and zebu. *J. Reprod. Fert.*, **77**:1-36.

60. Doherty, M. K., McLean, L. and Beynon, R. J. (2007). Avian proteomics: advances, challenges and new technologies. *Cytogenet Genome Res.*, **117**:358-369.

61. Dowling, D.F. (1956). An experimental study of heat tolerance of cattle. *Aust. J. Agric. Res.*, **7**: 469-481.

62. EAAP Technical Series No. 7. (2003). Interactions between Climate and Animal Production.

63. El-Darawany, A. A. (1999). Tunica dartos thermoregulatory index in Bull and Ram in Egypt. *Ind J Anim Sci .,* **69**(8):560–563.

64. Falvey, L. and Chantalakana, C. (1999). Smallholder dairying in the tropics. ILRI, Nairobi, Kenya.

65. FAO. (1989). Milking, milk production hygiene and udder health. FAO Animal Production and Health paper 78. Rome, Italy.

66. FAO. (1997). The water buffalo, FAO Animal production and Health series 4, Rome, Italy.

67. FAO. (1998). Secondary Guidelines for Development of National Farm Animal Genetic Resources Management Plans—Measurement of Domestic Animal Diversity (MoDAD), FAO.

68. FAO. (2007). The State of the World's Animal Genetic Resources for Food and Agriculture, edited by Barbara Rischkowsky & Dafydd Pilling. Rome, Italy.

69. FAO. (2011). Rearing young ruminants on milk replacers and starter feeds. FAO Animal Production and Health Manual No. 13. Rome, Italy.

70. FAOSTAT. (2010). Food and agriculture organization statistical database. http:// faostat. fao.org, browsed in January, 2013.

71. Fatet, A., Pellicer-Rubioa, M. T. and Leboeufb, B. 2011. Reproductive cycle of goats. *Animal Reproduction Science*, **124** :211–219.

72. Felius, M., Koolmees, P. A., Theunissen, B., European Cattle Genetic Diversity Consortium. and Lenstra, J. A. (2011). On the Breeds of Cattle-Historic and Current Classifications. *Diversity*, **3**:660-692.

73. Foley, R.C., Bath, D.L., Dickinson, F.N. and Tucker, H.A. (1973). Dairy Cattle: Principles, Practices, Problems, Profits. Lea & Febiger, Philadelphia.

74. Food and Agriculture Organization (FAO). (1983). Land Evaluation of Rainfed Agriculture. *Soil Bull.*, **52**: 237p.

75. Foster, H.L., Small, J.D. and Fov, J.G. (eds) (1983). The Mouse in biomedical research Vol. I-IV. Experimental biology and oncology, Academic Press, New York, U.S.A.

76. Fowler, M. (2008). Restraint and handling of wild and domestic animals. Wiley-Blackwell. Iowa, USA.

77. Frank, T. and Shutt, M. A. (1898). Barnyard manure its nature, functions, composition, fermentation; preservation and application, Bulletin no. 31. Department of Agriculture, Ottawa, Canada.

78. Galal, S. (2005). Biodiversity in goats. *Small Ruminant Research,* **60**:75-81.

79. Garg, D. N. (1987). Sources for ancient literature on veterinary sciences. *Indian Journal of History of Sciences*, **22**(2): 103-110.

80. Ghotge, N.S. (2000). Livestock farming systems- sustainable options for the new millennium: In Challenges of Agriculture in the 21st Century; (ed.) G. M. Pillai: pp. 153-177.

81. Gillespie, J. R. and Flanders, F. B. (2004). Modern Livestock and Poultry Production. Delmar Cengage Learning, NY, USA.

82. Giovanni, de F. (1986). Buffalo for Meat Production. 3rd World Congress on Genetics Applied to Livestock Production. Paper 8. *http://digitalcommons.unl.edu/ wcgalp/8*.

83. Goetsch, A. L., Gipson, T. A., Askar, A. R. and Puchala, R. (2010). Feeding behaviour of goats. *J. Anim. Sci.*, **88**:361-373.

84. Goetsch, A. L., Zeng, S. S. and Gipson, T. A. (2011). Factors affecting goat milk production and quality. *Small Ruminant Research*, **101**:55-63.

85. Government of India. (2001). Prevention of cruelty to animals (Transport of animals on foot), New Delhi, India.

86. Government of India. (2006). Report of the working group on animal husbandry and dairying-11th Five year plan (2007- 12). Planning Commission, New Delhi, India.

87. Government of India. (2010). Basic animal husbandry statistics (AHS Series 12). Ministry of Agriculture, Department of Animal Husbandry, Dairying and Fisheries. Krishi Bhavan, New Delhi.

88. Government of India. (2012). Basic animal husbandry statistics (AHS Series 13). Ministry of Agriculture, Department of Animal Husbandry, Dairying and Fisheries. Krishi Bhavan, New Delhi.

89. Government of India. Wool, Section-V. Ministry of textiles. Fibre_Policy_Sub_Groups_Report_dir_mg_d_20100608_5. New Delhi, India.

90. Green, C.J. (1979). Animal anaesthesia. Laboratory animal handbooks 8. pp.214-215. Laboratory Animals Ltd., London, UK.

91. Groeneveld, L. F., Lenstra, J. A., Eding, H., Toro, M. A., Scherf, B., Pilling, D., Negrini, R., Finlay, E. K., Jianlin, H., Groeneveld, E., Weigend S. and The GLOBALDIV Consortium. (2010). Genetic diversity in farm animals – a review. *Animal Genetics,* **41** (Suppl. 1) : 6–31.

92. Grooves, C. (1981). Ancestors for the pigs: taxonomy and phylogeny of the genus *Sus.* Technical bulletin no-3. Department of prehistory, Research school of Pacific studies, Australian National University, Canberra, Australia.

93. Guide for the care and use of laboratory animals. (1985). US Deptt. of health and human services, National Institute of Health, Bethesda, Marryland-20205, U.S.A.

94. Guide to the Care and Use of Experimental Animals Canadian Council on Animal Care (CCAC) Vol. 1, 1980, 112 pp; Vol.2, 1984, 208 pp. Ottawa. Ontario : Canadian Council on Animal Care (Available from CCAC, 1105-151 Slater Street, Ottawa, Ontario KIP 5H3, Canada).

95. Gunter, H. and Peter, H. (2007). Meat processing technology for small- to medium-scale producers. RAP publication 2007/20, FAO regional office for Asia and the Pacific Bangkok.

96. Gupta, L. and Tapadia, A. L. (2011). Strategic livestock feeding for sustainable production. Narendra Publishing House, New Delhi.

97. Gupta, V.K. and Singh, R. (1991). Impact of mechanization on employment and farm productivity. *Productivity,* **32**(3): 484-489.

98. Haenlein, G. F. W. (2004). Goat milk in human nutrition. *Small Ruminant Research,* **51**: 155-163.

99. Hafez, E. S. E. (1969). The behaviour of domestic animals. Bailliere, Tindal & Cassell Ltd, USA.

100. Hafez, E.S.E. (ed.) (1970). Reproduction and breeding techniques for laboratory animals. Lea & Febiger, Philadelphia, U.S.A.

101. Hafez, E.S.E.(1968). Adaptation of domestic animals. Philadelphia, Lea and Febiger, Philadelphia, USA 61-73.

102. Hales, J.R., Fawcett, A. A., Bennett, J. W. and Needham, A. D. (1978). Thermal control of blood flow through capillaries and arteriovenous anastomoses in skin of sheep. *Pflug. Arch.*, **378**, 55–63.

103. Hamilton, C., Hansson, I. T., Ekman, U. Emanuelson, and K. Forslund. (2002). Health of cows, calves and young stock on organic dairy herds in Sweden. *Vet. Rec.* **150**:503–508.

104. Hffman, R.A., Robinson, P.E. and Magalhaes, H. (eds) (1968). The Golden Hamster, Its biology and use in medical research. Lowa State University Press, Ames, iA, U.S.A.

105. *http://agriexchange.apeda.gov.in/India%20Production/AgriIndia_Productions.aspx?categorycode=0401.*

106. *http://www.climateandfarming.org/pdfs/FactSheets/III.3Cattle.pdf (online resource by USDA Northeast Sustainable Agriculture Research and Education (NE-SARE) with contributions from University of Vermont, Cornell University, and Clean Air-Cool Planet).*

107. *http://www.dpi.vic.gov.au/agriculture/farming-management/fire-flood-other-emergencies/animals-in-heat/caring-for-animals-during-extreme-heat (from article titled 'Caring for Animals during Extreme Heat' at online resource of Department of Primary Industries, Victoria, Australia).*

108. *http://www.fao.org/ag/againfo/programmes/en/pplpi/docarc/wp48.pdf.*

109. *http://www.fil-idf.org.*

110. *http://www.imd.gov.in/doc/learn-meteorology.htm (from website of Indian Meteorological Department, Ministry of Earth Sciences, Govt. of India).*

111. *http://www.inseda.org/Additional%20material/CD%20%20Agriculture%20and%20Environment%20Education/52-Manuers%20&%20 Fertilizers%20%28MFS%29/.*

112. *http://www.inseda.org/Additional%20material/CD%20%20 Agriculture %20 and% 20 Environment%20Education/12Agricultural %20Meteorology %20(AMS)/Agroclimatic % 20Divisions%20of%20 India-521.doc (from online resource of Integrated Sustainable Energy and Ecological Development Association – INSEDA).*

113. *http://www.ipcc.ch/publications_and_data/ar4/wg1/en/faq-1-3.html (from online resource of Intergovernmental Panel on Climate Change).*

114. *http://www.weatheronline.co.uk/reports/wxfacts/The-Stevenson-Screen.htm (from online resource for Weather Online Ltd. - Meteorological Services of United Kingdom).*

115. Hutchinson, J. C. D. and Brown, G. D. (1969). Penetrance of cattle coats by radiation. *J. Appl. Physiol.*, **26**: 454–464.

116. IAEA. (2008). Guidelines for sustainable manure management in Asian livestock production systems.Vienna.

117. Ibrahim, H. (1998). Small Ruminant Production Techniques. ILRI Manual 3. ILRI International Livestock Research Institute, Nairobi, Kenya. 207 pp.

118. Indian Council of Agricultural Research. (2002). Handbook of Animal Husbandry. Publications and Information Division, Krishi Anusandhan Bhawan, Pusa, New Delhi.

119. Indian National Science Academy. (2000). Guidelines for care and use of animals in scientific research. New Delhi.

120. Indian Standards Institution. (1968). Code for transport of dogs and cats by rail, road and air. IS4746. New Delhi.

121. Indian Standards Institution. (1971). Code for care, management and housing of laboratory animals. IS-6384, Part-1: Monkeys; Part-II: Laboratory dogs; Part-III; Laboratory cats. New Delhi.

122. Indian Standards Institution. (1977). Code for transport of laboratory animals, IS-3907, Part-1. Transport of mice, rats, rabbits, guinea pigs, cotton rats and hamsters. New Delhi.

123. Indian Standards Institution. (1978). Code for breeding, care, management and housing of laboratory animals. IS 5701, Part 1 – Mice & Rat: Part II – Laboratory rabbit: Part III Guinea pig: Part IV-Hamster. New Delhi.

124. Jainudeen, M. R., Wahid, H. and Hafez, E.S.E. (2000). Sheep and goats. In: Reproduction in farm animals, 7th edition. Hafez, B. and E.S.E. Hafez (Eds.). Lippincott Williams & Wilkins, Philadelphia, Baltimore.

125. Kahrs, R.F. (1995). General disinfection guidelines. *Rev. sci. tech. Off. int. Epiz.,* **14** (1) : 105-122.

126. Kaith, N.S. and Bhardwaj, J.C. (2009). Organic manure - a present day need, ENVIS Bulletin: *Himalayan Ecology* **17**.

127. Kanakraj, P. (2006). Elements of Breeding and Breeds of Cattle and Buffaloes. Jaypee Brothers Medical Publishers (P) Ltd, New Delhi.

128. Kandeepan, G., Biswas, S. and Rajkumar, R. S. (2009). Buffalo as a potential food animal. *International Journal of Livestock Production,* **1** (1) : 001-005.

129. Karanjkar, L. M. and Patil, R. A. (2008). Draught ability of bullocks: A Review. *Indian Journal of Animal Sciences,* **78** (9) : 1002-1018.

130. Karim, S. A. and Shakyawar, D. B. Present status and future strategies of wool production, processing and product manufacture. National Seminar cum workshop on recent R&D Initiatives and Developmental Schemes of wool and woolens.

131. Khan, A. G. (1998). An overview on development of broiler breeding research in India. *Indian Journal of Animal Sciences,* **68** (8, special issue): 728-734.

132. Khan, B. B., Iqbal, A., Riaz, M., Yaqoob, M. and Younas, M. (2004). Livestock Management Manual I, Department of Livestock Management, Uni. Agri., Faisalabad, Pakistan.

133. Khanna, S. S. (1989). The agro-climatic approach. In: Survey of Indian agriculture. *The Hindu*, Madras, India. pp. 28-35.

134. Kijlstra, A. and Eijck, I.A.J.M. (2006). Animal health in organic livestock production systems: a review. *NJAS Wageningen J. life Sci.,* **54** (1) : 71-84.

135. Kirk, R.W. (1970). Dogs In: reproduction and breeding techniques for laboratory animals (E.S.E. Hafez, ea.) Lea and Febiger, Philadelphia, PA, USA pp.224-236.

136. Kolay, A. K. 2007. Manures and fertilizers. Atlantic Publishers and Distributors (P) LTD, New Delhi, India.

137. Kumar, A. and Singh, D. K. (2008). Livestock production systems in India: An appraisal across agro-ecological regions. *Indian Journal of Agri. Econ.,* **63** (4):577-597.

138. Kumar, S. (2007). Commercial goat farming in India: An emerging agri-business opportunity. *Agricultural Economics Research Review,* 20 (Conference Issue):503-520.

139. Kumaravelu, N. and Murallidharan, R. (2008). Solid waste utilization in dairy farms. Dairy Year Book.

140. Kyriazakis, I. and Whittemore, C. T. (2006). Whittemore's Science and practice of pig production, 3rd edition. Blackwell Publishing Ltd, United Kingdom.

141. Lee, D. H. K. & Phillips, R. W. (1948). Assessment of the adaptability of livestock to climatic stress. *J. Anim. Sci.,* **7**: 391–425.

142. Lin, H., Jiao, H. C., Buyse, J. and Decuypere, E. (2006). Strategies for preventing heat stress in poultry. *World's Poultry Science Journal.* **62**: 71-85.

143. Linton, R. G. (1921). Veterinary Hygiene. Lorimer & Chalmers, Edinburgh.

144. Long, C. (1961). Biochemist's handbook, E. & F. N. Spon, Ltd, London.

145. Mandal, A. B., Elangovan, A. V. and Tyagi, P. K. (2005). Poultry nutrition for economic egg and meat production: A review. *Indian Journal of Animal Sciences,* **75** (10): 1215-1226.

146. Marai, I. F. M. and Habeeb, A. A. M. (1998). Adaptation of *Bos taurus* cattle under hot climate conditions. *Ann Arid Zone.* **37**(3):253–281.

147. Marai, I. F. M.; El-Darawany, A. A.; Abou-fandoud, E. I. (2006). Tunica dartos index as a parameter of adaptability of rams in sub-tropical conditions of Egypt. *Animal Science.***77**: 487-494.

148. Mazras, G. (1994). Traditional veterinary medicine in India. *Res. Sci. tech. off. int. Epiz.,* **13**(2): 443-451.

149. McGlone, J. and Pond, W. G. (2003). Pig production biology and adaptations. Thomson Delmar Learning, USA.

150. Mellen, I. M. (1952). The Natural History of the Pig. Exposition Press, New York, NY.

151. Mirkena ,T., Duguma,G., Haile, A., Tibbo , M., Okeyo, A., Wurzinger, M. M. And Sölkner, J. (2010). Genetics of adaptation in domestic farm animals: A review. *Livestock Science,* **132**:1–12.

152. Mishra, R. (1978). Management practices for goats. NDRI, Karnal, Haryana. Pub no.147.

153. Misra, A. K. and Singh, D. (1993). Rearing of calf. A scientific approach. *Indian Dairyman,* **64**: 526-529.

154. Misra, R. V. and Hesse, P. R. (1982). Comparative analyses of organic manures, FAO/UNDP Regional Project RAS/75/004, Project Field Document 24, 97p.

155. Mohanty, S. and Rajendran, K. (2003). 2020 Vision for Indian Poultry Industry. *International Journal of Poultry Science,* **2** (2): 139-143.

156. Morse, H.C.(1981). The Laboratory Mouse pp. 1-16. Academic Press. New York, USA.

157. Moulik, T. K. and Hariharan, V. (1986). Sheep rearing and wool production in India: problems and prospects, CMA Monograph No. 123.

158. Mudgal, V. D., Singhal, K. K. and Sharma, D. D. (2003). Advances in dairy animal production, 2nd edition. International Book Distributing Co., Lucknow, India.

159. Mukasa-Mugerwa, E. (1989). A review of reproductive performance of female Bos indicus (zebu) cattle. ILCA Monograph 6. ILCA, Addis Ababa, Ethiopia.

160. Muys, D. and Westenbrink, G. (2004). Pig keeping in the tropics. Agrodok-series No. Agromisa Foundation, Wageningen.

161. NABARD. Sheep farming for meat and wool. Technical Services Department and produced by Public Relations Division National Bank for Agriculture and Rural Development at Silver Point Press Pvt. Ltd., Mumbai, India.

162. Nagpal, S. K., Pankaj, P K., Ray B. and Talware, M. K. (2005). Shelter management for dairy animals: a review. *Indian Journal of Animal Sciences,* **75** (10): 1199-1214.

163. National Research Council (NRC). 1981. Nutrient requirements of goats: Angora, dairy and meat goats in temperate and tropical countries. No. 15, National Academy of Sciences, Washington, D.C., U.S.A., vii + 91pp.

164. National Research Council. (1996). Guide for the care and use of laboratory animals. National Academy Press. Wasington DC, U. S.A.

165. National Research Council. (2011). Guide to the care and use of experimental animals. 8th Edition. National Academic Press, Wasington DC, U.S.A.

166. NBAGR. (2009). New breeds of cattle and buffalo. *Indian Journal of Animal Sciences,* **80** (10): 1045-55.

167. Niemann,H., Rath, D. and Wrenzycki, C. (2003). Advances in Biotechnology: New Tools in Future Pig Production for Agriculture and Biomedicine. *Reprod Dom Anim.,* **38**:82–89.

168. NRC. (1971). A Guide to Environmental Research on Animals. Natl. Acad. Sci., Washington, DC. USA.

169. Online resource of electronic reference of The Merck Veterinary Manual.

170. Parekh, H. K.B., Srivastava, P.N. and Sarkhel, B.C. (2005). Textbook on Animal Genetics and Breeding, Kalyani Publishers, New Delhi.

171. Park, Y. W., Juarez, M., Ramos, M. and Haenlein, G. F. W. (2007). Physioco-chemical characteristics of goat and sheep milk. *Small Ruminant Research,* **68**: 88-113.

172. Parker, R. O. (2008). Equine science. Delmar Cengage learning. New York, USA.

173. Patnayak, B. C. (1992). Nutrition and feed resources for small ruminants. Small ruminant production: systems for sustainability. Proceedings from II International workshop for the PVO and technical personnel. The agency for international development bureau for research and development, office of agriculture. In collaboration with V International Conference on Goats 28-29 February 1992 New Delhi, India.

174. Pauw, R., Mack, S. and Maki-Hokkonen, J. (2003). Development of Animal Identification and Recording Systems for Developing Countries. ICAR, Villa del Ragno, Via Nomentana 134, 00162 Rome, Italy.

175. Payne, W. J. A. & Hodges, J. (1997). Tropical cattle: origins, breeds and breeding policies. Blackwell Science: Oxford.

176. Payne, W. J.A. and Wilson, R. T. (1999). An Introduction to Animal Husbandry in the Tropics. Blackwell Science Ltd, London, UK.

177. Peplow, A.M., Peplow, P.V. and Hafez, E.S.E. (1974). Parameters of Reproduction. In: CRC Handbook of laboratory animal science. CRC Press, Cleveland, Ohio, USA.

178. Phillips, C. J.C. (2010). Principles of cattle production. CABI. UK.

179. Pica-Ciamarra, U. and Otte, J. (2009). Poultry, Food Security and Poverty in India: Looking Beyond the Farm-Gate. Pro-poor livestock policy initiative, a living from livestock, RR Nr. 09-02. www.fao.org/ag/pplpi.html.

180. Pol, M. and Ruegg, P. L. (2007). Treatment practices and quantification of antimicrobial usage in conventional and organic dairy farms in Wisconsin. *J. Dairy Sci.*, **90**:249–261.

181. Pollott, G. and Wilson, R. (2009). Sheep and goat for diverse products and profits. Rural Infrastructure and Agro-Industries Division. Food and Agriculture Organization of the United Nations, Rome, Italy.

182. Porter, V. (1993). Pigs: A Handbook to the Breeds of the World. Cornell University Press, Ithaca, NY.

183. Porter, V. 2002. Mason's World Dictionary of Livestock Breeds, Types and Varieties. CABI, UK.

184. Prasad, J. (1997). Principles and Practices of Dairy Farm Management. Kalyani Publishers, New Delhi.

185. Price, E. O. (2002). Animal domestication and behavior. CABI, UK.

186. Rai, B., Singh, M. K. and Singh, S. K. (2005). Goats for meat, milk and fibre: a review. *Indian journal of Animal Sciences,* **75** (3):349-355.

187. Ranjan, S.K. (1999). Animal Nutrition and Feeding Practices. Vikas Publishing House, Delhi.

188. Ranjhan, S. K. (1999). Dairy feeding systems. *In* Falvey, L. and Chantalakana, C. Smallholder dairying in the tropics. ILRI, Nairobi, Kenya.

189. Ranjhan, S. K. (2003). Animal nutrition in tropics (5th edition) Vikas Publishing House Private Limited, New Delhi-110014.

190. Ranjhan, S.K. (2007). Buffalo as a social animal for humanity. *Ital. J. Anim. Sci.*,**6** (Suppl. 2): 30-38.

191. Rao, K. K. and Nagarcenkar, R. (1977). Potentialities of the buffalo. *World Review of Animal Production,* **13**:53-62.

192. Rautaray, S. K. (1985). Output of a pair of bullocks during tillage operation in summer. Central Institute of Agricultural Engineering, Bhopal.

193. Rege, J. E. O. and Gibson, J. P. (2003). Animal genetic resources and economic development: issues in relation to economic valuation. *Ecol. Econ.*, **45**:319–330.

194. Reksen, O., Tverdal, A. and Ropstad, E. (1999). A comparative study of reproductive performance in organic and conventional dairy husbandry. *J. Dairy Sci.*, **82**:2605–2610.

195. Rhoad, A .O. (1940). A Method of assaying genetic differences in the adaptability of cattle to tropical and subtropical climates. *Emp. J. Exp. Agric.*, **8**: 190–198.

196. Rhoad, A. O. (1944). The Iberia heat tolerance test for cattle. *Trop. Agr.* (Trinidad), **21** (9) : 162.

197. Roots, C. (2007). Domestication. Greenwood press Westport, Connecticut, London.

198. Roy, S. R., Negi, A. K. and Gupta, H. (1972). Crossbred bullocks vs. indigenous bullocks for drought purpose under West Bengal conditions. *Indian Dairyman,* **24**: 66-70.

199. Ruegg P. L. (2009). Management of mastitis on organic and conventional dairy farms. *J. Anim. Sci..* **87**:43-55.

200. Sanisbury, D. and Sanisbury, P. (1988). Livestock health and housing. Bailliere Tindal, England.

201. Saran, S., Sasidhar, P. V. K., Sastry, K. V. H. and Singh, R. (2005). Indian poultry industry: current scenario and future prospects: a review. *Indian Journal of Animal Sciences,* **75** (8): 992-998.

202. Sastry, N. S. R. (1993). Animal husbandry in different agro-climatic regions of India. Report of a consultancy mission, sponsored by Swiss Development Corporation, New Delhi, India.

203. Sastry, N. S. R. and Thomas, C. K. (2005). Livestock Production Management. Kalyani Publishers, New Delhi, India.

204. Sen, K.C., Ray, S.N. and Ranjhan, S. K. (1978). Chemical Composition of Indian Feeds Feeding of farm Animals. ICAR (Indian Council of Agricultural Research), New Delhi, India.

205. Sengar, O.P.S. (1980). Indian research on protein and energy requirements of goats. *J Dairy Sci.,* **63**:1655-1670.

206. Seré, C. S. H. and Groenewold, J. (1995). World Livestock Production Systems: Current Status, Issues and Trends. FAO Production and Health paper.

207. Sharda, D. P. (2005). Swine production. Indian Council of Agricultural Research. New Delhi.

208. Sharma, R. D., Kumar, and Sridhar, R. (1987). Historical background and analysis of scientific content of ancient Indian literature on practices for the treatment of diseases of domestic animals. *Indian Journal of History of science,* **22**(2):158-163.

209. Shelton, M. (1978). Reproduction and Breeding of Goats. *J Dairy Sci.,* **61**:994-1010.

210. Shethi, R. K. (2000). In the Souvenir of 4th Asian Buffalo Congress at New Delhi, India from Feb. 25-28.

211. Short, D.J. and Woodnott, D.P. (ads.). (1969). IAT Manual of Laboratory Animal Practice and Techniques.2nd Edition. Crosby Lockwood and Sons Ltd; London, Sw7 England.

212. Silanikove, N. (2000). The physiological basis of adaptation in goats to harsh environments. *Small Ruminant Research,* **35**:181-193.

213. Singh, C. M. (2002). The insignia of the Veterinary Council of India adopted from Emperor Ashoka's edict. In: Third Convocation of National Academy of Veterinary Sciences (India) and National Symposium on Historical Overview on Veterinary Sciences and Animal Husbandry in Ancient India (Vedic and Ashokan Period),

16–17 April 2002, Indian Veterinary Research Institute, Izatnagar, Uttar Pradesh, India. pp. 1–3.

214. Singh, G. (1999). Draught animal energy research in India. Proceedings of an ATNESA Workshop, South Africa.

215. Singh, G. 1994. Weight matrix of Indian cattle and their draught power. *Indian Journal of Agricultural Engineering*, **4** (3-4):100-106.

216. Singh, P. (2006). Final report of the working group on agro-climatic zonal planning including agriculture development in north-eastern India for XI five year plan (2007-12), Volume I-Main report. Government of India, New Delhi.

217. Singh, R. A. (2009). Poultry Production. Kalyani Publishers, New Delhi.

218. Singh, S. N. and Senger, O.P.S. (1978). Meat and milk potentialities of Indian goats. R.B.S. College, Bichpuri, Agra, India. Dep.Anim.Husbandry Dairying Tech Report.

219. Singh,V. K., Suresh, A., Gupta, D. C. and Jakhmola, R. C. (2005). Common property resources rural livelihood and small ruminants in India: A review. *Indian journal of Animal Sciences*, **75** (8): 1027-1036.

220. Smart, O. N., A. Michael, O. and Bamgbose, M. A. (2007). The unwanted creatures in poultry houses- a review. *Agric. Rev.*, **28** (3):157-168.

221. Smit, B., Mc Nabb, D. and Smihers, J. (1996). Agricultural adaptation to climatic variation. *Climatic change.* **33** : 7-29.

222. Solaiman, S. G. (2010). Goat science and production. Wiley-Blackwell, State Avenue, Ames, Iowa, USA.

223. Solarz, A.K. (1970). Seh, Inur In: The Beagle as an Experimental Dog pp. 453-468 (A.C.Anderson, ea.) Iowa State University Press, Ames, IA, USA.

224. Solorio, S., anchez, F.J. and Solorio S.B. (2002). Integrating fodder trees in to animal production systems in the tropics. *Tropical and Subtropical Agro-ecosystems*, **1**:1-11.

225. Somvanshi, R. (2006). Veterinary Medicine and animal keeping in ancient India. *Asian Agri-History*, **10**(2): 133-146.

226. Speedy, A. and Pugliese, P.L. (1992). Legumes tress and other fodder trees as protein sources for livestock. FAO Animal Production and Health Paper 102, pp. 339.

227. Srivastava, N. S. L. (1989). Research on Draught Animal Power in India. Draught Animals in Rural Development Proceedings of an International Research Symposium, Cipanas, Indonesia.

228. Srivastava, N. S. L. (1993). Status of research on engineering aspects of draught animal power. *Journal of Rural Energy*, **2**(2, 4):15- 32.

229. Starke, J.S. and Pretorius, A.G. (1955). Dentition of sheep as an indication of age. *Fmg S. Afr.*, **30** : 538.

230. Steinfeld,H. Wassenaar, T. and Jutzi, S. (2006). Livestock production systems in developing countries: status, drivers, trends. *Rev. sci. tech. Off. int. Epiz.*, **25** (2) : 505-516.

231. Taneja, V. K. (1999). Dairy breeds and selection. In Falvey, L. and Chantalakana, C. Smallholder dairying in the tropics. ILRI, Nairobi, Kenya.

232. Taverner, M.R. and Dunkin, A. C. (1996). Pig production. World animal science, C-10, Elsevier science B V, Amsterdam, The Netherland.

233. Taylor, R.E. and Field, T.G. (2001). Scientific farm animal production. Prentice-Hall, Inc., Upper saddle River, New Jersey.

234. Thamizhselvi, R. K. and Rao, S. V. N. (2010). Is contract broiler farming exploitative to small farmers. *Indian Journal of Animal Sciences,* **80** (12): 1243-50.

235. The Laboratory Guinea Pig. (2007). Laboratory Animals Centre. National University of Singapore.

236. Thiele, H. and Pottgüter, R. (2008). Management recommendations for laying hens in deep litter, perchery and free range systems. *Lohman Infirmation,* **43** (1), pp53.

237. Thom, E. C. (1959). The Discomfort Index. Weatherwise **12:** 57-60.

238. Thomas, C. K. and Sastry, N. S. R. (2000). Dairy Bovine Production. Kalyani Publishers. New Delhi.

239. Thomas, R and George, S. K. (2005). Organic Livestock Farming for meat production. *Vet. World.* **4**: 76-80.

240. Timon, V. M. and Hanrahan, J. P. (1985). Small ruminant production in the developing countries- Proceedings of an Expert Consultation held in Sofia, Bulgaria. FAO Animal Production and Health paper 58.

241. Tomar, S. S. (2004). Textbook of Animal Breeding, Kalyani Publishers, New Delhi.

242. Tony, S. (2000). Practical techniques for dairy farmers. University of Minnesota. St. Paul, MN.

243. Towne, C. W. and E. N. Wentworth. (1950). Pigs from Cave to Cornbelt. University of Oklahoma Press, Norman, OK.

244. Universities Federation for Animals Welfare. (1987). The UFAW handbook on the care and management of laboratory animals. 6th edn. Churchill Livingstone, New York, USA.

245. Upadhyay, R. C. and Madan, M. L. (1982). Work performance of Hariana and crossbred bullocks during summer. National seminar on animal energy utilization, Agriculture University, Coimbtore.

246. Upadhyay, R. C. and Madan, M. L. (1985). Studies on blood acid base status and muscle metabolism in working bullocks. *Animal Production,* **40**: 11-16.

247. USDA. (2004). India's Poultry Sector: Development and prospects market and trade economics division, economic research service, WRS-04-03. Washington, DC, USA.

248. Van der Westhuysen, J. M. (2005). Marketing goat fibres. *Small Ruminant Research,* **60**: 215-218.

249. Walkden-Brown, S. W. and Restall, B. J. (1996). Environmental and social factors affecting Reproduction. 6th Intl. Conference on Goats. 6-11 May, 1996, Beijing, China, Vol: 2.

250. Weerd, H. A. van de Jon and Day, E.L. (2009). A review of environmental enrichment for pigs housed in intensive housing systems. *Applied Animal Behaviour Science,* **116**:1-20.

251. Weilborth. S.H. Flatt, R.E. and Kraus, A.L. (1974). The biology of the laboratory rabbit. Academic Press, New York, USA.

252. Wells, R. (1997). Organic poultry production. Final report to MAFF, Contract ref.: CSA 3699. Welsh institute of rural studies university of Wales, Aberystwyth.

253. Wildeus, S. (1995). Successful reproductive management. Proceeding of the Virgina State Dairy Goat Association, April1 1995, Blacksburg, VA, pp: 21-26.

254. Willis, N. G. (2003). Animal carcass disposal. Conf. *OIE,* 149-159.

255. Wilson, R. T. (1995).Livestock production systems. The Tropical Agricultural Series. CTA. Macmillan Education Limited, London, U. K.

256. *www. lhu.emu.ee*

257. *www.cattletoday.com*

258. *www.fao.org/docrep/015/i2433e/i2433e07.pdf (from article titled 'rural structures in the tropics design and development : livestock housing' at online resource of FAO).*

259. *www.mla.com.au/files/02daccf7-a8ef.../heatload-in-feedlot-cattle.pdf (online resource of Meat and Livestock Australia).*

260. *www.rci.com.au/en/tibs/cattle/heatcow.pdf (from article titled 'Heat Stress in Dairy Cattle ' at online resource of rural chemical industries (Aust.) Pty Ltd).*

261. Yamamoto, S. (1983). The assessment of thermal environment for farm animals. *In:* New strategies for improving animal production for human welfare. *Proceedings of the 5th World Association for Animal Production* (WCAP), Japan. Society Zootech. Sci., Tokyo, Japan. pp. 197-204.

262. Yeck, R. G. and Kibler, H. H. (1958). Predicting Heat Tolerance from Calf Vaporization Rates. *J. Anim. Sci.,* **17**: 1228–1229.

263. Zeuner, F.E. (1963). A History of Domesticated Animals. Hutchinson: London.

FSC
www.fsc.org
MIX
Papier | Fördert
gute Waldnutzung
FSC® C083411

Zeitfracht Medien GmbH
Ferdinand-Jühlke-Straße 7
99095 Erfurt, Deutschland
produktsicherheit@kolibri360.de